Power Grid Resilience Against Natural Disasters

Power Grid Resilience Against Natural Disasters

Preparedness, Response, and Recovery

Shunbo Lei
Chong Wang
Yunhe Hou

This edition first published 2023
© 2023 John Wiley & Sons Ltd

The right of Shunbo Lei, Chong Wang, and Yunhe Hou to be identified as the authors of this work has been asserted in accordance with law.

Registered Office(s)
John Wiley & Sons, Inc., 111 River Street, Hoboken, NJ 07030, USA
John Wiley & Sons Ltd, The Atrium, Southern Gate, Chichester, West Sussex, PO19 8SQ, UK

For details of our global editorial offices, customer services, and more information about Wiley products visit us at www.wiley.com.

Wiley also publishes its books in a variety of electronic formats and by print-on-demand. Some content that appears in standard print versions of this book may not be available in other formats.

Library of Congress Cataloging-in-Publication Data Applied for:

Hardback ISBN: 9781119801474

Cover Design: Wiley
Cover Image: © MNBB Studio/Shutterstock

Set in 9.5/12.5pt STIXTwoText by Straive, Chennai, India
Printed and bound by CPI Group (UK) Ltd, Croydon, CR0 4YY

C9781119801474_151122

To

Jinhui, Qiuying, and Shunxia
      – Shunbo Lei

Xingxing, Xueheng, Yan, and Hui
      – Chong Wang

Lijuan and Jesse
      – Yunhe Hou

Contents

About the Authors *xv*
Preface *xvii*
Acknowledgments *xxiii*

Part I   Introduction *1*

1   Introduction *3*
1.1   Power Grid and Natural Disasters   *3*
1.2   Power Grid Resilience   *4*
1.2.1   Definitions   *4*
1.2.2   Importance and Benefits   *6*
1.2.2.1 Dealing with Weather-Related Disastrous Events   *6*
1.2.2.2 Facilitating the Integration of Renewable Energy Sources   *7*
1.2.2.3 Dealing with Cybersecurity-Related Events   *8*
1.2.3   Challenges   *9*
1.3   Resilience Enhancement Against Disasters   *12*
1.3.1   Preparedness Prior to Disasters   *12*
1.3.1.1 Component-Level Resilience Enhancement   *13*
1.3.1.2 System-Level Resilience Enhancement   *14*
1.3.2   Response as Disasters Unfold   *14*
1.3.2.1 System State Acquisition   *15*
1.3.2.2 Controlled Separation   *16*
1.3.3   Recovery After Disasters   *17*
1.3.3.1 Conventional Recovery Process   *17*
1.3.3.2 Microgrids for Electric Service Recovery   *18*
1.3.3.3 Distribution Grid Topology Reconfiguration   *18*
1.4   Coordination and Co-Optimization   *20*
1.5   Focus of This Book   *22*
1.6   Summary   *23*
         References   *23*

Part II   Preparedness Prior to a Natural Disaster *35*

2   **Preventive Maintenance to Enhance Grid Reliability** *37*
2.1   Component- and System-Level Deterioration Model *37*
2.1.1   Component-Level Deterioration Transition Probability *38*
2.1.2   System-Level Deterioration Transition Probability *40*
2.1.3   Mathematical Model without Harsh External Conditions *40*
2.2   Preventive Maintenance in Consideration of Disasters *41*
2.2.1   Potential Disasters Influencing Preventive Maintenance *41*
2.2.2   Preventive Maintenance Model with Disasters Influences *42*
2.2.2.1 Probabilistic Model of Repair Delays Caused By Harsh External
         Conditions *42*
2.2.2.2 Activity Vectors Corresponding to Repair Delays *42*
2.2.2.3 Expected Cost *43*
2.3   Solution Algorithms *44*
2.3.1   Backward Induction *44*
2.3.2   Search Space Reduction Method *44*
2.4   Case Studies *45*
2.4.1   Data Description *45*
2.4.2   Case I: Verification of the Proposed Model *45*
2.4.2.1 Verifying the Model Using Monte Carlo Simulations *46*
2.4.2.2 Selection of Optimal Maintenance Activities *47*
2.4.2.3 Influences of Harsh External Conditions on Maintenance *48*
2.4.3   Case II: Results Simulating the Zhejiang Electric Power Grid *48*
2.5   Summary and Conclusions *51*
         Nomenclature *52*
         References *53*

3   **Preallocating Emergency Resources to Enhance Grid
         Survivability** *55*
3.1   Emergency Resources of Grids against Disasters *55*
3.2   Mobile Emergency Generators and Grid Survivability *58*
3.2.1   Microgrid Formation *59*
3.2.2   Preallocation and Real-Time Allocation *59*
3.2.3   Coordination with Conventional Restoration Procedures *60*
3.3   Preallocation Optimization of Mobile Emergency Generators *61*
3.3.1   A Two-Stage Stochastic Optimization Model *61*
3.3.2   Availability of Mobile Emergency Generators *66*
3.3.3   Connection of Mobile Emergency Generators *66*
3.3.4   Coordination of Multiple Flexibility in Microgrids *67*

3.4   Solution Algorithms   *67*

3.4.1   Scenario Generation and Reduction   *68*

3.4.2   Dijkstra's Shortest-Path Algorithm   *69*

3.4.3   Scenario Decomposition Algorithm   *69*

3.5   Case Studies   *70*

3.5.1   Test System Introduction   *70*

3.5.2   Demonstration of the Proposed Dispatch Method   *71*

3.5.3   Capacity Utilization Rate   *73*

3.5.4   Importance of Considering Traffic Issue and Preallocation   *75*

3.5.5   Computational Efficiency   *76*

3.6   Summary and Conclusions   *77*

       Nomenclature   *78*

       References   *80*

4   **Grid Automation Enabling Prompt Restoration**   *85*

4.1   Smart Grid and Automation Systems   *85*

4.2   Distribution System Automation and Restoration   *87*

4.3   Prompt Restoration with Remote-Controlled Switches   *89*

4.4   Remote-Controlled Switch Allocation Models   *91*

4.4.1   Minimizing Customer Interruption Cost   *91*

4.4.2   Minimizing System Average Interruption Duration Index   *93*

4.4.3   Maximizing System Restoration Capability   *94*

4.5   Solution Method   *95*

4.5.1   Practical Candidate Restoration Strategies   *95*

4.5.2   Model Transformation   *99*

4.5.3   Linearization and Simplification Techniques   *100*

4.5.4   Overall Solution Process   *100*

4.6   Case Studies   *102*

4.6.1   Illustration on a Small Test System   *102*

4.6.1.1   Results of the CIC-oriented Model   *102*

4.6.1.2   Results of the SAIDI-oriented Model   *103*

4.6.1.3   Results of the RL-oriented Model   *105*

4.6.1.4   Comparisons   *105*

4.6.2   Results on a Large Test System   *106*

4.7   Impacts of Remote-Controlled Switch Malfunction   *109*

4.8   Consideration of Distributed Generations   *110*

4.9   Summary and Conclusions   *111*

       Nomenclature of RCS-Restoration Models   *112*

       Nomenclature of RCS Allocation Models   *113*

       References   *113*

Part III   Response as a Natural Disaster Unfolds   *119*

5   **Security Region-Based Operational Point Analysis for**
           Resilience Enhancement   *121*
5.1   Resilience-Oriented Operational Strategies   *121*
5.2   Security Region during an Unfolding Disaster   *123*
5.2.1   Sequential Security Region   *123*
5.2.2   Uncertain Varying System Topology Changes   *125*
5.3   Operational Point Analysis Resilience Enhancement   *126*
5.3.1   Sequential Security Region   *126*
5.3.2   Sequential Security Region with Uncertain Varying Topology
           Changes   *127*
5.3.3   Mapping System Topology Changes   *129*
5.3.4   Bilevel Optimization Model   *130*
5.3.5   Solution Process   *131*
5.4   Case Studies   *132*
5.5   Summary and Conclusions   *138*
           Nomenclature   *138*
           References   *140*

6   **Proactive Resilience Enhancement Strategy for Transmission**
           Systems   *143*
6.1   Proactive Strategy Against Extreme Weather Events   *143*
6.2   System States Caused by Unfolding Disasters   *145*
6.2.1   Component Failure Rate   *146*
6.2.2   System States on Disasters' Trajectories   *146*
6.2.3   Transition Probabilities Between Different System States   *147*
6.3   Sequentially Proactive Operation Strategy   *148*
6.3.1   Sequential Decision Processes   *148*
6.3.2   Sequentially Proactive Operation Strategy Constraints   *148*
6.3.3   Linear Scalarization of the Model   *150*
6.3.4   Case Studies   *152*
6.3.4.1   IEEE 30-Bus System   *152*
6.3.4.2   A Practical Power Grid System   *156*
6.4   Summary and Conclusions   *159*
           Nomenclature   *160*
           References   *162*

7   **Markov Decision Process-Based Resilience Enhancement for**
           Distribution Systems   *165*
7.1   Real-Time Response Against Unfolding Disasters   *165*
7.2   Disasters' Influences on Distribution Systems   *167*

7.2.1   Markov States on Disasters' Trajectories   *167*
7.2.2   Transition Probability Between Markov States   *169*
7.3   Markov Decision Processes-Based Optimization Model   *169*
7.3.1   Markov Decision Processes-based Recursive Model   *169*
7.3.2   Operational Constraints   *170*
7.3.2.1 Radiality Constraint   *170*
7.3.2.2 Repair Constraint   *170*
7.3.2.3 Power Flow Constraint   *171*
7.3.2.4 Power Balance Constraint   *171*
7.3.2.5 Line Capacity Constraint   *171*
7.3.2.6 Voltage Constraint   *172*
7.4   Solution Algorithms – Approximate Dynamic Programming   *172*
7.4.1   Solution Challenges   *172*
7.4.2   Post-decision States   *174*
7.4.3   Forward Dynamic Algorithm   *174*
7.4.4   Proposed Model Reformulation   *175*
7.4.5   Iteration Process   *177*
7.5   Case Studies   *177*
7.5.1   IEEE 33-Bus System   *177*
7.5.1.1 Data Description   *177*
7.5.1.2 Estimated Values of Post-Decision States   *178*
7.5.1.3 Dispatch Strategies with Estimated Values of Post-Decision States   *180*
7.5.2   IEEE 123-Bus System   *181*
7.5.2.1 Data Description   *181*
7.5.2.2 Simulated Results   *181*
7.6   Summary and Conclusions   *183*
         Nomenclature   *184*
         References   *186*

Part IV   Recovery After a Natural Disaster   *189*

**8   Microgrids with Flexible Boundaries for Service
           Restoration**   *191*
8.1   Using Microgrids in Service Restoration   *191*
8.2   Dynamically Formed Microgrids   *194*
8.2.1   Flexible Boundaries in Microgrid Formation Optimization   *194*
8.2.2   Radiality Constraints and Topological Flexibility   *195*
8.3   Mathematical Formulation of Radiality Constraints   *198*
8.3.1   Loop-Eliminating Model   *200*
8.3.2   Path-Based Model   *200*

8.3.3   Single-Commodity Flow-Based Model   *200*
8.3.4   Parent–Child Node Relation-Based Model   *201*
8.3.5   Primal and Dual Graph-Based Model   *201*
8.3.6   Spanning Forest-Based Model   *201*
8.4   Adaptive Microgrid Formation for Service Restoration   *202*
8.4.1   Formulation and Validity   *202*
8.4.2   Tightness and Compactness   *205*
8.4.3   Applicability and Application   *207*
8.5   Case Studies   *211*
8.5.1   Illustration on a Small Test System   *211*
8.5.2   Results on a Large Test System   *215*
8.5.3   LinDistFlow Model Accuracy   *219*
8.6   Summary and Conclusions   *219*
8.A   Appendix   *220*
8.A.1   Proof of Theorem 8.1   *220*
8.A.2   Proof of Proposition 8.1   *220*
          Nomenclature of Spanning Tree Constraints   *221*
          Nomenclature of MG Formation Model   *221*
          References   *222*

**9   Microgrids with Mobile Power Sources for Service
          Restoration**   *227*
9.1   Grid Survivability and Recovery with Mobile Power Sources   *227*
9.2   Routing and Scheduling Mobile Power Sources in Microgrids   *230*
9.3   Mobile Power Sources and Supporting Facilities   *233*
9.3.1   Availability   *233*
9.3.2   Grid-Forming Functions   *234*
9.3.3   Cost-Effectiveness   *234*
9.4   A Two-Stage Dispatch Framework   *235*
9.4.1   Proactive Pre-Dispatch   *235*
9.4.2   Dynamic Routing and Scheduling   *239*
9.5   Solution Method   *243*
9.5.1   Column-and-Constraint Generation Algorithm   *243*
9.5.2   Linearization Techniques   *245*
9.6   Case Studies   *245*
9.6.1   Illustration on a Small Test System   *246*
9.6.1.1   Results of MPS Proactive Pre-positioning   *246*
9.6.1.2   Results of MPS Dynamic Dispatch   *247*
9.6.2   Results on a Large Test System   *251*
9.7   Summary and Conclusions   *255*
          Nomenclature   *255*
          References   *257*

10 **Co-Optimization of Grid Flexibilities in Recovery Logistics** *261*

10.1 Post-Disaster Recovery Logistics of Grids *261*
10.1.1 Power Infrastructure Recovery *262*
10.1.2 Microgrid-Based Service Restoration *263*
10.1.3 A Co-Optimization Approach *264*
10.2 Flexibility Resources in Grid Recovery Logistics *265*
10.2.1 Routing and Scheduling of Repair Crews *265*
10.2.2 Routing and Scheduling of Mobile Power Sources *268*
10.2.3 Grid Reconfiguration and Operation *271*
10.3 Co-Optimization of Flexibility Resources *277*
10.4 Solution Method *280*
10.4.1 Pre-assigning Minimal Repair Tasks *280*
10.4.2 Selecting Candidate Nodes to Connect Mobile Power Sources *281*
10.4.3 Linearization Techniques *283*
10.5 Case Studies *284*
10.5.1 Illustration on a Small Test System *284*
10.5.2 Results on a Large Test System *287*
10.5.3 Computational Efficiency *290*
10.5.4 LinDistFlow Model Accuracy *292*
10.6 Summary and Conclusions *293*
10.A Appendix *293*
10.A.1 Proof of Proposition 10.1 *293*
References *294*

Index *301*

About the Authors

Shunbo Lei received the BE degree in Electrical Engineering and Automation from Huazhong University of Science and Technology, Wuhan, China, in 2013, and the PhD degree in Electrical and Electronic Engineering from The University of Hong Kong, Hong Kong SAR, China, in 2017. He was a visiting scholar at Argonne National Laboratory, Lemont, IL, USA, from 2015 to 2017, a postdoctoral researcher with The University of Hong Kong from 2017 to 2019, and a research fellow with the University of Michigan, Ann Arbor, MI, USA, from 2019 to 2021. He is currently an assistant professor with the School of Science and Engineering, The Chinese University of Hong Kong, Shenzhen, China. He is the Secretary of the IEEE PES Loads Subcommittee, and the Chair of the IEEE PES Task Force on Flexible Grid-interactive Efficient Buildings to Enhance Electric Service Resilience. He is also an associate editor of the *IEEE Transactions on Smart Grid*, and a Young Editorial Board Member of the *Protection and Control of Modern Power Systems*. He is the awardee of the 2019–2021 IEEE Transactions on Smart Grid Top 5 Outstanding Papers Award, the 2022 IEEE PES General Meeting Best Conference Papers Award, and the Young Resilience Fellowship by the Netherlands' 4TU Centre for Resilience Engineering. His research interests include power systems, resilience, grid-interactive efficient buildings, optimization, and learning.

Chong Wang received the BE and MS degrees in Electrical Engineering from Hohai University, Nanjing, China, in 2009 and 2012, respectively, and the PhD degree in Electrical Engineering from The University of Hong Kong, Hong Kong SAR, China, in 2016. He was a postdoctoral researcher at The University of Hong Kong in 2016 and was a postdoctoral researcher at Iowa State University, Ames, IA, USA, from 2017 to 2018. He is currently a professor with the College of Energy and Electrical Engineering, Hohai University. His research interests include power system resilience, renewable integration, and integrated energy system modeling and operation.

Yunhe Hou received the BE and PhD degrees in Electrical Engineering from Huazhong University of Science and Technology, Wuhan, China, in 1999 and 2005, respectively. He was a postdoctoral research fellow at Tsinghua University, Beijing, China, from 2005 to 2007, and a postdoctoral researcher at Iowa State University, Ames, IA, USA, and the University College Dublin, Dublin, Ireland, from 2008 to 2009. He was also a visiting scientist at the Laboratory for Information and Decision Systems, Massachusetts Institute of Technology, Cambridge, MA, USA, in 2010. He has been a guest professor with Huazhong University of Science and Technology from 2017 and an Academic Adviser of the China Electric Power Research Institute from 2019. He joined the faculty of The University of Hong Kong, Hong Kong SAR, China, in 2009, where he is currently an associate professor with the Department of Electrical and Electronic Engineering. Dr. Hou is the chair of IEEE PES Task Force on "Power System Restoration with Renewable Energy Sources" under the PSDP/PSSC-Power System Stability Controls Subcommittee. Dr. Hou was an associate editor of the *IEEE Transactions on Smart Grid* from 2016 to 2021. Dr. Hou is currently an associate editor of the *IEEE Transactions Power Systems* and the *Journal of Modern Power Systems and Clean Energy*.

Preface

Motivation for This Book

Resilience indicates the capabilities of dealing with the unexpected and is one of the most important merits of a variety of systems. Climate change, which results in more frequent extreme weather events, and threats from malicious attacks, especially highlight the urgency of establishing a resilient power and energy system, as the resilience of many other critical infrastructures depends on it. Generally, decision-makers have to consider various challenges including those associated with the climate change, infrastructure interdependencies, and information uncertainties, in establishing social, economic, and environmental resilience.

Power grid resilience, specifically, characterizes its ability to prepare for and adapt to changing conditions, and withstand and recover rapidly from disruptions. Effective functioning and resilience of power grids are critical for the economic stability, etc. Recent years have witnessed more frequent and more severe power outages due to cyber-attacks (e.g. by hackers) and physical attacks (e.g. by extreme weather events). Those large-scale outages have resulted in significant economic loss and greatly increased the life risk of sensitive groups of people, thus highlighting the importance of improving grid resilience. In power grids, we must understand the size and scope of key parameters required to facilitate the establishment of resilience.

First, emergency preparedness is essential for resilient power grids anticipating a natural disaster, e.g. a hurricane. It greatly determines the ability of the grid to withstand and reduce the magnitude and duration of disruptive events. Effective disaster preparedness measures include preventive maintenance of critical system components, installation or upgradation of grid automation systems, and prepositioning or predispatching of emergency resources for future response. Other new preparedness strategies of engineering application values are being proposed and evaluated, too. Many existing studies are particularly devoted to optimizing disaster planning based on approximate dynamic programming, scenario decomposition-based stochastic programming, etc. While there have already been some disaster preparedness strategies adopted by grid operators for years, there are also some recently developed strategies that have been or are being transitioned from academia to industry. In the future, industry needs of new disaster preparedness strategies and more efficient algorithms will increase, especially for fully utilizing and coordinating different flexibility resources.

Second, natural disaster-induced damages to the power grid sequentially unfold themselves during such a catastrophic event, resulting in disturbances to system states that require operators to implement appropriate and real-time response. Effective and efficient mitigation response should take into full consideration the sequential and uncertain characteristics of power grid component damages. Currently, there are some approaches using optimization methods such as robust programming and scenario-based stochastic programming to deal with the above challenges. Some researchers have also been establishing Markov decision process-based approaches that map the stochastic disturbances with corresponding optimal mitigation strategies in consideration of complex system dynamics. Still, there are currently inadequate studies on efficient decision-making approaches for sequential responses against uncertain damages over time. In future years, there will be increasing industry needs of methods for real-time state-based operation and control against natural disaster-induced disturbances. Techniques and technologies of engineering application values will also start transitioning from academia to industry in the near future.

Third, postdisaster recovery of power grids, including the restoration of electric service and repairing of grid infrastructures, needs to be improved. Different from common single fault-induced outages, natural disaster-induced blackouts are typically wide-range and long-duration, requiring novel and efficient recovery strategies especially for distribution grids. There have been many studies on resilience-enhancing service restoration and infrastructure recovery of power grids utilizing and coordinating distributed renewables, automation systems, and interdependent infrastructures. Microgrid technologies specifically provide unique opportunities of leveraging grid flexibility resources to achieve prompt recovery of power grids. However, many critical issues, e.g. the management of flexible boundaries of microgrids, are not fully addressed. In future years, there will be increasingly more studies to fulfill industry needs of decision-making tools for the postdisaster operation of power grids, and the dispatch of repair crews and available mobile power sources. Holistic and integrated optimization approaches for disaster recovery logistic of power grids are also needed.

Therefore, we perceive the necessity of disseminating fundamental theories and methodologies that are useful in improving power grid resilience against natural disasters.

Organization of This Book

This book covers three important topics related to the planning and operational resilience of power grids against natural disasters, including (i) preparedness before a natural disaster, which increases both component- and system-level reliability, extends grid flexibility, and enhances system readiness for prompt and effective dispatch of emergency resources; (ii) response as a natural disaster unfolds, which analyzes and implements real-time state-based sequential strategies for mitigating the impact of transmission or distribution grid damages caused by the natural disaster; and (iii) postdisaster recovery, which establishes a systematic methodology for electric service restoration and infrastructure recovery exploiting microgrid technologies and utilizing a comprehensive set of flexibility resources.

To help readers have a better understanding of what we have done, we would like to provide a short review of the 10 chapters as follows:

Chapter 1 provides an overview on several critical issues of power grid resilience against natural disasters. The definition of power grid resilience is discussed. The benefits and challenges of improving grid resilience are elaborated on. Different stages involved in power grid resilience enhancement, i.e. before, during, and after a natural disaster, are also discussed. The importance of coordinating and co-optimizing measures of different stages and flexibility resources of different kinds is highlighted.

Chapter 2 explores a dynamic model for condition-based maintenance strategies considering harsh external conditions. A Markov process is used to describe the physical characteristics of component deterioration, and the effects of harsh external conditions are represented as probabilistic models. A cost-to-go model, including the maintenance cost and the system reliability cost, is constructed to optimize the maintenance strategies to enhance power grid resilience.

Chapter 3 uses truck-mounted mobile emergency generators (MEGs) as an example of emergency resources and proposes scheduling MEGs as distributed generations in distribution systems to recover critical loads by temporarily forming microgrids. A two-stage dispatch approach consisting of preallocation and real-time allocation is presented. Particularly, preallocation is conducted via a two-stage scenario-based stochastic optimization problem, in which the first-stage preallocation decisions are assessed by a number of second-stage real-time allocation subproblems associated with the considered scenarios.

Chapter 4 presents a new approach to grid automation planning that allocates remote-controlled switches (RCSs) to improve the restoration performance and optimize resilience and reliability benefits with an appropriate cost. Particularly, the optimal number and locations of to-be-upgraded switches can be determined by the approach for different objectives: (i) To maximize the reduction in customer interruption cost; (ii) To maximize the reduction in system average interruption duration index; or (iii) To maximize loads that can be recovered by the grid automation system with upgraded RCSs.

Chapter 5 studies the sequential steady-state security region (SSSR). SSSR-based resilience enhancement of power transmission systems, considering uncertain time-varying topology changes due to the extreme weather event, is investigated. In consideration of uncertain time-varying topology changes with SSSR, the resilience enhancement problem is constructed as a bilevel optimization model, which can be utilized by grid operators to find an optimal strategy against the most threatening weather-related scenario.

Chapter 6 proposes a proactive operation strategy to enhance power transmission system resilience during an unfolding extreme event. A Markov process model is used to represent the uncertain sequential transition of system states driven by the evolution of the extreme event. Transition probabilities depend on failure rates caused by the extreme event. For each state, a recursive value function with a current cost and a future cost is modeled in consideration of intertemporal constraints and operation constraints.

Chapter 7 proposes a state-based decision-making model to enhance distribution grid resilience throughout an unfolding event. The time-varying system topologies are modeled as Markov states, and the probabilities between different Markov states depend on the component failures caused by the unfolding event. A Markov decision process-based model is proposed to make state-based actions, i.e. system reconfiguration, at each decision time. An approximate dynamic programming method is employed to optimize the proposed model.

Chapter 8 presents a new method for formulating radiality constraints to fully include topological and related flexibilities in reconfiguration-related microgrid and distribution grid optimization problems, whose feasibility and optimality thus are extended and enhanced, respectively. The method's theoretical validity is certified by graph–theoretic analyses. As it involves integer variables, this chapter further analyzes the tightness and compactness issues. The new radiality constraints are particularly used for postdisaster microgrid formation that is involved in many resilience-enhancing service restoration and infrastructure recovery problems.

Chapter 9 presents a two-stage framework for resilient routing and scheduling of mobile power sources (MPSs). In the first stage, i.e. prior to the natural disaster, MPSs are pre-positioned in the distribution grid to enable prompt prerestoration of electric service. The first-stage decisions are obtained by a two-stage robust optimization model. In the second stage, i.e. after the disaster, MPSs are dynamically dispatched in the distribution grid to cooperate with conventional service recovery efforts, in order to improve system recovery. A novel mixed-integer programming model is formulated.

Chapter 10 proposes a resilient approach for disaster recovery logistics co-optimizing distribution grid restoration with the dispatch of MPSs and repair crews (RCs). A novel co-optimization model is constructed to coordinately route MPSs and RCs in the transportation network, schedule them in the distribution grid, and reconfigure the distribution grid for forming microgrids. Resilient recovery strategies therefore are obtained to improve service restoration, especially by dynamically forming microgrids that are energized by MPSs and topologized by network reconfiguration of the distribution grid and repairing actions of RCs.

Prerequisites and Usage of This Book

The prerequisite knowledge to read this book is an undergraduate level of understanding of power grid analysis and operation, basic knowledge of optimization methods including the Markov decision process, graph theory, and probability theory.

This book may be of interest to senior undergraduate students, postgraduate students, researchers from academia (in the fields of engineering, operations research, etc.), and specialists, engineers, planners, and operators from industry. We also hope that this book will be useful as a reference for advanced courses related to power and energy, industrial engineering, and operations research.

Summary

To summarize, this book provides a comprehensive and systematic introduction to resilience-enhancing planning and operation strategies of power grids against natural disasters. It covers three closely related important problems, i.e. preparedness prior to an extreme event, response as the event unfolds, and postevent recovery. State-of-the-art methods are introduced and illustrated in detail with examples for the readers to learn how to use them to address realistic problems and improve today's practices.

Shunbo Lei
Shenzhen, China

Chong Wang
Nanjing, China

Yunhe Hou
Hong Kong SAR, China

August 2022

Acknowledgments

This book summarizes some of our research on power grid resilience in recent years. Many people contributed to this book in various ways. We would like to thank Prof. Jianhui Wang from Southern Methodist University, Prof. Ping Ju and Prof. Feng Wu from Hohai University, Prof. Chen Chen from Xi'an Jiaotong University, Prof. Zhaoyu Wang from Iowa State University, Dr. Hui Zhou from the State Grid Zhejiang Electric Power Company, Prof. Tao Liu and Dr. Yue Song from the University of Hong Kong, Dr. Yupeng Li from Hong Kong Baptist University, for their contributions to some research presented in this book.

We also would like to thank Prof. Felix F. Wu from the University of California, Berkeley, Prof. David J. Hill from the University of New South Wales, Sydney, Prof. Ian A. Hiskens and Prof. Johanna L. Mathieu from the University of Michigan-Ann Arbor, Prof. Kai Strunz from the Technical University of Berlin, Prof. Mingbo Liu from South China University of Technology, Dr. Feng Qiu from Argonne National Laboratory, Prof. Feng Liu and Prof. Wei Wei from Tsinghua University, Dr. Miao Miao from the Department of Development and Planning, State Grid Qinghai Electric Power Company, Prof. Wei Sun from the University of Central Florida, Prof. Liang Liang from Harbin Institute of Technology (Shenzhen), and Dr. Chaoyi Peng from China Southern Power Grid Company, for valuable discussions and suggestions when conducting some of these research.

We thank Ms. Chenxi Hu and Mr. Qinfei Long for contributing some materials to Chapter 1 of this book. We also thank Dr. Wenqian Yin, Dr. Yujia Lauren Li and Ms. Yixuan Chen from the University of Hong Kong, Ms. Mingze Xu, Mr. Haoran Liu, Mr. Cheng Ma, Mr. Fengqi Lyu, Mr. Weimin Wu, Mr. Aichao Zhang, and Mr. Haochen Song from the Chinese University of Hong Kong, Shenzhen, for pointing out typos and checking the whole book.

In addition, we acknowledge the innovative research and practice contributed by others in this increasingly important field, and appreciate the staff at Wiley and IEEE Press for their assistance and help in preparing this book.

xxiv *Acknowledgments*

This book is supported in part by the National Natural Science Foundation of China (NSFC) under Grant 52177118 and Grant 51907050, in part by the Joint Research Fund in Smart Grid under cooperative agreement between the NSFC and State Grid Corporation of China under Grant U1966601, in part by the Research Grants Council of Hong Kong under Grant GRF17207818, in part by the University Development Fund (No. 01002140) of the Chinese University of Hong Kong, Shenzhen, and in part by the Shenzhen Institute of Artificial Intelligence and Robotics for Society. The authors really appreciate their support.

Shunbo Lei
Chong Wang
Yunhe Hou

Part I

Introduction

1

Introduction

1.1 Power Grid and Natural Disasters

The electric power grid is the largest and most complex machine in the world. It is employed to supply, transfer, and utilize electric power. Its history can be traced back to 1881, when the world's first power system was built at Godalming in England. In the twentieth century, electricity had gradually become one of the basic necessities in the modern society. The power grid has been performing effectively in satisfying the energy need and, meanwhile, has been causing adverse impacts on the natural environment. The associated carbon emissions also contribute to the climate change that has been causing more frequent natural disasters.

In Asia and the Pacific, the primary energy demand is estimated to have more than 2.4% increase each year by 2030, while typically the electricity demand has a higher increase, at about 3.4%. For the increased electricity demand, it is desired to have an efficient and reliable power supply. In recent years, the economy, society, and environment have also introduced new pressures or requirements on power grids, e.g. shifting from centralized to decentralized structures. Hence, long-term sustainability of power grids is of critical importance in developing the twenty-first century power grids, i.e. smart grids.

A power grid usually covers a wide geographical region, and many of its components in the system are exposed to the external environment, which makes the power grid vulnerable to natural disasters, e.g. wind storms, ice storms, thunderstorms, earthquakes, wildfires, hurricanes, and flooding [1–3]. In recent years, more frequent natural disasters have resulted in severe power outages that are large-scale and long-duration. For example after the Hurricane Sandy struck the East Coast of the United States in 2012, approximately 8.35 million customers were

Power Grid Resilience against Natural Disasters: Preparedness, Response, and Recovery, First Edition.
Shunbo Lei, Chong Wang, and Yunhe Hou.
© 2023 John Wiley & Sons Ltd. Published 2023 by John Wiley & Sons Ltd.

reported without power [4]. Some studies have indicated that the climate change leads to the increase in disastrous events. The global temperature rise has been considered as one of the important underlying causes of disastrous events with higher intensity and frequency [4, 5].

Disastrous event-related power outages have introduced tremendous economic losses and significant life risks, highlighting the importance of enhancing power grid resilience [6], which generally refers to the ability to withstand and rapidly recover from disruptive events [7, 8]. A natural disaster can inflict widespread and severe damages to the power grid, leaving numerous customers without power for days, sometimes even for over a week. One of the critical requirements on resilient power grids is that the system can effectively prepare for, response to and recover from natural disasters, as most social activities greatly depend on the reliable power supply [9].

Aside from external natural disasters (e.g. weather-related events), the cyber systems, which enable system operators to efficiently monitor and control the power grid, make the system vulnerable to cyber intrusions. Conventional strategies, e.g. common preventive and emergency measures, due to their little consideration of weather-related and cybersecurity-related events, fail to be resilient preparedness, response, or recovery strategies. Considering potential weather-related events attacking the physical system and cybersecurity-related events attaching the cyber system, resilient power grids have to be constructed.

In this chapter, the definition and importance of power grid resilience will be introduced. Then, challenges brought by different kinds of events that may jeopardize resilient power grid operation will be discussed, and the corresponding resilience enhancement strategies will also be discussed.

1.2 Power Grid Resilience

1.2.1 Definitions

As mentioned above, sustainable power grids have to balance economic growth and social progress, meanwhile, preserving the natural environment [10]. With more frequent severe power outages caused by natural disasters, power grid resilience is receiving much attention. Resilience can be generally understood as the ability of power grids to avoid or reduce failures and to recover quickly after failure occurrence [11]. Currently, there is not a unified definition of power grid resilience. Several definitions given by different organizations are shown as follows:

- In [12], the U.S. National Academies of Sciences, Engineering, and Medicine define resilience as "the ability to prepare and plan for, absorb, recover from and more successfully adapt to adverse events."

- In [13], the Cabinet Office of the United Kingdom refers to resilience as "the ability of assets, networks and systems to anticipate, absorb, adapt to and/or rapidly recover from a disruptive event."
- In [14], the U.S. President's National Infrastructure Advisory Council specifies resilience as "the ability to anticipate, absorb, adapt to, and/or rapidly recover from a potentially disruptive event."
- In [15], the IEEE Power and Energy Society Industry Technical Support Task Force prescribes resilience as "the ability to withstand and reduce the magnitude and/or duration of disruptive events, which includes the capability to anticipate, absorb, adapt to, and/or rapidly recover from such an event."
- In [8], the U.S. Electric Power Research Institute states that "grid resilience includes hardening, advanced capabilities, and recovery/reconstitution."

Although the resilience definitions given by different organizations are different, the key understanding that decision-makers should plan for, ride through, and recover from each potential disastrous event is consistent. According to the goals of enhancing power grid resilience, a sequence of resilience merits, including robustness, resourcefulness, rapid recovery, and adaptability, have been highlighted [16, 17]:

- *Robustness prior to an event.* This necessitates that the grid is capable of remaining to stand and operate in the face of extreme events. For instance, hardening a grid's critical structure, from a mid- or short-term perspective, can be performed to guarantee a strong system prior to an event. In addition, to ensure the robustness, investments, and maintenance scheduling of critical electric devices, from a long- or mid-term perspective, can be entailed prior to weather-related and cybersecurity-related disastrous events.
- *Resourcefulness during an event.* This requires skillful abilities to manage the power grid when a disruptive event unfolds. Effective and real-time strategies are expected to be implemented to mitigate the negative impacts. For instance, determining what should be conducted to mitigate the damages is a critical issue during a disastrous event. Furthermore, adequate resources for communications among different decision-makers are important in implementing established mitigation strategies.
- *Rapid recovery after an event.* This demands the ability to recover the power grid back to a normal state quickly after adverse events. For example, detailed recovery plans under the conditions of various blackouts should be established in time and adequate resources for implementing the recovery strategies should be guaranteed.
- *Adaptability to future events.* This denotes the ability to absorb new lessons from past events and generalize to new situations. Instead of case-by-case methods, more flexible measures and strategies are necessary to be fitted into various situations as well as to improve the power grid's capability of dealing with extreme events.

1.2.2 Importance and Benefits

As power grids are critical infrastructures for social and economic development [12], a power outage might cause severe consequences. The U.S. National Research Council [12] and the U.K. House of Lords [18] have emphasized the importance of resilient power and energy infrastructures. The North American Electric Reliability Corporation [14, 19, 20] and the U.S. Electric Power Research Institute [21] have further recognized the functionalities of power grid resilience. In general, enhancing power grid resilience can improve economic, social, and environmental sustainability.

Note that a reliable power grid is not necessarily resilient. Specifically, power grid reliability guarantees its operation under normal-state conditions, or in high-probability, low-impact events. On the other hand, a resilient power grid is capable of performing well in low-probability, high-impact events such as natural disasters [12]. Many resilience-oriented power grid planning and operation strategies have been proposed and studied in the literature. Nevertheless, many problems, including distribution grid automation to enhance the restoration capability and the utilization of mobile generation resources which involve the consideration of road networks, have not been addressed.

In the following, the importance and benefits of power grid resilience are briefly discussed from three perspectives.

1.2.2.1 Dealing with Weather-Related Disastrous Events

As mentioned above, electric power grids have the characteristic of wide geographical coverage, making grid components exposed to extreme weather events such as tornadoes, typhoons, windstorms, hurricanes, and blizzards. These disastrous weather events are major causes of power outages. In addition, the aging nature of electric devices also makes power grids more susceptible to extreme weather events. For instance, about 679 power outages were caused by weather events from 2003 to 2012 in the United States, and each event affected at least 50 000 customers [8].

Table 1.1 shows the number of blackouts between 1984 and 2006 in the United States [22]. As indicated, around 44% of the outage events were weather-related. Based on the analysis of the U.S. President's Executive Office, weather-related outages lead to about $25 billion in economic losses annually. In addition, weather-related outages have an increasing trend [8].

To sum up, outages induced by weather-related disastrous events happen more frequently all around the world, causing significant economic and safety damages to the human society. With the advancement of power grids, this type of extreme weather will incur greater risks to various security issues, making it an urgent task for each country to improve the power grid resilience against such disastrous events. As aforementioned, improving power grid resilience in terms of

Table 1.1 Outages in the United States for 1984–2006.

Cause	Percentage of events	Mean size in MW	Mean size in customers
Earthquake	0.8	1408	3 75 900
Hurricane/tropical storm	4.2	1309	7 82 695
Lightning	11.3	270	70 944
Wind/rain	14.8	793	1 85 199
Ice storm	5	1152	3 43 448
Tornado	2.8	367	1 15 439
Other cold weather	5.5	542	1 50 255
Fire	5.2	431	1 11 244
Intentional attack	1.6	340	24 572
Supply shortage	5.3	341	1 38 957
Other external causes	4.8	710	2 46 071
Equipment failure	29.7	379	57 140
Operator error	10.1	489	1 05 322
Voltage reduction	7.7	153	2 12 900
Volunteer reduction	5.9	190	1 34 543

the handling of weather-related disastrous events can significantly enhance the economic, social, and environmental sustainability.

1.2.2.2 Facilitating the Integration of Renewable Energy Sources

Renewable energy sources (RESs), including but not limited to wind turbines and photovoltaic panels, have been integrated into power grids worldwide at an increasing rate. Their effects in relieving the energy crisis concern are promising. However, the variable and uncertain natures of RESs have introduced new challenges to power grid planning and operation and brought about new issues to power grid resilience. Conventionally, the power grid consisted of controllable generators and semipredictable electric power demands. Thus, power grid operators could just adjust generation sources to accommodate admissible deviations of power demands. Now with the growing RES integration, such an operation paradigm becomes ineffective. New operating strategies utilizing smart grid technologies, e.g. advanced optimization methods, are needed, so that power grid resilience regarding the integration and utilization of RESs can be improved.

To generate operation strategies with both robustness and economy, stochastic optimization and robust optimization have been applied to cope with the uncertainty of wind power. With probability distribution information of uncertain parameters, stochastic optimization is a mature methodology to provide decisions against uncertainties. To name a few, stochastic unit commitment and stochastic economic dispatch models were proposed in [23, 24]. Nevertheless, an accurate probability distribution of unknown parameters can be quite difficult to identify [25]. In this regard, robust optimization is a promising alternative receiving much attention in recent years, partially because it does not require accurate distribution information of uncertainties. It evaluates the worst-case performance of a decision, resulting in an optimization solution that is robust against any possible scenario in the uncertainty set.

However, to build resilient power grids with high penetration of RESs, many critical problems, e.g. restoration of transmission grids with large-scale RESs, have not yet been addressed.

In general, RESs can empower more flexible strategies for improving power grid resilience, but we must also consider the impact of RESs' characteristics on the power grid. Reaching a fair balance between RES integration and resilience improvement in power grid is critical for relevant research. In fact, only with resilient strategies that sufficiently mitigate the adverse impacts of RESs, the economic, social, and environmental benefits of RESs can be fully utilized.

1.2.2.3 Dealing with Cybersecurity-Related Events

The former two parts are mainly about security and resilience of the physical system of a power grid. Apart from that, security of the cyber system is also an important part of power grid resilience. The reliable operation of modern power grids is fundamentally supported by cyber systems. The devices that monitor and control power grids are typical information and communications technologies-based systems. Those systems face the threats of cyber-attacks, which can undermine the control systems and endanger the secure operation of power grids.

In recent years, cyber-attacks have resulted in many security problems. The U.S. National Security Agency has reported that there have been some cyber intrusions to critical infrastructures and has emphasized the importance and benefits of improving the resilience of cyber systems. The following shows several reported incidents of cyber intrusions into existing information and communications technologies-based systems [26]:

- *BlackEnergy.* It was first reported in 2007, with critical energy infrastructures being its targets. In 2014, several information and communications technologies-based systems were infected by BlackEnergy. With BlackEnergy, cyber-attackers can deliver some plug-in modules for audio recording, keylogging, and grabbing screenshots, etc.
- *HAVEX.* For the early version of HAVEX, it was distributed through spear-phishing attacks or spam e-mails. After several revisions, HAVEX

has become a Trojan horse used to modify "legitimate" software in information and communications technologies-based systems including the supervisory control and data acquisition (SCADA) system, and supervisory control and data acquisition (SCADA) systems by adding additional instructions to codes.

- *Sandworm.* It is a Trojan horse, which is used to deliver malware on thumb drives.

As indicated by the aforementioned events, cyber-attacks cannot be neglected. With the development of cyber networks, cyber security of power grids faces new threats from attacks that are very stealthy and less expensive. Therefore, the cyber security in power grids is currently a hot topic in resilience research. For example in [27], a novel criterion for assessing the resilience of power supply to data centers was proposed to evaluate the system's capability of sustaining functionality during an outage. Enhancing the cyber system resilience of power grids not only mitigates the threats from cyber-attacks but also enables the cyber system to more effectively and more efficiently monitor and control power grids.

1.2.3 Challenges

Conventional strategies, e.g. common normal state-based preventive and emergency strategies in the planning and operation timescales, fail to consist of resilient measures against disastrous events, as they have little consideration of such low-probability, high-impact events either weather-related or cybersecurity-related. Since power grid resilience requires robustness, resourcefulness, rapid recovery, and adaptability at different stages, appropriate and sophisticated strategies of different stages should be implemented [28]. Furthermore, complicated characteristics of modern power grids, from the perspectives of the source, network, storage, and load, pose great challenges to the construction of resilient power grids.

In the following, some challenges to improving power grid resilience are briefly discussed:

- *Component reliability enhancement.* Electric power companies try to maximize their profits and maintain a good system technical behavior from the reliability perspective. As a weather-related disastrous event unfolds, the critical devices on the trajectory of the event are expected to have high reliability, which can reduce the probability of being in failure and therefore ensure system resilience [29]. In this regard, component reliability enhancement, e.g. maintenance scheduling in consideration of potential extreme events, is needed to guarantee good conditions of electric devices. Establishing a comprehensive maintenance plan considering potential natural disasters is not an easy task. The following challenges should be considered: First, weather-related events occur with uncertainties, and the influences of those events on the deterioration of each component are stochastic. Second, the dimension of enormous electric devices in power grids results in large-scale optimization problems. For short-term maintenance scheduling, many system operation constraints, including the $N-1$ security, ramping rates of generating units, power balance, and spinning reserve capacity requirements, should also be considered. The relevant optimization models are usually computationally intractable with large-scale systems [30]. Third, considering a fast growth of renewable energy in power grids, the influences of uncertainties of renewables on maintenance scheduling considering weather-related extreme events are critical issues. To tackle those problems, proactive operation strategies to enhance system resilience considering the uncertain sequential transitions of states are needed.

- *System state acquisition.* System state acquisition is a prerequisite for power grid operators to perform resilient, proactive, and emergency strategies, before the extreme event, during its unfolding, and after its occurrence [31]. Facing extreme events, a system is under continuous and severe disturbances. In this regard, the dynamic states of the system should be identified. Measurements from phasor measurement units (PMUs) are rapidly updated and can be employed to perform dynamic state estimation [32]. Performing dynamic state estimation should consider the following challenges. First, multiple control areas, resulting from the power industry's deregulation, should be considered. These interdependent areas usually do not share all information of their own control areas. Second, appropriate approaches for estimating dynamic states are needed. Currently, some approaches can be used for estimating dynamic states with the assumption of Gaussian distributions of measurements' noises [33, 34]. However, some measurements' noises are not satisfied with the assumption of Gaussian distributions. Third, a rapid computation speed for dynamic state estimation is required.

- *Multiple energy systems.* It is expected that power grids have more contributions to the sustainability and low-carbon development of energy sectors. In China, carbon neutrality is an essentially important target, and a series of policy measures have been implemented to reduce greenhouse gas emissions. By means of integrating different energy sources across different pathways, the multi-energy system provides a promising way to reduce carbon emissions. However, the multi-energy system has more complicated characteristics compared to traditional power grids, and in consequence needs sufficient and novel measures to guarantee high system resilience. In 2021, one severe power outage occurred in Texas, partly due to inappropriate operation of the multi-energy system in the face of extreme cold weather. In fact, different energy carriers of the multienergy system have different responses to disturbances, and this leads to difficulties in constructing coordinated strategies. Therefore, it is a challenge for multiple energy systems to construct systematic frameworks and techniques that enhance system resilience.

- *Renewable energy uncertainty.* With increasing concerns on possible energy shortage, worldwide efforts have been carried out to integrate enormous RESs into power grids [35]. Among different kinds of renewable energy, wind power attains the highest penetration in some countries, partially owing to the relatively mature wind turbine technologies. Although wind power is promising in easing the worries over the energy crisis, its variability and uncertainty have brought about great challenges to the reliable and economic operation of power grids. Other kinds of renewable energy also possess variability and uncertainty features, which also impose similar challenges and worsen relevant problems. Two critical issues need to be considered when a power grid is under the threat of a natural disaster, especially an extreme weather event. The first one is that the extreme weather event has a great impact on renewable generation, and the second one is that the grid in the face of the event may not be as strong as that under normal conditions. These two critical issues make conventional strategies improper to the resilient operation of high-renewable power grids against extreme weather events. In addition, RESs such as photovoltaic power and wind power are usually connected to the grid via power electronics devices with low inertia, which reduces the grid strength against disturbances. The abovementioned factors pose great challenges to resilient high-renewable grid operation in the face of natural disasters.
- *Cyber-physical systems.* Many research studies focus on power grid resilience from the perspective of physical systems. However, information and communications technologies have been playing important roles in physical system monitoring and control and have driven the conventional power system into the cyber-physical system. Many components in the cyber-physical system are directly exposed to external environment, and they are both vulnerable to natural disasters (including extreme weather events) from the perspectives of the security and resilience of information networks and physical networks. If a cyber-physical system is affected by an extreme weather event, unavailability of some parts of the cyber system might result in incomplete information, which in consequence can lead to failed state estimation and large control errors. Therefore, how to analyze and mitigate the impacts of disastrous events on cyber-physical systems is a critical challenge with regard to power grid resilience improvement.
- *Multi-area networks.* The development of power markets, multiple energy systems, and system expansion, etc. have been driving power grids into multi-area interconnected systems. Conventional centralized strategies usually cannot be directly employed for multi-area interconnected systems when considering specific jurisdictional mandates, extensive communication burdens,

and information privacy, etc. Boundary restrictions between different energy systems complicate the interaction problems and coordinated strategies and result in challenges to the improvement of system resilience.

Note that other than the abovementioned issues, many other challenges need to be addressed in building resilient power grids. Relevant discussions are included in relevant parts, where appropriate, of this book.

1.3 Resilience Enhancement Against Disasters

A power grid resides in different stages when it is exposed to natural disasters including extreme weather events. It is necessary to define these stages to enable systematic enhancements of power grid resilience against these events. In October 2010, the U.S. President's National Infrastructure Advisory Council released a report presenting a resilience structure with four features, based on the sequence of "prior to an event," "during an event," "after an event," and "postincident learning," respectively [14, 16, 17]. Building effective resilience enhancement strategies requires an understanding of preventive, real-time and recovery strategies, and an awareness of how the related actions impact grid planning and operation [36].

Specifically, prior to an event, "robustness" requires the grid to stay standing or keep operating in the face of the event that can be catastrophic. Strengthening and hardening the system is one of the acceptable preventive strategies. In general, maintaining and investing in critical infrastructure elements can improve grid robustness so that the system can withstand those extreme events. During an event, "resourcefulness" requires that the grid has sufficient capabilities of managing the event as it unfolds. In this stage, it is necessary to identify available strategies and prioritize what can be implemented to mitigate the impact of damages caused by the event. After an event, it is desired that we can get the system back to its normal state as quickly as possible. To this end, determining emergency and restoration approaches is important, and scheduling appropriate resources and right people to right places is also critical [37, 38].

1.3.1 Preparedness Prior to Disasters

Prior to natural disasters, assessments and preventive strategies need to be implemented to improve the power grid's capability of dealing with the events, such as state assessments using historical data-based models [39–41], and resilience enhancement strategies based on distributed generation (DG) allocation [42], microgrid technologies [43, 44], and switch placement [45]. Overall, methods of enhancing power grid resilience can be divided into component level and system level. A brief discussion is provided in the following:

1.3.1.1 Component-Level Resilience Enhancement

Good conditions of electrical devices are important for power grid resilience enhancement. Usually, maintenance activities are employed to mitigate the deterioration of grid components. However, such activities often increase the total operating cost of grids. To achieve an appropriate trade-off between grid resilience and operating cost, system operators need to develop a series of combined long-, mid-, and short-term maintenance activities for various components in the power grid.

Currently, different categories of maintenance scheduling methods, e.g. planned maintenance scheduling, condition-based maintenance scheduling, reliability-centered maintenance scheduling, and optimization-based maintenance scheduling, can be implemented to improve grid resilience:

- *Planned maintenance scheduling.* In planned maintenance scheduling, each maintenance event is prescheduled, and all future maintenance events are preprogrammed. Different maintenance events implemented on different devices are programmed separately according to legislation or manufacturer recommendations. Its advantages include the easiness of scheduling and programming maintenance events, evenly distributed costs, and low costs of instruments used for supervision of devices. It surely also has disadvantages, such as requirements for ongoing labor costs and training investment, and very expensive operation because of frequent changes of parts.

- *Condition-based maintenance scheduling.* Condition-based maintenance scheduling means that maintenance tasks are only performed when it is necessary according to equipment conditions [46]. One prerequisite is having the capability of monitoring the equipment's health, i.e. condition monitoring. The developments in sensor technologies, signal processing, and online diagnosis empower system operators to have access to more accurate conditions of grid components, e.g. generating units [47] and transformers. Among current research, the Markov model-based methodology is a promising tool for improving condition-based maintenance scheduling. Existing studies have shown that Markov models are useful for establishing effective and efficient maintenance scheduling [48–56].

- *Reliability-centered maintenance scheduling.* Reliability-centered maintenance scheduling is a well-organized method by which the maintenance processes aim to improve system reliability. It was introduced into the power engineering field after its successful deployment in the aerospace and aircraft industry in the 1960s [57]. Relevant studies focused on voltage regulators [58], circuit breakers [59], overhead lines [60], underground systems [61], and power transformers [62]. The advantages of reliability-centered maintenance scheduling is that it has the ability to minimize the frequency of overhauls and increase the reliability of system components [63, 64].

- *Optimization-based maintenance scheduling.* Optimization-based maintenance scheduling refers to methods that formulate the maintenance scheduling problem as an optimization model and solve it by optimization algorithms. For example, using integer programming or mixed-integer programming methods, the statuses of electric devices are modeled as binary decision variables. When a device is scheduled for maintenance, the value of its associated variable regarding the device status is 0; otherwise, the value is 1. As the operating conditions of power grids should be satisfied under maintenance events, security-constrained optimal maintenance scheduling, associated with unit commitment constraints and $N - 1$ security constraints, should be considered [65, 66].

1.3.1.2 System-Level Resilience Enhancement

Apart from component-level resilience enhancement, the system-level improvement of grid resilience is also essential. Prior to disastrous events, outage evaluation and proactive preparedness are the main methods to make power grids less susceptible to damages caused by disasters.

With sufficiently accurate outage predictions, power grid operators can pre-allocate repair personnel and components for the following restoration after the disaster. Other proactive strategies can also be implemented based on available predictions. Current studies regarding outage predictions mainly focus on data-based statistical models [39–41, 67, 68]. Though performing well in some specific service areas, they may suffer from poor scalability and generalizability.

Regarding proactive strategies, the extension and hardening of power grids, such as adding DGs, adding redundant lines and hardening existing electric devices [42], are effective approaches recommended by the U.S. Department of Energy. Decision-makers also have to determine how to allocate and mobilize available resources to guarantee timely responses and recovery efforts against possible damages [69–71], so as to improve the grid resilience.

1.3.2 Response as Disasters Unfold

During disastrous events, power grid operators need to monitor the condition of each system component and the overall system's operating state, schedule repair groups to secure critical facilities, and dispatch existing system components such as generators, relay protection devices, and transformers to maintain voltage and frequency balances. Specifically, some real-time strategies are expected to be performed to mitigate the negative impacts of disaster-induced damages. Relevant research includes state acquisition [72], controlled separation [73], and microgrid sectionalization [74], etc. The benefits of these measures are noticeable. That is, they lower the blackout risk while maintaining regular system performance. A brief discussion on major measures is provided in the following.

1.3.2.1 System State Acquisition

Acquiring states of the power grid, by state estimation, etc. is a prerequisite for monitoring and controlling the system. Without it, responsive strategies as the extreme event unfolds cannot be established during the event. In general, measurements such as transmission line power flow in the SCADA system have slow update rates, and conventional state estimation methods based on steady-state models of a system can only attain the static states. On the other hand, measurements from PMUs are rapidly updated and can be employed to perform dynamic state estimation. In addition, to achieve smart features such as demand response in smart grids, more accurate models and more rapid algorithms are critically needed for state estimation of distribution grids. Selected categories of state estimation are briefly discussed as follows:

- *Static state estimation.* Static state estimation is mainly to determine the system's steady-state status, e.g. bus voltages, based on measurements at each short-term time slot. Specifically, it translates telemetered data into a reliable estimate of transmission and distribution network topology and status. For relevant analyses, lots of algorithms have been designed for solving static state estimation problems, such as the weighted least squares algorithm [75], parallel computation algorithm [76], and interior-point based algorithm [77], etc. However, natural disasters including extreme weather events are considered as severe disturbances where static state estimation methods might only have quite limited usefulness and effectiveness.
- *Dynamic state estimation.* With increasing disturbances in the power grid, e.g. from wind power and solar power, dynamic characteristics of the system are crucial for operators to implement control actions and ensure system stability. As PMU measurements are updated rapidly, their measurements can be employed for dynamic state estimation, concerned with the tracking of system states with more often and fast changes. The extended Kalman filter is one of the effective methods for conducting dynamic state estimation [78–80]. Based on a relevant IEEE standard [81], PMU can provide measurements at the rate of up to 120 samples per second, while in practice, the rates of PMU measurements may be much lower, e.g. 30 samples per second. To estimate dynamics accurately under the condition of lower measurement rates, algorithms based on the extended particle filter [82–84] were proposed, which can improve the robustness of dynamic state estimation.

- *Multi-area state estimation.* Multi-area state estimation primarily tackles the issue of performing efficient state estimation on large power grids. One objective is to decrease the computation time, which helps take advantage of real-time measurements gathered within multiple areas across the system. Specifically, multi-area state estimation is based on a selected type of decomposition-coordination framework, which takes use of weaker geographical or measurement connections across regions, in conjunction with well-established solution approaches. With more PMUs in power grids, there have been many research studies focusing on multiarea state estimation [85, 86]. In general, there are two computation architectures for multiarea state estimation, i.e. the hierarchical architecture [72, 87] and the decentralized architecture [88, 89].

1.3.2.2 Controlled Separation

Currently, power girds in near areas are often connected together to construct an interconnected power system, which ensures efficiency, enhances reliability, and improves resilience against some issues. However, under the condition of some rare circumstances, e.g. extreme weather events and man-made incorrect operations, different groups of generators in an interconnected system may be unsynchronized. If emergency controls [90, 91] cannot terminate the unsynchronization among different groups of generators, the controlled separation strategy, i.e. disconnecting some lines to divide an interconnected system into several separated grids, might need to be performed to prevent cascading failures and blackouts.

For an interconnected power system, coherent groups of generators usually exist [92–94]. Generators in a coherent group are usually synchronized, and these generators should be included in one separated system. The grouping of generators after large disturbances can be considered as an NP-hard partition problem from the perspective of mathematics. Many methods, e.g. geometric methods [95–97], combinatorial methods [98, 99], and spectral methods [100, 101], can be used.

Then for controlled separation, various strategies have been proposed to solve this power system partitioning problem. When selecting the optimal splitting strategy, the minimum load-generation imbalance is a critical criterion, which partly determines the splitting lines. This criterion aims to ensure that the frequency of each separated system is within the acceptable range. Most of the relevant research studies have not considered the constraints of reactive power balance. Reactive power determines the voltage profile, and its imbalance may result in voltage instability [102]. For example insufficient reactive power in the Idaho area of the United States on 2 July 1996 resulted in a blackout [103]. Therefore, it is necessary to include both real power balance and reactive power balance when establishing a splitting strategy.

1.3.3   Recovery After Disasters

Even though many preventive preparedness and proactive response actions can be performed prior to and during a disastrous event, it is generally impossible to completely avoid outages. When outages occur after the event, it is necessary to recover the electric service as quickly as possible to improve the system resilience.

The recovery of a power grid after a partial or total failure is a complicated procedure. Many aspects must be considered, including the system's operational condition, equipment availability, and restoration time, etc. A conventional power grid recovery plan usually includes three stages, i.e. preparation, system restoration, and load restoration [104–106] (especially for the transmission grid). However, there are unique characteristics associated with outages caused by natural disasters, leading to different requirements on the recovery strategies. New approaches, such as microgrid-based restoration strategies [107] and decentralized restoration schemes [108], are needed. Moreover, power grid recovery is a multi-objective, multi-stage, multi-variable and multi-constraint optimization problem with nonlinearity and uncertainty. Effective and efficient algorithms are needed to find the optimal recovery plan.

In the following, several issues on postdisaster grid recovery are briefly discussed.

1.3.3.1   Conventional Recovery Process

How to quickly recover a power grid following a partial or complete blackout is a significantly challenging task. There have already been many research studies focusing on conventional power grid recovery methods. As mentioned above, a typical power grid recovery process includes three stages, i.e. preparation, system restoration, and load restoration:

- *Preparation stage.* The preparation stage is mainly to assess the grid's state and identify critical loads and initial cranking sources.
- *System restoration stage.* In the system restoration stage, the main objective is to establish a strong bulk power network by restarting appropriate blackstart and nonblackstart generating units associated with appropriate transmission lines and some critical loads [105, 106].
- *Load restoration stage.* In the load restoration stage, the main objective is to restore as many loads as possible. This stage is performed after a sufficiently strong bulk power network is established to maintain the system frequency and voltage profiles. Many approaches, e.g. expert systems [109, 110], fuzzy logic [111, 112], heuristic approaches [113, 114], and mathematical programming [115, 116], have been employed for decision-making in load restoration.

Note that the above recovery approaches are typically adopted by transmission grids. However, the effectiveness of conventional restoration approaches for post-disaster recovery might be limited. More proactive preparedness measures prior to the extreme event are needed. Moreover, as the grid might have been separated into subgrids by the damages induced by the disaster, the nesting of system restoration and load restoration is much more complicated. In general, novel and effective recovery methods need to be developed for enhancing grid resilience against disastrous events.

1.3.3.2 Microgrids for Electric Service Recovery

For integrating distributed energy resources, etc., microgrid technologies have been extensively adopted in power grids. In this regard, using microgrids for electric service recovery is becoming critically important and has received increasing attention.

Reference [117] studied voltage and frequency controls in the blackstart restoration of microgrids. In [118], an algorithm for identifying the automatic switching time was proposed for restoring microgrids. Reference [119] modeled microgrids as virtual feeders and applied spanning tree search algorithms to find the strategy maximizing restored loads and minimizing the number of switching actions. Reference [74] proposed to optimally sectionalize a distribution grid into networked self-adequate microgrids for continuously providing reliable power supply for the maximum loads. In [120], based on the continuous operating time concept, the availability of microgrids for critical load restoration and the service time were evaluated to enhance power grid resilience. More detailed literature reviews can be found in [121] and [120], etc.

The above works assume the microgrids to be installed beforehand, which may not be available at the current stage in many places. Followed by [122] and [123], etc., reference [121] is the first to temporarily form microgrids with DGs to continue supplying critical loads after a disastrous event. This microgrid formation strategy can be further extended as the dynamic microgrid formation strategy, which dynamically changes the boundaries of temporarily formed microgrids according to the system's state changes. Moreover, instead of fixed DGs, mobile power sources can also be applied in microgrid formation for electric service recovery.

As mentioned above, novel power grid recovery approaches including the above ones are needed by systems endangered by natural disasters. Also note that both transmission and distribution grids may adopt the above system recovery methods.

1.3.3.3 Distribution Grid Topology Reconfiguration

Topology reconfiguration (or called network reconfiguration) is an essential measure of power grids for many different objectives. For example, the above microgrid formation-based restoration strategy needs to conduct topology reconfiguration. As distribution grids are typically built with quite smaller levels of redundancy compared with transmission grids, topology reconfiguration is a critically important and must-used strategy in electric service recovery of power distribution systems.

Both static network and dynamic network reconfigurations have been investigated by the research community. For instance, the microgrid formation problem studied in [121–123] is essentially a static network reconfiguration. As for dynamic network reconfiguration of distribution grids, it often relies on the real-time operation of remote-controlled switches. The effects of hourly distribution network reconfiguration on power loss reduction and operation cost minimization were investigated in [120, 124–127], some of which considered DGs. Specifically, reference [124] studied segmented-time reconfiguration coupled with DGs' reactive power control to minimize the distribution grid operation cost. In [127], hourly reconfiguration in the presence of RESs was studied based on mixed-integer, second-order cone programming to minimize daily network losses. In [120], to minimize power losses, a hierarchical decentralized agent-based dynamic network reconfiguration methodology was presented.

Some other studies applied dynamic network reconfiguration to DG integration in a more straightforward manner, i.e. minimizing DG curtailment or maximizing DG penetration. In [128], it is demonstrated that using hourly distribution network reconfiguration can significantly reduce both wind and solar DG curtailments for systems with high DG penetration. In [129], where DG penetration maximization was studied, results showed that distribution network reconfiguration can mitigate the over-voltage problem caused by the increased DG penetration. In [130], with a mixed-integer nonlinear multiperiod optimal power flow model, both static and dynamic reconfigurations were adopted to improve DG hosting capacity, and the results suggested that integrating larger amounts of DGs can be achieved by a small number of line switching actions. To reduce network losses in grid-connected operation and load curtailments in islanded operation, microgrid optimal scheduling with dynamic network reconfiguration was studied in [131]. More detailed literature reviews on distribution grid dynamic network reconfiguration can be found in [130, 132, 133] etc.

In summary, the effectiveness of distribution network dynamic reconfiguration in enhancing distributed renewable energy integration has been studied extensively, mainly from two aspects, i.e. using it to explicitly minimize the curtailment of existing DGs or maximize the penetration of DGs to be accommodated, and using it to mitigate DGs' negative impacts on the considered objectives and constraints.

However, the utilization of more flexible and more adaptive distribution network reconfiguration for postdisaster recovery of distribution grids has received somewhat less attention. For example distribution network reconfiguration greatly relies on the deployment of remote-controlled switches, which are highly expensive, and their massive deployment is not likely in the near future. Therefore, minimizing installation and maintenance costs of remote-controlled switches while achieving certain reconfiguration requirements are an important research topic. It has been studied mostly for the purpose of reliability improvement [134–136], but quite less explored considering the recovery of distribution grids after extreme events.

Also note that distribution grids mostly have to operate in a radial topology. Therefore, radiality constraints need to be considered in most relevant optimization problems. However, existing formulations of radiality constraints might implicitly impose restrictions on the flexibility that can be considered by the optimization models. To deal with such current shortcomings, a new formulation of radiality constraints for reconfiguration-related optimization problems that fully enables the topological and related flexibility of distribution grids is necessary.

1.4 Coordination and Co-Optimization

Coordination and co-optimization have become essential issues for power grid resilience enhancement against natural disasters for many reasons. To name a few, first, the damages caused by disastrous events are typically large-scale and long-duration, so the available resources need to be coordinated and co-optimized in their utilization. Second, different from common single-fault outage scenarios, the outages induced by extreme events can lead to significant economic losses, so more proactive preparedness measures, more timely response actions, and more effective recovery efforts need to be coordinated and co-optimized for mitigating the impacts. Third, a disastrous event can cause damages and outages to different subgrids in an interconnected power system, so the coordination and co-optimization among different areas are essentially necessary.

Selected issues, some aforementioned, regarding the coordination and co-optimization in power grid resilience enhancement against disastrous events, are briefly discussed as follows:

- *Different stages.* As mentioned above, resilience enhancement strategies are implemented in three different stages, i.e. prior to the event, as the event unfolds, and after the event. Coordination and co-optimization among different stages are critically necessary. First, preparedness measures before the event need to consider the flexibility needs of the power grid during and after the event. Second, response actions as the event unfolds might have to take into account the actions' impacts on the system's postdisaster state. Third, some postdisaster recovery efforts might result in permanently installed flexibility resources, etc. Therefore, the effectiveness of those efforts in the prior-event preparedness measures of future disasters needs to be evaluated.

- *Various flexibility resources.* In preparing for, responding to and recovering from a disastrous event, various flexibility resources might be available. Although the power grid operator understands the potentially significant impacts of such extreme events and provides lots of emergency resources, they still need to be coordinated and co-optimized, so that they can be fully utilized. For example, for distribution grid restoration and recovery, the dispatch of flexibility provided by mobile power sources, repair crews, remote-controlled switches, and DGs, etc. needs to be coordinated and co-optimized. Their utilization also needs to be coordinated and co-optimized among different stages mentioned above.

- *Transmission and distribution grids.* Both transmission and distribution grids are endangered by extreme events. That is, disaster-induced damages happen in both of them. Moreover, flexibility resources disperse in both of them. Therefore, resilience enhancement strategies often need to coordinate and co-optimize both levels of power grids. For example, if the transmission grid experiences a partial or complete blackout due to the disaster, cranking power provided by distribution grids can greatly reduce the transmission grid's restoration time. In fact, whether a bottom-up approach, a top-down approach, or a mixed one should be used is an important issue in the postdisaster recovery of transmission and distribution grids.

- *Multiple areas.* As mentioned above, a natural disaster typically impacts a large geographical area covering an interconnected power system. Therefore, in different stages and when using various flexibility resources, multi-area issues need to be dealt with. First, the interimpacts among different areas need to be mitigated. For example, the blackout of a subgrid might result in the loss of some tie-line power, which further leads to negative influences on other subgrids. Second, the interassistance among different areas needs to be utilized. For example repair crews of different subgrids can be coordinated and co-optimized, so that the recovery of the interconnected power system is enhanced, which improves the resilience of individual subgrids.

- *Cyber and physical systems.* With the development and applications of smart grid technologies, power grids have become a systematic integration of cyber and physical systems. Disastrous events not only impact physical parts of a power grid but also affect its cyber parts. In many cases, power grid resilience enhancement approaches have to consider both cyber and physical systems. For example, in prior-event preparedness, other than hardening the power lines, etc., the communications system also needs to be strengthened, as state acquisition is critically important in responding to and recovering from the extreme event. In fact, when recovering a power grid, in some cases, some physical parts need to be repaired first to restore the power supply to critical cyber systems, while in some other cases, several cyber components have to be repaired first to restore the communications with important physical parts.

- *Interdependent critical infrastructures.* The interdependencies among different critical infrastructures make the resilience enhancement of power grids more complicated. For example, transportation networks (including road networks) and power networks are highly coupled, partly due to the increasing penetration of electric vehicles. In improving grid resilience, the power demands of transportation systems, the use of road networks to deliver flexibility resources, and the coordinated charging/discharging of electric vehicles, etc. have to be considered or dealt with. Other essential interdependencies, e.g. between natural gas networks and power grids, also need to be studied for resilience objectives so that the involved flexibility in different stages and areas can be fully coordinated and co-optimized.

1.5 Focus of This Book

Researchers and engineers in the electric power industry around the world have been conducting diversified studies or practices to attain the aforementioned components of power grid resilience, e.g. applying information and communications technologies in power grid controls. Among many worth-investigating topics on power grid resilience against disastrous events, this book concentrates on three critical issues, i.e. preparedness prior to disasters, response as disasters unfold, and recovery after disasters.

Specifically, this book summarizes the authors' selected research accomplishments related to the challenges or issues mentioned in the above sections, so as to provide applicable methods and some insights for the development of resilient power grids.

The main objectives of the research involved in this book include the following:

- *Preparedness prior to a natural disaster.*
  - Establish an optimal and coordinated preventive maintenance strategy that considers harsh external conditions (Chapter 2).
  - Construct a two-stage dispatch framework for the preallocation and real-time allocation of emergency resources to improve grid survivability (Chapter 3).
  - Develop a new approach to distribution grid automation that allocates remote-controlled switches empowering prompt restoration (Chapter 4).

- *Response as a natural disaster unfolds.*
  - Provide a sequential steady-state security region-based method to describe grid operability impacted by sequential extreme events (Chapter 5).
  - Propose a proactive operation strategy to enhance grid resilience during an unfolding extreme event considering uncertain sequential transitions of grid states (Chapter 6).
  - Develop a Markov decision process-based approach for distribution grids' real-time response throughout an unfolding event (Chapter 7).
- *Recovery after a natural disaster.*
  - Establish a new formulation of distribution grid topology constraints enabling restoration methods that use dynamically formed microgrids with flexible and adaptive boundaries (Chapter 8).
  - Construct a resilient routing and scheduling method for mobile power sources used in microgrids to enhance both survivability and recovery (Chapter 9).
  - Propose a novel co-optimization approach that coordinates the utilization of various flexibility resources in postdisaster recovery logistics (Chapter 10).

1.6 Summary

This chapter provides a general overview of resilience research and practice in power grids that have been endangered by natural disasters. The resilience concept is clarified based upon its definitions, importance and benefits, and challenges. Then, some relevant studies on grid resilience enhancement against disasters are discussed, divided by stages of resilience objectives, i.e. preparedness prior to disasters, response as disasters unfold, and recovery after disasters. Additionally, selected coordination and co-optimization issues in enhancing grid resilience are discussed. Last but not least, the focus of this work is clarified.

References

1 I. Abdin, Y.-P. Fang, and E. Zio, "A modeling and optimization framework for power systems design with operational flexibility and resilience against extreme heat waves and drought events," *Renewable and Sustainable Energy Reviews*, vol. 112, pp. 706–719, 2019.

2 D. K. Mishra, M. J. Ghadi, A. Azizivahed, L. Li, and J. Zhang, "A review on resilience studies in active distribution systems," *Renewable and Sustainable Energy Reviews*, vol. 135, p. 110201, 2021.

3 Y. Wang, A. O. Rousis, and G. Strbac, "On microgrids and resilience: A comprehensive review on modeling and operational strategies," *Renewable and Sustainable Energy Reviews*, vol. 134, p. 110313, 2020.

4 The North American Electric Reliability Corporation, *Hurricane Sandy Event Analysis Report*, 2014.

5 N. Martin and J. Rice, "Power outages, climate events and renewable energy: Reviewing energy storage policy and regulatory options for Australia," *Renewable and Sustainable Energy Reviews*, vol. 137, p. 110617, 2021.

6 L. Molyneaux, C. Brown, L. Wagner, and J. Foster, "Measuring resilience in energy systems: Insights from a range of disciplines," *Renewable and Sustainable Energy Reviews*, vol. 59, pp. 1068–1079, 2016.

7 U.S. Federal Emergency Management Agency, *Mitigation Assessment Team Report: Hurricane Sandy in New Jersey and New York*, 2013.

8 U.S. Executive Office of the President and Department of Energy's Office of Electricity and Energy Reliability, "Economic benefits of increasing electric grid resilience to weather outages," 2013.

9 R. J. Campbell and S. Lowry, "Weather-related power outages and electric system resiliency," Congressional Research Service, Library of Congress, Washington, DC, 2012.

10 S. Chu and A. Majumdar, "Opportunities and challenges for a sustainable energy future," *Nature*, vol. 488, no. 7411, pp. 294–303, 2012.

11 Z. Bie, Y. Lin, G. Li, and F. Li, "Battling the extreme: A study on the power system resilience," *Proceedings of the IEEE*, vol. 105, no. 7, pp. 1253–1266, 2017.

12 National Academy of Sciences U.S.A., *National Research Council: Disaster Resilience: A National Imperative*. Washington, DC, US.A: National Academies Press, 2012.

13 Cabinet Office. Natural Hazards and Infrastructure. [Online]. Available: https://www.gov.uk/government/uploads/system/uploads/attachment_data/file/6134 2/natural-hazards-infrastructure.pdf.

14 National Infrastructure Advisory Council. A Framework for Establishing Critical Infrastructure Resilience Goals. [Online]. Available: https://www.dhs .gov/xlibrary/assets/niac/niac-a-framework-for-establishing-criticalinfra-structure-resilience-goals-2010-10-19.pdf.

15 A. Stankovic, "The definition and quantification of resilience," *IEEE PES Industry Technical Support Task Force: Piscataway, NJ, USA*, pp. 1–4, 2018.

16 L. Das, S. Munikoti, B. Natarajan, and B. Srinivasan, "Measuring smart grid resilience: Methods, challenges and opportunities," *Renewable and Sustainable Energy Reviews*, vol. 130, p. 109918, 2020.

17 S. N. Emenike and G. Falcone, "A review on energy supply chain resilience through optimization," *Renewable and Sustainable Energy Reviews*, vol. 134, p. 110088, 2020.

18 U.K. House of Lords, "The resilience of the electricity system," 2015.

19 North American Electric Reliability Corporation, "High-impact, low-frequency event risk to the north American bulk power system," 2010.

20 The North American Electric Reliability Corporation, *Electricity Sub-sector Coordinating Council: Critical Infrastructure Strategic Roadmap*, 2010.

21 U.S. Electric Power Research Institute, *Enhancing Distribution Resiliency: Opportunities for Applying Innovative Technologies*, 2013.

22 P. Hines, J. Apt, and S. Talukdar, "Trends in the history of large blackouts in the United States," in *2008 IEEE Power and Energy Society General Meeting-Conversion and Delivery of Electrical Energy in the 21st Century*, pp. 1–8, IEEE, 2008.

23 Q. Wang, J. Wang, and Y. Guan, "Stochastic unit commitment with uncertain demand response," *IEEE Transactions on Power Systems*, vol. 28, no. 1, pp. 562–563, 2012.

24 Y.-Y. Lee and R. Baldick, "A frequency-constrained stochastic economic dispatch model," *IEEE Transactions on Power Systems*, vol. 28, no. 3, pp. 2301–2312, 2013.

25 R. Billinton and Z. Pan, "Incorporating reliability index probability distributions in financial risk assessment with performance based regulation," *Electric Power Components and Systems*, vol. 33, no. 6, pp. 685–697, 2005.

26 Congressional Research Service Report, *Cybersecurity Issues for the Bulk Power System*, 2015.

27 U. Müller and K. Strunz, "Resilience of data centre power system: Modelling of sustained operation under outage, definition of metrics, and application," *The Journal of Engineering*, vol. 2019, no. 12, pp. 8419–8427, 2019.

28 K. Sun, Y. Hou, W. Sun, and J. Qi, *Power System Control under Cascading Failures: Understanding, Mitigation, and System Restoration*. John Wiley & Sons, 2019.

29 R.-P. Liu, S. Lei, C. Peng, W. Sun, and Y. Hou, "Data-based resilience enhancement strategies for electric-gas systems against sequential extreme weather events," *IEEE Transactions on Smart Grid*, vol. 11, no. 6, pp. 5383–5395, 2020.

30 Y. Fu, M. Shahidehpour, and Z. Li, "Long-term security-constrained unit commitment: Hybrid Dantzig-Wolfe decomposition and subgradient approach," *IEEE Transactions on Power Systems*, vol. 20, no. 4, pp. 2093–2106, 2005.

31 A. Abur and A. G. Exposito, *Power System State Estimation: Theory and Implementation*. CRC Press, 2004.

32 N. Zhou, D. Meng, Z. Huang, and G. Welch, "Dynamic state estimation of a synchronous machine using PMU data: A comparative study," *IEEE Transactions on Smart Grid*, vol. 6, no. 1, pp. 450–460, 2014.

33 G. Valverde and V. Terzija, "Unscented Kalman filter for power system dynamic state estimation," *IET Generation, Transmission and Distribution*, vol. 5, no. 1, pp. 29–37, 2011.

34 K. Emami, T. Fernando, H. H.-C. Iu, H. Trinh, and K. P. Wong, "Particle filter approach to dynamic state estimation of generators in power systems," *IEEE Transactions on Power Systems*, vol. 30, no. 5, pp. 2665–2675, 2014.

35 L. E. Jones, *Renewable Energy Integration: Practical Management of Variability, Uncertainty, and Flexibility in Power Grids*. Academic Press, 2017.

36 B. Chiu and A. Bose, "Resilience Framework, Methods, and Metrics for the Electricity Sector, IEEE Power & Energy Society, Technical Report PES-TR83, Piscataway, NJ, USA, 2020.

37 A. N. Tari, M. S. Sepasian, and M. T. Kenari, "Resilience assessment and improvement of distribution networks against extreme weather events," *International Journal of Electrical Power & Energy Systems*, vol. 125, p. 106414, 2021.

38 A. Sharifi and Y. Yamagata, "Principles and criteria for assessing urban energy resilience: A literature review," *Renewable and Sustainable Energy Reviews*, vol. 60, pp. 1654–1677, 2016.

39 H. Liu, R. A. Davidson, D. V. Rosowsky, and J. R. Stedinger, "Negative binomial regression of electric power outages in hurricanes," *Journal of Infrastructure Systems*, vol. 11, no. 4, pp. 258–267, 2005.

40 S. D. Guikema, R. Nateghi, S. M. Quiring, A. Staid, A. C. Reilly, and M. Gao, "Predicting hurricane power outages to support storm response planning," *IEEE Access*, vol. 2, pp. 1364–1373, 2014.

41 R. Nateghi, S. D. Guikema, and S. M. Quiring, "Forecasting hurricane-induced power outage durations," *Natural Hazards*, vol. 74, no. 3, pp. 1795–1811, 2014.

42 W. Yuan, J. Wang, F. Qiu, C. Chen, C. Kang, and B. Zeng, "Robust optimization-based resilient distribution network planning against natural disasters," *IEEE Transactions on Smart Grid*, vol. 7, no. 6, pp. 2817–2826, 2016.

43 M. H. Amirioun, F. Aminifar, and H. Lesani, "Resilience-oriented proactive management of microgrids against windstorms," *IEEE Transactions on Power Systems*, vol. 33, no. 4, pp. 4275–4284, 2018.

44 M. H. Amirioun, F. Aminifar, and H. Lesani, "Towards proactive scheduling of microgrids against extreme floods," *IEEE Transactions on Smart Grid*, vol. 9, no. 4, pp. 3900–3902, 2018.

45 M. Zare-Bahramabadi, A. Abbaspour, M. Fotuhi-Firuzabad, and M. Moeini-Aghtaie, "Resilience-based framework for switch placement problem in power distribution systems," *IET Generation, Transmission and Distribution*, vol. 12, no. 5, pp. 1223–1230, 2018.

46 D. Zhang, W. Li, and X. Xiong, "Overhead line preventive maintenance strategy based on condition monitoring and system reliability assessment," *IEEE Transactions on Power Systems*, vol. 29, no. 4, pp. 1839–1846, 2014.

47 C. J. Crabtree, D. Zappalá, and P. J. Tavner, "Survey of commercially available condition monitoring systems for wind turbines," 2014.

48 J. Endrenyi, G. Anders, and A. Leite da Silva, "Probabilistic evaluation of the effect of maintenance on reliability. An application [to power systems]," *IEEE Transactions on Power Systems*, vol. 13, no. 2, pp. 576–583, 1998.

49 S. K. Abeygunawardane, P. Jirutitijaroen, and H. Xu, "Adaptive maintenance policies for aging devices using a Markov decision process," *IEEE Transactions on Power Systems*, vol. 28, no. 3, pp. 3194–3203, 2013.

50 P. Jirutitijaroen and C. Singh, "The effect of transformer maintenance parameters on reliability and cost: A probabilistic model," *Electric Power Systems Research*, vol. 72, no. 3, pp. 213–224, 2004.

51 M. E. El-Hawary, *Electric Power Applications of Fuzzy Systems*. Wiley-IEEE Press, 1998.

52 H. Ge and S. Asgarpoor, "Reliability evaluation of equipment and substations with fuzzy Markov processes," *IEEE Transactions on Power Systems*, vol. 25, no. 3, pp. 1319–1328, 2010.

53 D. Mohanta, P. Sadhu, and R. Chakrabarti, "Fuzzy Markov model for determination of fuzzy state probabilities of generating units including the effect of maintenance scheduling," *IEEE Transactions on Power Systems*, vol. 20, no. 4, pp. 2117–2124, 2005.

54 O. Duque and D. Morinigo, "A fuzzy Markov model including optimization techniques to reduce uncertainty," in *Proceedings of the 12th IEEE Mediterranean Electrotechnical Conference (IEEE Cat. No.04CH37521)*, vol. 3, pp. 841–844, 2004.

55 M. L. Littman, "A tutorial on partially observable Markov decision processes," *Journal of Mathematical Psychology*, vol. 53, no. 3, pp. 119–125, 2009.

56 E. Byon and Y. Ding, "Season-dependent condition-based maintenance for a wind turbine using a partially observed Markov decision process," *IEEE Transactions on Power Systems*, vol. 25, no. 4, pp. 1823–1834, 2010.

57 A. M. Smith and G. R. Hinchcliffe, *RCM–Gateway to World Class Maintenance*. Elsevier, 2003.

58 D. W. Holladay, B. D. Dallman, and C. H. Grigg, "Reliability centered maintenance study on voltage regulators," in *ESMO 2006 - 2006 IEEE 11th International Conference on Transmission & Distribution Construction, Operation and Live-Line Maintenance*, 2006.

59 T. Chan, C.-C. Liu, and J.-W. Choe, "Implementation of reliability-centered maintenance for circuit breakers," in *IEEE Power Engineering Society General Meeting, 2005*, vol. 1, pp. 684–690, 2005.

60 J. Goodfellow, "Applying reliability centered maintenance (RCM) to overhead electric utility distribution systems," in *2000 Power Engineering Society Summer Meeting (Cat. No.00CH37134)*, vol. 1, pp. 566–569, 2000.

61 W. Reder and D. Flaten, "Reliability centered maintenance for distribution underground systems," in *2000 Power Engineering Society Summer Meeting (Cat. No.00CH37134)*, vol. 1, pp. 551–556, 2000.

62 S. Costa and A. Brandao, "Optimization of the maintenance of oil filled power transformers based on reliability criteria," in *2004 IEEE/PES Transmision and Distribution Conference and Exposition: Latin America (IEEE Cat. No. 04EX956)*, pp. 297–301, 2004.

63 P. Dehghanian, M. Fotuhi-Firuzabad, F. Aminifar, and R. Billinton, "A comprehensive scheme for reliability centered maintenance in power distribution systemsPart I: Methodology," *IEEE Transactions on Power Delivery*, vol. 28, no. 2, pp. 761–770, 2013.

64 P. Dehghanian, M. Fotuhi-Firuzabad, F. Aminifar, and R. Billinton, "A comprehensive scheme for reliability-centered maintenance in power distribution systemsPart II: Numerical analysis," *IEEE Transactions on Power Delivery*, vol. 28, no. 2, pp. 771–778, 2013.

65 Y. Fu, M. Shahidehpour, and Z. Li, "Security-constrained optimal coordination of generation and transmission maintenance outage scheduling," *IEEE Transactions on Power Systems*, vol. 22, no. 3, pp. 1302–1313, 2007.

66 Y. Wang, H. Zhong, Q. Xia, D. S. Kirschen, and C. Kang, "An approach for integrated generation and transmission maintenance scheduling considering $N-1$ contingencies," *IEEE Transactions on Power Systems*, vol. 31, no. 3, pp. 2225–2233, 2016.

67 R. Nateghi, S. D. Guikema, and S. M. Quiring, "Comparison and validation of statistical methods for predicting power outage durations in the event of hurricanes," *Risk Analysis: An International Journal*, vol. 31, no. 12, pp. 1897–1906, 2011.

68 S. D. Guikema, S. M. Quiring, and S.-R. Han, "Prestorm estimation of hurricane damage to electric power distribution systems," *Risk Analysis: An International Journal*, vol. 30, no. 12, pp. 1744–1752, 2010.

69 E. Yamangil, R. Bent, and S. Backhaus, "Designing resilient electrical distribution grids," *arXiv preprint arXiv:1409.4477*, 2014.

70 A. Arab, A. Khodaei, S. K. Khator, K. Ding, V. A. Emesih, and Z. Han, "Stochastic pre-hurricane restoration planning for electric power systems infrastructure," *IEEE Transactions on Smart Grid*, vol. 6, no. 2, pp. 1046–1054, 2015.

71 C. Coffrin, P. Van Hentenryck, and R. Bent, "Strategic stockpiling of power system supplies for disaster recovery," in *2011 IEEE Power and Energy Society General Meeting*, pp. 1–8, 2011.

72 A. El-Keib, J. Nieplocha, H. Singh, and D. Maratukulam, "A decomposed state estimation technique suitable for parallel processor implementation," *IEEE Transactions on Power Systems*, vol. 7, no. 3, pp. 1088–1097, 1992.

73 M. Panteli, D. N. Trakas, P. Mancarella, and N. D. Hatziargyriou, "Boosting the power grid resilience to extreme weather events using defensive islanding," *IEEE Transactions on Smart Grid*, vol. 7, no. 6, pp. 2913–2922, 2016.

74 Z. Wang and J. Wang, "Self-healing resilient distribution systems based on sectionalization into microgrids," *IEEE Transactions on Power Systems*, vol. 30, no. 6, pp. 3139–3149, 2015.

75 F. C. Schweppe and J. Wildes, "Power system static-state estimation, Part I: Exact model," *IEEE Transactions on Power Apparatus and Systems*, vol. PAS-89, no. 1, pp. 120–125, 1970.

76 H. Sasaki, K. Aoki, and R. Yokoyama, "A parallel computation algorithm for static state estimation by means of matrix inversion lemma," *IEEE Transactions on Power Systems*, vol. 2, no. 3, pp. 624–631, 1987.

77 H. Wei, H. Sasaki, J. Kubokawa, and R. Yokoyama, "An interior point method for power system weighted nonlinear L_1 norm static state estimation," *IEEE Transactions on Power Systems*, vol. 13, no. 2, pp. 617–623, 1998.

78 Z. Huang, P. Du, D. Kosterev, and B. Yang, "Application of extended Kalman filter techniques for dynamic model parameter calibration," in *2009 IEEE Power & Energy Society General Meeting*, pp. 1–8, 2009.

79 L. Fan and Y. Wehbe, "Extended Kalman filtering based real-time dynamic state and parameter estimation using PMU data," *Electric Power Systems Research*, vol. 103, pp. 168–177, 2013.

80 H. Tebianian and B. Jeyasurya, "Dynamic state estimation in power systems: Modeling, and challenges," *Electric Power Systems Research*, vol. 121, pp. 109–114, 2015.

81 "IEEE draft standard for synchrophasor measurements for power systems," *IEEE PC37.118.1/D4.2, May 2011*, pp. 1–56, 2011.

82 M. K. Pitt and N. Shephard, "Filtering via simulation: Auxiliary particle filters," *Journal of the American Statistical Association*, vol. 94, no. 446, pp. 590–599, 1999.

83 M. Arulampalam, S. Maskell, N. Gordon, and T. Clapp, "A tutorial on particle filters for online nonlinear/non-Gaussian Bayesian tracking," *IEEE Transactions on Signal Processing*, vol. 50, no. 2, pp. 174–188, 2002.

84 N. Zhou, D. Meng, and S. Lu, "Estimation of the dynamic states of synchronous machines using an extended particle filter," *IEEE Transactions on Power Systems*, vol. 28, no. 4, pp. 4152–4161, 2013.

85 W. Jiang, V. Vittal, and G. T. Heydt, "Diakoptic state estimation using phasor measurement units," *IEEE Transactions on Power Systems*, vol. 23, no. 4, pp. 1580–1589, 2008.

86 L. Yan, Z. Xiaoxin, and Z. Jingyang, "A new algorithm for distributed power system state estimation based on PMUs," in *2006 International Conference on Power System Technology*, pp. 1–6, 2006.

87 K. Clements, "A multi-area approach to state estimation in power system networks," in *IEEE PES Summer Meeting*, 1972.

88 C. W. Brice and R. K. Cavin, "Multiprocessor static state estimation," *IEEE Transactions on Power Apparatus and Systems*, vol. PAS-101, no. 2, pp. 302–308, 1982.

89 S.-Y. Bin and C.-H. Lin, "An implementable distributed state estimator and distributed bad data processing schemes for electric power systems," *IEEE Transactions on Power Systems*, vol. 9, no. 3, pp. 1277–1284, 1994.

90 T. Overbye and R. Klump, "Determination of emergency power system voltage control actions," *IEEE Transactions on Power Systems*, vol. 13, no. 1, pp. 205–210, 1998.

91 D. Sutanto and W. Lachs, "Power system emergency control against voltage collapse," in *2000 IEEE Power Engineering Society Winter Meeting. Conference Proceedings (Cat. No.00CH37077)*, vol. 2, pp. 1501–1505, 2000.

92 R. Podmore, "Identification of coherent generators for dynamic equivalents," *IEEE Transactions on Power Apparatus and Systems*, vol. PAS-97, no. 4, pp. 1344–1354, 1978.

93 J. Thapar, V. Vittal, W. Kliemann, and A. Fouad, "Application of the normal form of vector fields to predict interarea separation in power systems," *IEEE Transactions on Power Systems*, vol. 12, no. 2, pp. 844–850, 1997.

94 V. Vittal, W. Kliemann, Y.-X. Ni, D. Chapman, A. Silk, and D. Sobajic, "Determination of generator groupings for an islanding scheme in the Manitoba hydro system using the method of normal forms," *IEEE Transactions on Power Systems*, vol. 13, no. 4, pp. 1345–1351, 1998.

95 Berger and Bokhari, "A partitioning strategy for nonuniform problems on multiprocessors," *IEEE Transactions on Computers*, vol. C-36, no. 5, pp. 570–580, 1987.

96 H. D. Simon, "Partitioning of unstructured problems for parallel processing," *Computing Systems in Engineering*, vol. 2, no. 2–3, pp. 135–148, 1991.

97 C.-W. Ou, S. Ranka, and G. Fox, "Fast and parallel mapping algorithms for irregular problems," *The Journal of Supercomputing*, vol. 10, no. 2, pp. 119–140, 1996.

98 P. Sadayappan and F. Ercal, "Nearest-neighbor mapping of finite element graphs onto processor meshes," *IEEE Transactions on Computers*, vol. C-36, no. 12, pp. 1408–1424, 1987.

99 B. W. Kernighan and S. Lin, "An efficient heuristic procedure for partitioning graphs," *The Bell System Technical Journal*, vol. 49, no. 2, pp. 291–307, 1970.

100 B. Hendrickson and R. Leland, "An improved spectral graph partitioning algorithm for mapping parallel computations," *SIAM Journal on Scientific Computing*, vol. 16, no. 2, pp. 452–469, 1995.

101 A. Pothen, H. D. Simon, and K.-P. Liou, "Partitioning sparse matrices with eigenvectors of graphs," *SIAM Journal on Matrix Analysis and Applications*, vol. 11, no. 3, pp. 430–452, 1990.

102 J. Li, C.-C. Liu, and K. P. Schneider, "Controlled partitioning of a power network considering real and reactive power balance," *IEEE Transactions on Smart Grid*, vol. 1, no. 3, pp. 261–269, 2010.

103 V. Venkatasubramanian and Y. Li, "Analysis of 1996 Western American electric blackouts," *Bulk Power System Dynamics and Control-VI*, Cortina d'Ampezzo, Italy, pp. 22–27, 2004.

104 M. Adibi and L. Fink, "Overcoming restoration challenges associated with major power system disturbances - Restoration from cascading failures," *IEEE Power and Energy Magazine*, vol. 4, no. 5, pp. 68–77, 2006.

105 Y. Hou, C.-C. Liu, K. Sun, P. Zhang, S. Liu, and D. Mizumura, "Computation of milestones for decision support during system restoration," *IEEE Transactions on Power Systems*, vol. 26, no. 3, pp. 1399–1409, 2011.

106 W. Sun, C.-C. Liu, and L. Zhang, "Optimal generator start-up strategy for bulk power system restoration," *IEEE Transactions on Power Systems*, vol. 26, no. 3, pp. 1357–1366, 2011.

107 A. Castillo, "Microgrid provision of blackstart in disaster recovery for power system restoration," in *2013 IEEE International Conference on Smart Grid Communications (SmartGridComm)*, pp. 534–539, 2013.

108 F. Ren, M. Zhang, D. Soetanto, and X. Su, "Conceptual design of a multi-agent system for interconnected power systems restoration," *IEEE Transactions on Power Systems*, vol. 27, no. 2, pp. 732–740, 2012.

109 C.-C. Liu, S. Lee, and S. Venkata, "An expert system operational aid for restoration and loss reduction of distribution systems," *IEEE Transactions on Power Systems*, vol. 3, no. 2, pp. 619–626, 1988.

110 C.-S. Chen, C.-H. Lin, and H.-Y. Tsai, "A rule-based expert system with colored petri net models for distribution system service restoration," *IEEE Transactions on Power Systems*, vol. 17, no. 4, pp. 1073–1080, 2002.

111 S.-I. Lim, S.-J. Lee, M.-S. Choi, D.-J. Lim, and B.-N. Ha, "Service restoration methodology for multiple fault case in distribution systems," *IEEE Transactions on Power Systems*, vol. 21, no. 4, pp. 1638–1644, 2006.

112 S.-J. Lee, S.-I. Lim, and B.-S. Ahn, "Service restoration of primary distribution systems based on fuzzy evaluation of multi-criteria," *IEEE Transactions on Power Systems*, vol. 13, no. 3, pp. 1156–1163, 1998.

113 A. Morelato and A. Monticelli, "Heuristic search approach to distribution system restoration," *IEEE Transactions on Power Delivery*, vol. 4, no. 4, pp. 2235–2241, 1989.

114 S. Toune, H. Fudo, T. Genji, Y. Fukuyama, and Y. Nakanishi, "Comparative study of modern heuristic algorithms to service restoration in distribution systems," *IEEE Transactions on Power Delivery*, vol. 17, no. 1, pp. 173–181, 2002.

115 R. Perez-Guerrero, G. T. Heydt, N. J. Jack, B. K. Keel, and A. R. Castelhano, "Optimal restoration of distribution systems using dynamic programming," *IEEE Transactions on Power Delivery*, vol. 23, no. 3, pp. 1589–1596, 2008.

116 R. E. Perez-Guerrero and G. T. Heydt, "Viewing the distribution restoration problem as the dual of the unit commitment problem," *IEEE Transactions on Power Systems*, vol. 23, no. 2, pp. 807–808, 2008.

117 C. L. Moreira, F. O. Resende, and J. A. Peas Lopes, "Using low voltage microgrids for service restoration," *IEEE Transactions on Power Systems*, vol. 22, no. 1, pp. 395–403, 2007.

118 E. Zare and M. Shahabi, "Microgrid restoration after major faults in main grid with automatic and constant time switching," *International Journal of Intelligent Systems and Applications*, vol. 5, no. 10, p. 50, 2013.

119 J. Li, X.-Y. Ma, C.-C. Liu, and K. P. Schneider, "Distribution system restoration with microgrids using spanning tree search," *IEEE Transactions on Power Systems*, vol. 29, no. 6, pp. 3021–3029, 2014.

120 H. Gao, Y. Chen, Y. Xu and C.-C. Liu, "Resilience-oriented critical load restoration using microgrids in distribution systems," *IEEE Transactions on Smart Grid*, vol. 7, no. 6, pp. 2837–2848, 2016.

121 C. Chen, J. Wang, F. Qiu, and D. Zhao, "Resilient distribution system by microgrids formation after natural disasters," *IEEE Transactions on Smart Grid*, vol. 7, no. 2, pp. 958–966, 2016.

122 T. Ding, Y. Lin, G. Li, and Z. Bie, "A new model for resilient distribution systems by microgrids formation," *IEEE Transactions on Power Systems*, vol. 32, no. 5, pp. 4145–4147, 2017.

123 K. S. A. Sedzro, A. J. Lamadrid, and L. F. Zuluaga, "Allocation of resources using a microgrid formation approach for resilient electric grids," *IEEE Transactions on Power Systems*, vol. 33, no. 3, pp. 2633–2643, 2018.

124 S. Chen, W. Hu, and Z. Chen, "Comprehensive cost minimization in distribution networks using segmented-time feeder reconfiguration and reactive power control of distributed generators," *IEEE Transactions on Power Systems*, vol. 31, no. 2, pp. 983–993, 2016.

125 E. Lopez, H. Opazo, L. Garcia, and P. Bastard, "Online reconfiguration considering variability demand: Applications to real networks," *IEEE Transactions on Power Systems*, vol. 19, no. 1, pp. 549–553, 2004.

126 D. Bernardon, A. Mello, L. Pfitscher, L. Canha, A. Abaide, and A. Ferreira, "Real-time reconfiguration of distribution network with distributed generation," *Electric Power Systems Research*, vol. 107, pp. 59–67, 2014.

127 M. R. Dorostkar-Ghamsari, M. Fotuhi-Firuzabad, M. Lehtonen, and A. Safdarian, "Value of distribution network reconfiguration in presence of renewable energy resources," *IEEE Transactions on Power Systems*, vol. 31, no. 3, pp. 1879–1888, 2016.

128 C. Lueken, P. M. Carvalho, and J. Apt, "Distribution grid reconfiguration reduces power losses and helps integrate renewables," *Energy Policy*, vol. 48, pp. 260–273, 2012.

129 P. C. Ramaswamy, P. Vingerhoets, and G. Deconinck, "Reconfiguring distribution grids for more integration of distributed generation," in *22nd International Conference and Exhibition on Electricity Distribution (CIRED 2013)*, pp. 1–4, 2013.

130 F. Capitanescu, L. F. Ochoa, H. Margossian, and N. D. Hatziargyriou, "Assessing the potential of network reconfiguration to improve distributed generation hosting capacity in active distribution systems," *IEEE Transactions on Power Systems*, vol. 30, no. 1, pp. 346–356, 2015.

131 A. Kavousi-Fard, A. Zare, and A. Khodaei, "Effective dynamic scheduling of reconfigurable microgrids," *IEEE Transactions on Power Systems*, vol. 33, no. 5, pp. 5519–5530, 2018.

132 T. Thakur and Jaswanti, "Study and characterization of power distribution network reconfiguration," in *2006 IEEE/PES Transmission & Distribution Conference and Exposition: Latin America*, pp. 1–6, 2006.

133 S. Lei, Y. Hou, F. Qiu, and J. Yan, "Identification of critical switches for integrating renewable distributed generation by dynamic network reconfiguration," *IEEE Transactions on Sustainable Energy*, vol. 9, no. 1, pp. 420–432, 2018.

134 P. Carvalho, L. Ferreira, and A. da Silva, "A decomposition approach to optimal remote controlled switch allocation in distribution systems," *IEEE Transactions on Power Delivery*, vol. 20, no. 2, pp. 1031–1036, 2005.

135 D. P. Bernardon, M. Sperandio, V. J. Garcia, L. N. Canha, A. d. R. Abaide, and E. F. B. Daza, "AHP decision-making algorithm to allocate remotely controlled switches in distribution networks," *IEEE Transactions on Power Delivery*, vol. 26, no. 3, pp. 1884–1892, 2011.

136 L. S. de Assis, J. F. V. Gonz'ález, F. L. Usberti, C. Lyra, and F. Von Zuben, "Optimal allocation of remote controlled switches in radial distribution systems," in *2012 IEEE Power and Energy Society General Meeting*, pp. 1–8, 2012.

Part II

Preparedness Prior to a Natural Disaster

2

Preventive Maintenance to Enhance Grid Reliability

Chapter 2 investigates a dynamic model for condition-based maintenance strategies considering harsh external conditions. A Markov process is used to describe the physical characteristics of component deterioration, and the effects of harsh external conditions are represented as probabilistic models. The impacts of different maintenance strategies on an individual component and the entire system are involved. A cost-to-go model, including the maintenance cost and the system reliability cost, is constructed to optimize the maintenance strategies, and a backward induction algorithm associated with a search space reduction approach is used to solve the model. The proposed model is validated by two systems, and the results demonstrate that this optimal maintenance strategy model in consideration of harsh external conditions provides insight for scheduling appropriate maintenance activities.

2.1 Component- and System-Level Deterioration Model

Power system operators try to maximize profits and maintain acceptable reliability levels when the system is under different conditions. One of the effective means is to implement coordinated maintenance on system components to mitigate deterioration of components and in consequence reduce operational costs and potential fault losses. However, an additional cost would be caused by such maintenance activities. To achieve a trade-off between the system reliability and the operational cost, system operators need to develop a long-term series of combined maintenance activities on various components from the perspective of the entire system. The importance of maintenance scheduling for aging components over a given time horizon is already well recognized. Using the advantages of reliability-centered maintenance (RCM), e.g. minimizing the frequency of

Power Grid Resilience against Natural Disasters: Preparedness, Response, and Recovery, First Edition.
Shunbo Lei, Chong Wang, and Yunhe Hou.
© 2023 John Wiley & Sons Ltd. Published 2023 by John Wiley & Sons Ltd.

overhauls and increasing the reliability of components, Dehghanian et al. [1] and [2] proposed a practical framework by which the RCM procedure could be implemented in power distribution systems. Considering the influence of maintenance activities implemented on a component, Silva et al. [3] solved the problem of generator maintenance scheduling with network constraints by using Benders decomposition. Reference [4] studied the impact of a power market on the maintenance scheduling of a generating unit. However, maintenance scheduling should also depend on the different operating states of a component at different time intervals. A Markov model was introduced to represent the deterioration processes of an individual component [5, 6], while [7] enhanced the Markov model by using state diagrams. Due to the various aging rates of components in a power system, different maintenance activities on multiple components should be involved in maintenance scheduling. References [8] and [9] proposed models that included several components with different stochastic deterioration states. References [10] and [11] developed a two-stage maintenance management model that incorporated joint midterm and short-term maintenance. The proposed maintenance model considered both network constraints and maintenance constraints.

In practice, the impacts of harsh external conditions on the maintenance scheduling should be considered. Compliance application notices concerning protection system maintenance [12], provided by the North American Electric Reliability Corporation (NERC), verifies that maintenance activities should consider extreme weather conditions, scheduling conflicts, reliability issues, etc.

The NERC Report on the 2011 southwest cold weather event [13] illustrates the importance of extreme weather conditions on maintenance scheduling. IEEE Standard 516-2009 [14] dictates that live work, e.g. maintenance activities on electrical devices, should not be performed under adverse weather conditions such as storms, lightning activity, high humidity, and heavy snow. IEEE Standard 3007.2-2010 [15] emphasizes that maintenance activities on electrical devices such as switches and generators should not be performed under unfavorable weather conditions. In this case, the maintenance activities should enhance the system reliability to avoid severe damage caused by repair delays due to harsh external conditions. Therefore, it is necessary to schedule the maintenance activities to improve the performance of the system over the entire time horizon, in consideration of harsh external conditions.

2.1.1 Component-Level Deterioration Transition Probability

A component might be in failure at any time interval due to component deterioration with a certain probability. The stochastic state transitions of components such as transformers and lines are represented by a Markov model. A component

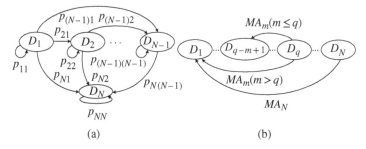

Figure 2.1 (a) Transitions between different states without maintenance activity. (b) Transitions between different states with maintenance activity.

could have a failure state and several normal operating states. For the state D_k, we assume that a larger value of k is associated with a worse state, while D_N represents the failure state. Figure 2.1a illustrates the transition diagram of a component without any maintenance activities. Mathematically, $\mathbf{P}_{i,0}$ can be written as (2.1) follows:

$$\mathbf{P}_{i,0} = \left[p_{jk}^{(i)} \right]_{N \times N} \tag{2.1}$$

In practice, the conditions of each component can be divided into several levels based on industrial standards, and each level can be considered as a Markov state. A system monitoring and condition tracking infrastructure can be used to monitor the states of all components to determine a specific component's state.

Maintenance activities are employed to mitigate the negative effects caused by deterioration processes. For a component with N states, there are N activities, i.e. $MA_1, MA_2, \ldots, MA_N$, which could be performed on the component. For example, state D_q $(1 \leq q \leq N-1)$ can be driven to state D_{q-m+1} through maintenance activity MA_m when $m \leq q$. When $m > q$, state D_q $(1 \leq q \leq N)$ can be driven to state D_1, as shown in Figure 2.1b. In addition, a repair activity, represented by MA_N, can be implemented on a component in failure and drives the state D_N to the state D_1. The replacement of a failed component is considered as a repair activity with a different cost.

For the component i, the state transition matrix under the activity MA_m $(1 \leq m \leq N)$ can be expressed as follows:

$$\mathbf{P}_{i,m} = \left[p_{jk}^{(i)} \right]_{N \times N} = \begin{cases} 1, & j = [k-m]_+ + 1, \ k = 1, \ldots, N \\ 0, & else \end{cases} \tag{2.2}$$

where $[\bullet]_+$ is defined as $\max(\bullet, 0)$. Based on the state transitions, the component can either retain its present state or return to a better state after a maintenance activity. In other words, maintenance activities reduce the future failure rates for components.

Using transition matrix (2.2), the transition probabilities for any states in adjacent time intervals can be expressed as a Markov chain.

Because only one activity can be selected in a given time interval, $N + 1$ elements are included in $a_{i,t}$ where one element is "1" and the others are "0." The first element of $a_{i,t}$ means that no activities are implemented on the ith component. Meanwhile, the second element to the $(N + 1)$th element of $a_{i,t}$ represent N activities. $S_{i,t}$ represents the ith component in state D_k at time t, and the corresponding value is "1" and "0" elsewhere. According to the definition of $S_{i,t}$ and $a_{i,t}$, the transition probability from $S_{i,t}$ to $S_{i,(t+1)}$ under $a_{i,t}$ can be represented as follows:

$$P(\mathbf{S}_{i,(t+1)}|\mathbf{S}_{i,t}, \mathbf{a}_{i,t}) = (\mathbf{S}_{i,(t+1)})^{\mathrm{T}} \cdot \mathbf{P}_{i,m} \cdot \mathbf{S}_{i,t} \tag{2.3}$$

where $\mathbf{P}_{i,m}$ is defined as (2.1) and (2.2).

2.1.2 System-Level Deterioration Transition Probability

From the system-level perspective, the maintenance activity and the state can be described by the activity matrix $\mathbf{A}_t$ and the state matrix $\mathbf{\Omega}_t$, respectively.

These two matrices consist of the activity vector $a_{i,t}$ and the state vector $S_{i,t}$ and can be expressed as follows:

$$\mathbf{\Omega}_t = [\mathbf{S}_{1,t}, \ldots, \mathbf{S}_{i,t}, \ldots, \mathbf{S}_{N_c,t}] \tag{2.4}$$

$$\mathbf{A}_t = [\mathbf{a}_{1,t}, \ldots, \mathbf{a}_{i,t}, \ldots, \mathbf{a}_{N_c,t}] \tag{2.5}$$

When considering external conditions and operational conditions, the deterioration processes for adjacent components are not independent. Therefore, changing the transition probabilities for a component can directly alter the dependency of the deterioration process of other components. Furthermore, different components depend on each other because of operational constraints. The conditional probability of $\mathbf{\Omega}_{t+1}$ under states $\mathbf{\Omega}_t$ and $\mathbf{A}_t$ can be described as follows:

$$P(\mathbf{\Omega}_{t+1}|\mathbf{\Omega}_t, \mathbf{A}_t) = \prod_{i \in \Theta_c} P(\mathbf{S}_{i,(t+1)}|\mathbf{S}_{i,t}, \mathbf{a}_{i,t}) \tag{2.6}$$

2.1.3 Mathematical Model without Harsh External Conditions

This section presents a recursive model for maintenance optimization without consideration of harsh external conditions. The expected cost-to-go at the tth time interval consists of three parts: (i) the successive cost, $C_{S,t}$; (ii) the cost due to the loss of load, $C_{L,t}$; (iii) the cost of maintenance activities, $C_{A,t}$.

For these three costs, $C_{A,t}$ depends on maintenance activities, and $C_{L,t}$ depends on maintenance activities and the load level of the system. $C_{L,t}$ can be optimized by utilizing optimal power flow with the objective of minimizing the loss of load. Note that $C_{A,t}$ and $C_{L,t}$ can be expressed as the functions of the maintenance activity matrix $\mathbf{A}_t$ and can be written as $C_{A,t}(\mathbf{A}_t)$ and $C_{L,t}(\mathbf{A}_t)$. The maintenance activities determine the successive cost $C_{S,t}$, which can be written as $C_{S,t}(\mathbf{\Omega}_t, \mathbf{A}_t)$. Therefore, the expected cost-to-go $v_t(\mathbf{\Omega}_t, \mathbf{A}_t)$ at the tth time interval can be expressed as follows:

$$v_t(\mathbf{\Omega}_t, \mathbf{A}_t) = C_{A,t}(\mathbf{A}_t) + C_{L,t}(\mathbf{A}_t) + C_{S,t}(\mathbf{\Omega}_t, \mathbf{A}_t) \tag{2.7}$$

$$C_{S,t}(\mathbf{\Omega}_t, \mathbf{A}_t) = \sum_{\mathbf{\Omega}_{t+1} \in \Theta_N \cup \Theta_F} \left[v_{t+1}^*(\mathbf{\Omega}_{t+1}) \cdot P(\mathbf{\Omega}_{t+1} | \mathbf{\Omega}_t, \mathbf{A}_t) \right] \tag{2.8}$$

The objective is to minimize the expected cost-to-go for the given state $\mathbf{\Omega}_t$ by finding an activity matrix $\mathbf{A}_t$ for the components, that is

$$v_t^*(\mathbf{\Omega}_t) = \min \left\{ v_t(\mathbf{\Omega}_t, \mathbf{A}_t), \ \mathbf{A}_t \in \Theta_A \right\}, \quad \mathbf{\Omega}_t \in \Theta_N \tag{2.9}$$

2.2 Preventive Maintenance in Consideration of Disasters

2.2.1 Potential Disasters Influencing Preventive Maintenance

Recent years have seen exceptionally harsh weather conditions occurring in many different areas of the world. Some of these events are listed in Table 2.1. These rare events could lead to very high economic losses, and the severe impacts of these extreme weather events on the power systems have brought the concept of system resilience to the attention of many governments. The US government emphasizes the importance of resilient grids to counter the effects of the increased frequency and intensity of severe weather. As such, appropriate maintenance scheduling in consideration of severe weather and climate events is

Table 2.1 Harsh weather.

Time	Location	Event
Aug 2005	United States	Hurricane Katrina
Jan 2008	China	Snow Storm
Feb 2011	United States	Snow Storm
Oct 2012	United States and Canada	Hurricane Sandy
Feb 2014	Slovenia and Australia	Heat Wave

a significant step in building a resilient grid. Given the difficulty in determining the occurrence of harsh extreme conditions, many research studies [16, 17] have focused on models and methodologies, using historical data provided by weather observations.

2.2.2 Preventive Maintenance Model with Disasters Influences

Repair activities on failed components are usually affected by harsh external conditions, and these kinds of impacts will be included in the model. The system would suffer severe damage when harsh external conditions delay the repair activities. Therefore, maintenance activities over time intervals on different components should be coordinated to ensure high-system reliability during harsh external conditions in consideration of system operational constraints, maintenance costs, and deterioration processes. To include the impacts of harsh external conditions, an expected cost in the proposed recursive model is used to quantify the influence of harsh external conditions on repair delays, and the uncertainties of repair delays caused by harsh external conditions are modeled as probabilities in the expected cost. These probabilities can be determined based on historical data.

2.2.2.1 Probabilistic Model of Repair Delays Caused By Harsh External Conditions

Given $M(M < N_c)$ fault components in a certain future time interval, there will be $2M$ combinations of repair scenarios when considering potential repair delays caused by harsh external conditions. This is written as follows:

$$v_t^*(\mathbf{\Omega}_t) = \min\left\{v_t(\mathbf{\Omega}_t, \mathbf{A}_t), \ \mathbf{A}_t \in \Theta_A\right\}, \quad \mathbf{\Omega}_t \in \Theta_N \tag{2.10}$$

where $C_M^r, r \in \{0, 1, \dots, M\}$ denotes that repair activities on r components will be delayed due to harsh external conditions. The general formula of the probability of one scenario that the repair activities on certain components are delayed is represented as follows:

$$P_t^{(E)}(\mathbf{B}) = \prod_{l=1}^{M}\left[\mathbf{B}(l) \cdot P_{l,t}^{(R)} + (1 - \mathbf{B}(l)) \cdot \left(1 - P_{l,t}^{(R)}\right)\right] \tag{2.11}$$

where $\mathbf{B}(l)$ equals 0 when the repair on the lth fault component is delayed, otherwise, $\mathbf{B}(l)$ equals 1. Usually, harsh weather conditions affect power systems over an entire area, and it is assumed that the impacts of harsh weather conditions on adjacent components are the same.

2.2.2.2 Activity Vectors Corresponding to Repair Delays

Different repair delays would have different activity vectors. The element representing the repair activity is 1, and the other elements are 0 when the repair on the lth fault component is not delayed at the tth time interval. That is

$$\mathbf{A}_t^{(E)}(N+1, l') = 1 \tag{2.12}$$

$$\mathbf{A}_t^{(E)}(n, l') = 0, \quad n = \{1, 2, \ldots, N\} \tag{2.13}$$

where l' is the serial number of the lth fault component in all components. The element in the activity vector representing no activity should be 1 when the repair on the lth fault component is delayed at the tth time interval. That is

$$\mathbf{A}_t^{(E)}(N+1, l') = 1 \tag{2.14}$$

$$\mathbf{A}_t^{(E)}(n, \ l') = 0, \quad n = \{2, 3, \ldots, N+1\} \tag{2.15}$$

It is assumed that there are no maintenance activities on a component if the component is in failure. Therefore, the activity vectors for the ith normal operating component can be represented as follows:

$$\mathbf{A}_t^{(E)}(1, \ i) = 1, \quad i \neq l' \tag{2.16}$$

$$\mathbf{A}_t^{(E)}(n, \ i) = 0, \quad i \neq l', \ n = \{2, 3, \ldots, N+1\} \tag{2.17}$$

2.2.2.3 Expected Cost

Based on the model of the potential repair delay scenarios, the expected cost in consideration of failure states of components can be established as follows:

$$C_{R,t}^{(E)}(\mathbf{B}) = \sum_{l=1}^{M} \left[B(l) \cdot c_{R,l} \right] \tag{2.18}$$

$$C_{S,t}^{(E)}\left(\mathbf{\Omega}_t, \mathbf{A}_t^{(E)}\right) = \sum_{\mathbf{\Omega}_{t+1} \in \Theta_N \cup \Theta_F} \left[v_{t+1}^*(\mathbf{\Omega}_{t+1}) \cdot P\left(\mathbf{\Omega}_{t+1} | \mathbf{\Omega}_t, \mathbf{A}_t^{(E)}\right) \right] \tag{2.19}$$

$$v_t^*(\mathbf{\Omega}_t) = C_{L,t}^{(E)}(\mathbf{\Omega}_t) + \sum_{B(1)=0}^{1} \cdots \sum_{B(M)=0}^{1} \left[P_t^{(E)}(\mathbf{B}) \times \left(C_{R,t}^{(E)}(\mathbf{B}) + C_{S,t}^{(E)}\left(\mathbf{\Omega}_t, \mathbf{A}_t^{(E)}\right) \right) \right] \tag{2.20}$$

where (2.18) denotes the repair cost of a possible repaired scenario, (2.19) is the successive cost with a possible repaired scenario, where $\sum \cdots \sum[\bullet]$ indicates all possible repair delay scenarios, and (2.20) calculates the expected cost with potential delays of repair activities on failure components. Because harsh external conditions are assumed to affect repair activities, (2.18)–(2.20) associated with (2.7)–(2.9) are employed to achieve the optimal scheduling of maintenance activities for non-failure states. Since the proposed model is a recursive model with the costs of various maintenance activities on different components, the recursive characteristic is capable of handling maintenance alternatives with dissimilar time requirements.

2.3 Solution Algorithms

Backward induction [18] associated with a search space reduction method is employed to solve the proposed dynamic model.

2.3.1 Backward Induction

Backward induction is a deduction process operating backward from the end of a problem to determine a sequence of optimal activities. At each time interval, Bellman's equations are used to optimize the expected cost-to-go of one state with a certain activity. At each time interval, the action resulting in the minimal expected cost-to-go for each state is the optimal strategy. This process continues backward until all time intervals are covered.

2.3.2 Search Space Reduction Method

When system states and time intervals increase, the size of the search space grows dramatically and in consequence results in a computationally intensive task. The search space is reduced by means of a search space reduction method. The key point of the proposed search space reduction method is to neglect state transitions with tiny probabilities. In realistic systems, the deterioration probability between different states is usually small, especially in consideration of advanced manufacturing technologies and materials. In this case, there will be a small probability that many components deteriorate simultaneously in one time interval. When calculating the expected cost-to-go of each state, state transitions with minor probabilities are neglected to reduce the huge search space.

The steps to solve the proposed model are as follows:

Step 1. Using the deterioration process of each component, generate the sets Θ_N and Θ_F.

Step 2. Repeat for $t = T, T - 1, \ldots, 1$.

Step 2.1. For $\Omega_t \in \Theta_F$, calculate the probability and the corresponding active matrix using (2.10)–(2.17).

Step 2.2. Calculate the repair cost and the successive cost using (2.18) and (2.19) in conjunction with the search space reduction method. Then, get the expected cost for $\Omega_t \in \Theta_F$ using (2.20).

Step 2.3. For $\Omega_t \in \Theta_N$ with activity $\mathbf{A}_t$, use (2.7) and (2.8) to calculate the corresponding expected cost-to-go. Again, the search space reduction method is employed.

Step 2.4. For all possible activities, calculate all expected costs and find the optimal strategy for $\Omega_t \in \Theta_N$ using (2.9).

Step 2.5. Set $t = t - 1$ and return to Step 2.1. For the Tth time interval, the successive cost of each state is not considered.

2.4 Case Studies

Two test systems are used to validate the proposed model. The proposed model is verified by Case I, and the influences of harsh external conditions on the maintenance scheduling are also discussed. Case II presents simulation results for a practical power system. For the sake of exposition, we focus on the maintenance of transformers. However, it is worth pointing out that the proposed model also determines appropriate maintenance scheduling for other components, e.g. transmission and distribution lines.

2.4.1 Data Description

The following data are required in the model:

* Probability of repair delays caused by harsh external conditions.
* Transition probabilities between different states, e.g. deterioration states and failure states.
* Costs of different activities on different components.
* Costs of loss of load when components are out of service due to failure.

Furthermore, the following parameters are needed for simulation.

* Three states, i.e. the good state (D_1), the deteriorated state (D_2), and the failure state (D_3), are considered.
* Three activities, i.e. the minor maintenance activity (M_1), the major maintenance activity (M_2), and the repair activity (M_3), are included.
* 52 weeks for the maintenance scheduling using a one-week time interval.

2.4.2 Case I: Verification of the Proposed Model

The IEEE 30-bus system is used as the test system, and Table 2.2 shows the repair costs and the maintenance activity costs for the transformers in the system, where T_1, T_2, T_3, and T_4 represent the transformers at bus 6-10, bus 6-9, bus 27-28,

Table 2.2 Costs (10^3 \$) of Transformers.

	T_1	T_2	T_3	T_4
M_1	0.8	1.0	1.0	1.0
M_2	6.5	6.4	6.0	6.5
Repair	14	15	16	15.5

Table 2.3 Unrepaired Probabilities over All Time Intervals.

Weeks	Unrepaired probability
1–3	0.12
4–24, 38–41	0.05
25–32, 42–47	0.10
33–37, 48–52	0.15

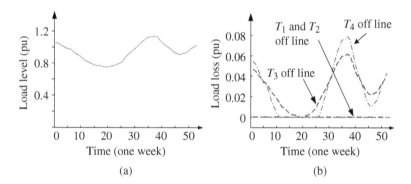

Figure 2.2 (a) Load curve. (b) Load losses with different offline transformers.

and bus 4-12, respectively. Table 2.3 shows the unrepaired probabilities, and it is assumed that unrepaired probabilities of different transformers at the same time interval are the same. Figure 2.2a,b shows the load curve over 52 weeks and the load losses with regard to different offline transformers. For all transformers, the probabilities from D_1 to D_2 and from D_1 to D_3 are 0.055 and 0.005, respectively. The probability from D_2 to D_3 is 0.015.

2.4.2.1 Verifying the Model Using Monte Carlo Simulations
The Monte Carlo method is used to verify the proposed model.

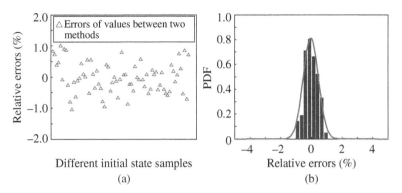

Figure 2.3 (a) Errors of expected costs based on the proposed model and Monte Carlo simulations. (b) PDF of the relative errors.

Take an initial state with all transformers in the state D_1 as an example. Based on 5000 Monte Carlo simulations, the expected cost is 20 218.4 \$, which is close to the expected cost of 20 164.4 \$ calculated by the proposed model. Figure 2.3a shows the relative errors of the expected costs based on the proposed model and the Monte Carlo simulations, and (b) shows the probability density function (PDF) of the relative errors. The tiny errors indicate that the proposed model and its solution are correct.

2.4.2.2 Selection of Optimal Maintenance Activities

The optimal maintenance activities are determined based on observed states. Figure 2.4 shows that the optimal maintenance activities are adjusted dynamically based on the observed states of the transformers at each time interval. For example, the optimal strategy should be a minor maintenance activity on T_4 if T_2 is in state D_2 and T_1, T_3, and T_4 are in D_1 at the 26th time interval. The optimal strategy should be a major maintenance activity if T_1 and T_3 deteriorate and other

Figure 2.4 Sample paths of states with regard to the costs of two scenarios.

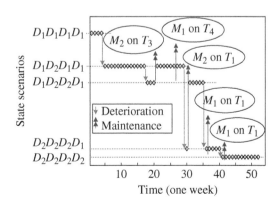

transformers retain their previous states, i.e. T_1, T_2, and T_3 are in state D_2 and T_4 is in D_1 in the 30th time interval.

2.4.2.3 Influences of Harsh External Conditions on Maintenance

The expected costs are impacted by harsh external conditions. The expected costs based on the proposed optimal maintenance strategies are smaller than the strategies without consideration of harsh external conditions. The reason for the cost reduction is that future effects due to the non-repair of components caused by harsh external conditions are taken into account for the proposed maintenance scheduling. For example, when T_1, T_2, and T_3 are in D_2 states and T_4 is in D_1 state in the first time interval, the expected cost of the proposed optimal maintenance strategy in consideration of harsh external conditions is 18 582.8 \$, while the expected cost of the strategy without consideration of harsh external conditions is 20 010.4 \$. For all states at the first time interval, 6.13–9.25% of the expected costs can be reduced while considering harsh external conditions.

Since the expected costs are influenced by harsh external conditions, the maintenance strategies are correspondingly influenced by harsh external conditions. Two sample paths of states with and without harsh external conditions are presented in Figure 2.5a,b, respectively. The states of some transformers deteriorate in the 28th time interval. For the proposed maintenance scheduling in this chapter, M_2 is implemented on T_4, and T_4 reaches to a good state in the next time interval. In this case, it has a small probability of failing in the future. However, if harsh external conditions are not considered, no maintenance activities are implemented. In this case, T_4 has a larger probability for failure in the future.

To provide an overview of the influences of harsh external conditions on maintenance scheduling, Figure 2.6 shows the distribution of inconsistent maintenance strategies with and without harsh external conditions, and Figure 2.7a–d shows the distributions of inconsistent maintenance strategies on T_1, T_2, T_3, and T_4 with and without harsh external conditions. The results show that most of inconsistent strategies are performed on T_1, T_2, and T_4, rather than on T_3. One of important reasons is that a higher cost of loss of load would be caused by disconnecting T_3 in most time intervals. In addition, inconsistent maintenance activity on T_4 mainly occurs around the 11–28th time intervals. One reason is no loss of load during these periods, and another reason is to enhance the reliability during peak periods, avoiding potential repair delays.

2.4.3 Case II: Results Simulating the Zhejiang Electric Power Grid

In this section, one practical system, i.e. a subnetwork of Zhejiang Electric Power Grid containing six transformers, is used to validate the dynamic optimal maintenance scheduling. The probabilities from D_1 to D_2, from D_1 to D_3, and from D_2

Figure 2.5 (a) Sample path of states with external conditions. (b) Sample path of states without external conditions.

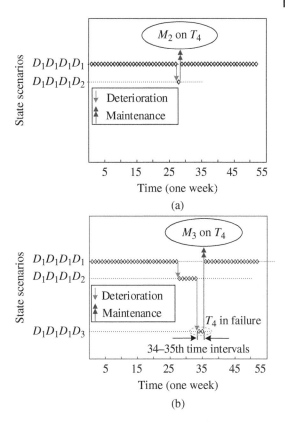

(a)

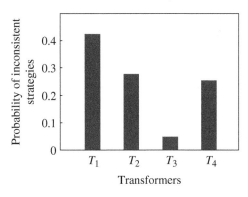

(b)

Figure 2.6 Distribution of inconsistent strategies on T_1, T_2, T_3, and T_4.

to D_3 are 0.055, 0.005, and 0.015, respectively. With the established maintenance scheduling, system operators first observe the states of all components at each time interval with monitoring systems, then choose the optimal activities that should be performed corresponding to the observed states.

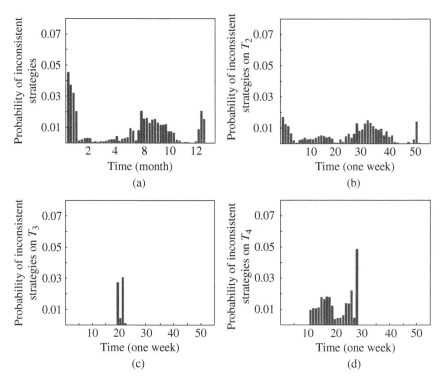

Figure 2.7 Distribution of inconsistent maintenance strategies on (a) T_1, (b) T_2, (c) T_3, and (d) T_4 over the whole 52-week time interval.

Comparisons between the expected costs caused by the proposed maintenance activities (ECPA) and the expected costs caused by given maintenance activities with regular time intervals (ECGA) are presented in Figure 2.8a. Expected costs include loss of load costs, maintenance activity costs, and successive costs. Figure 2.8b shows the improvement on expected costs and system reliability when using the ECPA. When the initial states of all transformers are in D_2, ECPA is 23% lower. When the initial states of all transformers are all in a good state, i.e. D_1, 65% of the expected cost can be reduced. For these two states, we also see that the reliability can also be enhanced. For other states, improvements on expected costs and the reliability (loss of loads) are presented in Figure 2.8b. The results show that the proposed strategy achieves a better balance between the reliability level and maintenance scheduling.

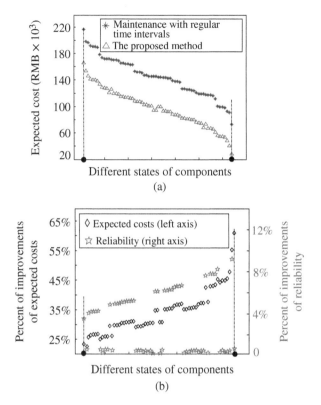

Figure 2.8 (a) Expected costs for proposed maintenance activities and given maintenance activities with regular time intervals. (b) Improvements in expected costs and reliability.

2.5 Summary and Conclusions

This chapter establishes a sequentially coordinated condition-based maintenance scheduling by using a Markov-based model in consideration of the influence of harsh external conditions. A better performance can be achieved by coordinating the maintenance activities of multiple components over a time horizon. The proposed model is solved by the backward induction with the search space reduction approach. The major findings from the case studies are listed as follows: (i) The harsh external conditions are critical for the maintenance scheduling. When considering influences of harsh external conditions, 6.13–9.25% of the expected cost of the maintenance scheduling can be reduced in the case studies. (ii) Sequential coordination of the multiple components over a time horizon can significantly enhance the performance of the system. It is believed that with the development of the condition monitoring technology and the external conditions prediction, the proposed methods is promising for realistic systems.

Nomenclature

N	number of states for a component
N_c	number of components
t	the tth time interval
i	the ith component
Θ_D	set of states, $\Theta_D = \{D_k, 1 \leq k \leq N\}$
D_k	one state of a component, $k = \{1, 2, \ldots, N\}$
MA_m	The mth activity, $m = \{1, 2, \ldots, N\}$
$\mathbf{P}_{i,0}$	transition matrix of component i with deterioration processes
$\mathbf{P}_{i,m}$	transition matrix of component i with maintenance activity MA_m.
$p_{jk}^{(i)}$	transition probability of component i from state D_k to D_j
$\mathbf{a}_{i,t}$	activity vector of component i at the tth time interval
$\mathbf{S}_{i,t}$	state vector of component i at the tth time interval
$\mathbf{\Omega}_t$	state matrix for multiple components at the tth time interval
$\mathbf{A}_t$	activity matrix for multiple components at the tth time interval
$P(\cdot)$	probability of going from one state to another for a given set of activities.
Θ_C	set of all components
$C_{A,t}(\cdot)$	cost caused by activities at the tth time interval
$C_{L,t}(\cdot)$	cost caused by the loss of load at the tth time interval
$C_{S,t}(\cdot)$	successive cost at the tth time interval
$v_t(\cdot)$	expected cost-to-go at the tth time interval
$v_t^*(\cdot)$	minimum expected cost-to-go at the tth time interval
Θ_N	set of state matrices with all components in their normal operating states
Θ_F	set of state matrices with any components in failure states
Θ_A	set of activity matrices with all components in normal operating states
M	number of fault components
l	the lth fault component
l'	serial number of the lth fault component in all components
$\mathbf{B}$	an $M \times 1$ vector, where the value of each element in $\mathbf{B}$ can be 1 or 0
$P_t^{(E)}(\cdot)$	probability that certain fault components can be repaired and others cannot be repaired

$P_{l,t}^{(R)}$	probability that the lth fault component can be repaired at the tth time interval
$\mathbf{A}_t^{(E)}$	activity matrix for multiple components at the tth time interval with harsh external conditions
$C_{L,t}^{(E)}(\cdot)$	cost caused by the loss of load at the tth time interval considering harsh external conditions
$C_{R,t}^{(E)}(\cdot)$	repair cost at the tth time interval considering harsh external conditions
$C_{S,t}^{(E)}(\cdot)$	successive cost at the tth time interval considering harsh external conditions
$c_{R,l}$	repair cost for the lth fault component

References

1 P. Dehghanian, M. Fotuhi-Firuzabad, F. Aminifar, and R. Billinton, "A comprehensive scheme for reliability centered maintenance in power distribution systems–part I: Methodology," *IEEE Transactions on Power Delivery*, vol. 28, no. 2, pp. 761–770, 2013.

2 P. Dehghanian, M. Fotuhi-Firuzabad, F. Aminifar, and R. Billinton, "A comprehensive scheme for reliability-centered maintenance in power distribution systems–part II: Numerical analysis," *IEEE Transactions on Power Delivery*, vol. 28, no. 2, pp. 771–778, 2013.

3 E. Silva, M. Schilling, and M. Rafael, "Generation maintenance scheduling considering transmission constraints," *IEEE Transactions on Power Systems*, vol. 15, no. 2, pp. 838–843, 2000.

4 A. Conejo, R. Garcia-Bertrand, and M. Diaz-Salazar, "Generation maintenance scheduling in restructured power systems," *IEEE Transactions on Power Systems*, vol. 20, no. 2, pp. 984–992, 2005.

5 Y. Fu, M. Shahidehpour, and Z. Li, "Security-constrained optimal coordination of generation and transmission maintenance outage scheduling," *IEEE Transactions on Power Systems*, vol. 22, no. 3, pp. 1302–1313, 2007.

6 J. Endrenyi, G. Anders, and A. Leite da Silva, "Probabilistic evaluation of the effect of maintenance on reliability. an application [to power systems]," *IEEE Transactions on Power Systems*, vol. 13, no. 2, pp. 576–583, 1998.

7 T. M. Welte, "Using state diagrams for modeling maintenance of deteriorating systems," *IEEE Transactions on Power Systems*, vol. 24, no. 1, pp. 58–66, 2009.

8 F. Yang and C. S. Chang, "Multiobjective evolutionary optimization of maintenance schedules and extents for composite power systems," *IEEE Transactions on Power Systems*, vol. 24, no. 4, pp. 1694–1702, 2009.

9 J.-H. Heo, M.-K. Kim, G.-P. Park, Y. T. Yoon, J. K. Park, S.-S. Lee, and D.-H. Kim, "A reliability-centered approach to an optimal maintenance strategy in transmission systems using a genetic algorithm," *IEEE Transactions on Power Delivery*, vol. 26, no. 4, pp. 2171–2179, 2011.

10 A. Abiri-Jahromi, M. Parvania, F. Bouffard, and M. Fotuhi-Firuzabad, "A two-stage framework for power transformer asset maintenance management–part I: Models and formulations," *IEEE Transactions on Power Systems*, vol. 28, no. 2, pp. 1395–1403, 2013.

11 A. Abiri-Jahromi, M. Parvania, F. Bouffard, and M. Fotuhi-Firuzabad, "A two-stage framework for power transformer asset maintenance management–part II: Validation results," *IEEE Transactions on Power Systems*, vol. 28, no. 2, pp. 1404–1414, 2013.

12 NERC's report, *Compliance application notice-protection system maintenance and testing evidence.* http://www.nerc.com.

13 NERC's report, *FERC/NERC staff report on the 2011 Southwest cold weather event.* http://www.nerc.com.

14 IEEE Standard 516-2009, *Guide for maintenance methods on energized power lines.* 2009.

15 IEEE Standard 3007.2-2010, *Recommended practice for the maintenance of industrial and commercial power systems.* 2010.

16 G. Li, P. Zhang, P. B. Luh, W. Li, Z. Bie, C. Serna, and Z. Zhao, "Risk analysis for distribution systems in the Northeast U.S. under wind storms," *IEEE Transactions on Power Systems*, vol. 29, no. 2, pp. 889–898, 2014.

17 B. Ruszczak and M. Tomaszewski, "Extreme value analysis of wet snow loads on power lines," *IEEE Transactions on Power Systems*, vol. 30, no. 1, pp. 457–462, 2015.

18 M. Puterman, *Markov Decision Process.* New York: Wiley, 1994.

3

Preallocating Emergency Resources to Enhance Grid Survivability

Emergency resources are critical flexibility for resilient response of grids against natural disasters. Nonetheless, typically they are insufficiently utilized. For their improved utilization and higher grid survivability, this chapter uses truck-mounted mobile emergency generators (MEGs) as an example, and proposes scheduling MEGs as distributed generations (DGs) in distribution systems (DSs) to recover critical loads by temporarily forming microgrids (MGs). As MEGs' travel time on road networks (RNs) can significantly influence critical loads' outage duration, a two-stage dispatch approach consisting of preallocation and real-time allocation is presented, and the traffic issue is taken into account in the problem of vehicle routing. Preallocation places MEGs in staging locations before the natural disaster, and real-time allocation dispatches MEGs from staging locations into DSs to recover critical loads by MG formation after the natural disaster strikes. Particularly, using an objective that minimizes the expected outage duration weighted by the loads' priorities and demand sizes, preallocation is conducted via a two-stage scenario-based stochastic optimization problem, in which the first-stage preallocation decisions are assessed by a number of second-stage real-time allocation subproblems corresponding to the considered scenarios of DS and RN damages or congestion. A scenario decomposition algorithm is used to solve this problem. Illustrative cases are presented to demonstrate the effectiveness of the proposed dispatch approach and algorithm.

3.1 Emergency Resources of Grids against Disasters

Many severe power outages have been caused by natural disasters in recent years. For instance, in 2012, around 8.35 million people were reported losing power, after the Hurricane Sandy struck the US East Coast [1]. Power outages due to extreme weather events have resulted in significant economic loss and life risk, highlighting the importance of improving resilience for electric power

Power Grid Resilience against Natural Disasters: Preparedness, Response, and Recovery, First Edition.
Shunbo Lei, Chong Wang, and Yunhe Hou.
© 2023 John Wiley & Sons Ltd. Published 2023 by John Wiley & Sons Ltd.

grids [2]. Prompt and efficient response for electric service recovery is one of the critical requirements on a resilient power grid with a high-survivability level, since most of the fast recovery activities greatly rely on a reliable power supply [3]. However, widespread and severe damages to power grids can be caused by a natural disaster, leaving lots of customers without power for days, sometimes even for over a week. For a higher-survivability level by more prompt recovery, resilient response strategies are critically needed when natural disasters threaten communities.

Emergency resources, with MEGs as the focus of this chapter, are critical flexibility of power grids for fast and survivability-enhancing electric service restoration, especially when customers do not have access to power from the main grid, which is usually the case after the striking of a natural disaster [4–6]. MEGs are truck-mounted emergency generators having the merits of large capacity (up to several MVA) and mobility. When sustained damages result in prolonged electric service outages, MEGs are one of the most efficient emergency response resources. However, typically they are underutilized. For instance, 400 industrial-size truck-mounted emergency response generators were prepared by the Federal Emergency Management Agency (FEMA) before the striking of Hurricane Sandy. However, even three days after Sandy made landfall, only a fraction of them were providing power [7].

Effective utilization of emergency resources including MEGs, i.e. fast and efficient response, is greatly hindered by several challenges that are interrelated. First, assessments are needed to evaluate each outage area's needs for MEGs and emergency resources of different types. Currently, such assessments can take from hours to days after the striking of a natural disaster. Second, appropriate matching between unserved loads and MEGs and other emergency resources is required to assure the resources to be rationally utilized. Such a matching can be very difficult when considering multiple factors such as load priorities, operation modes, and grid damages. Third, RN congestion and damages might prevent the timely dispatch of emergency resources including MEGs, since natural disasters can result in very inefficient traffic that is a complicated yet needed consideration in the dispatch of MEGs and other emergency resources.

Specifically, the research on utilizing MEGs, i.e. dispatching MEGs for resilient and survivability-enhancing response against natural disasters, is insufficient. Reference [8] presents a detailed system design and experimental studies for MEGs using fuel cells. Reference [9] presents a MEG design and test results based on an integrated controller for both single-phase and three-phase distribution line backup. Reference [10] investigates the risks in grid protection due to the use of MEGs. Reference [11] briefly introduces various MEG designs and their backup roles in the British telecommunications network during emergency situations. In

Japan, telecommunications companies are also using MEGs to ensure the surviv-ability of telecommunications services, especially after an earthquake [12]. How-ever, references [11] or [12] do not present dispatch strategies, and power grids are surely very different from telecommunications networks.

References [13] and [14] study the optimal dispatch of MEGs. The sum of MEG investment and maintenance cost and customers' outage cost is minimized using a genetic algorithm in [13]. In [14], loads are first classified and prioritized based on multiple safety factors, and then the load loss is minimized using a hierarchical dispatch strategy. However, the models or methods presented in [13] and [14] have limited effectiveness. First, they do not appropriately consider or model the conse-quences of natural disasters (e.g. DS damages, RN damages or congestion). Second, they do not include a detailed formulation for the operation of power grids.

In this chapter, it is proposed to dispatch MEGs as DGs to some distribution grid nodes for restoring critical loads by forming multiple temporary MGs. Ref-erence [4] proposes and studies a novel electric service restoration method that forms MGs energized by existing DGs. One significant difference is that in [4], DGs are preinstalled resources, while the method in this chapter involves the optimal MEG placement. Note that the optimal DG placement problem, which generally assumes normal operation states for DSs and aims to improve service reliability, enhance voltage profiles, reduce power losses, etc. [15], is also quite different from the problem considered here.

This chapter proposes to schedule and dispatch emergency resources in a two-stage framework that includes preallocation and real-time allocation. As indicated in Figure 3.1, prior to a natural disaster, according to the damage forecast of DSs and RNs, preallocation decisions are made via a scenario-based two-stage stochastic optimization model that minimizes the expected outage duration of loads considering their priorities and demand sizes. Then MEGs are preallocated (i.e. placed) in staging locations, so as to attain prompt real-time allocation in the future for improving grid survivability. The real-time allocation is optimized after the natural disaster strikes. Then MEGs are sent from staging

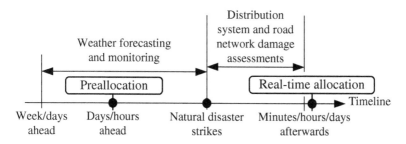

Figure 3.1 Timing of preallocation and real-time allocation.

locations to selected DS nodes for forming MGs and picking up critical loads. Thus, emergency resources can be much more effectively utilized for resilient and survivability-enhancing response of grids against natural disasters.

3.2 Mobile Emergency Generators and Grid Survivability

Mostly, the dispatch of emergency resources is to allocate them in the power grid for recovering critical loads under contingency situations. A MEG might serve a single customer, e.g. a government building or a hospital. However, facing a natural disaster, MEGs will have to play a role more important than that. First, most large and critical loads that are small in number have backup power from self-installed emergency generators or alternate feeders. And note that before a natural disaster, many MEGs will be prepared. Therefore, a considerable number of MEGs can be used to serve many small yet critical or less-critical loads. Second, MEGs can be vital power sources for many customers to survive for days or even for over a week after the striking of a natural disaster, since it often causes prolonged outages. Besides, as a response against natural disasters, MEG dispatch has the characteristics as follows:

- A partial or complete loss of power supply from the main grid to DSs is often caused by natural disasters. Broken DS feeders or laterals, substation faults, and transmission system outages, etc., are the major reasons. Conventional restoration strategies for timely electric service restoration of isolated outage regions can be hindered. Those are the main areas of interest for the dispatch of MEGs and other emergency resources. However, for areas that still have access to the main grid power, as their full restoration may be prohibited by operational constraints, they should also be considered in MEG dispatch.
- Prior to a natural disaster, sustained DS component damages, the resulting prolonged power outages to customers, and RN damages or congestion are very uncertain. After the natural disaster strikes, assessments of these may be finished in minutes or hours, or even days. Then, the matching between critical loads and MEGs is implemented to instruct MEG dispatch. Considering the grid damages, requirements on the utilization efficiency of MEGs, DS operational constraints, desired timely restoration, and survivability of critical loads, etc., such a matching is a task quite difficult and complicated.
- Including MEGs that are truck-mounted emergency-response generators, emergency resources will have to travel on RNs to the allocated locations. For critical loads to be picked up by MEGs, their outage duration and survivability level are greatly impacted by the travel time of MEGs. Some critical loads may also require

a MEG to arrive before their backup power runs out. Note that RNs can also be vulnerable to natural disasters. The traffic issue therefore is a critical factor in the dispatch of emergency resources. Proactive predispatch strategies (i.e. pre-allocation in this chapter) are needed, and the vehicle routing (VR) of MEGs needs to be considered, so as to reduce the critical loads' outage duration.

The typical MEG dispatch practice can be quite ineffective. First, major dispatch efforts are implemented after a natural disaster strikes, though MEGs are prepared to be ready in advance. It is preferred that proactive measures are taken. Second, the assessment of needs for MEGs and the matching between critical loads and MEGs are not rationally conducted. After the outage and damage assessment, sub-districts request MEGs based on some statistical data, and then match loads and MEGs based on limited experience. Third, as a critical factor noted previously, the traffic issue is not considered appropriately. Thus, MEGs' capability to improve grid survivability is typically underutilized.

3.2.1 Microgrid Formation

This chapter proposes to dispatch MEGs as DGs to some DS nodes. MEGs then form multiple MGs that operate in islanded mode to recover critical loads. The objective is minimizing the outage duration, considering the loads' sizes and priorities. The survivability of critical loads is thus enhanced.

Main decision variables are as follows:

- Allocation of MEGs: To allocate MEGs to candidate nodes in distribution grids.
- Formation of MGs powered by MEGs: To close or open each branch to form MGs, and to pick up or not to pick up each load.

Major constraints or conditions to be considered are as follows:

- MEGs' different capacities.
- Critical loads' priorities.
- Damages of grid components.
- RN damages or congestion.
- The radial topology requirements, operational constraints, etc.

Also note that, constrained by firmware or equipment requirements, only a fraction of DS nodes will be feasible as candidate nodes for allocating and connecting MEGs.

3.2.2 Preallocation and Real-Time Allocation

The proposed restoration strategy of forming MGs with MEGs is achieved via a two-stage dispatch approach, comprising preallocation and real-time allocation, to improve electric service survivability. The proposed two-stage dispatch methodology is applicable to the utilization of different emergency resources of power grids.

Prior to the natural disaster, i.e. in the first stage, preallocation is implemented. It is a common proactive measure of electric utilities to preallocate resources. They allocate resources (including restoration equipment and repair crews) in their staging locations before the natural disaster, to assure the earliest possible response after the natural disaster strikes [16]. We believe that preallocating MEGs, i.e. placing MEGs in staging locations for earliest response in the future, is also needed for the following interrelated reasons:

- Improving the survivability level depends on the earliest possible electric service restoration.
- The travel time of MEGs to allocated locations can be reduced.
- The impact of RN damages or congestion can be mitigated.

After the natural disaster strikes, i.e. in the second stage, the optimal real-time allocation is conducted. From staging locations, MEGs are sent to allocated places. MEGs are connected to the grid and form MGs upon arrival, to restore critical loads.

3.2.3 Coordination with Conventional Restoration Procedures

Here, the conceptual resilience curve in [17] is used for clearer explanations. In Figure 3.2, R is a metric indicating the grid resilience level, and t_{MEG} the timing of MEGs or other emergency resources participating in grid restoration.

Associated with an event, a distribution grid has the following states: event progress $t_{\mathrm{e}} \sim t_{\mathrm{pe}}$, postevent degraded state $t_{\mathrm{pe}} \sim t_{\mathrm{r}}$, restorative state $t_{\mathrm{r}} \sim t_{pr}$, postrestoration state $t_{pr} \sim t_{\mathrm{ir}}$, and infrastructure recovery $t_{\mathrm{ir}} \sim t_{\mathrm{pir}}$. The time window $t_{\mathrm{r}} \sim t_{\mathrm{ir}}$ is the concern of conventional distribution grid restoration strategies

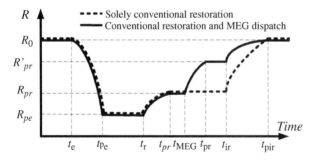

Figure 3.2 Conceptual resilience curves associated with an event.

that can be produced by multiagent systems [18], expert systems [17] and other optimization approaches [19]. Nevertheless, conventional strategies may only have limited effect for a system attacked by a natural disaster. That is, the improvement from R_{pe} to R_{pr} might be small. In such cases, emergency resources such as MEGs are desirable for flexibility to improve the grid resilience level from R_{pr} to R'_{pr} in the postrestoration state. With quick restoration enabled by emergency resources, the grid's survivability level is improved.

The concerned regions of emergency resources scheduling are mainly the isolated outage areas and some regions that cannot be fully restored by the surviving power from the main grid. Specifically, the concerned time window of dispatching MEGs is generally the same as that of conventional distribution grid restoration, while their concerned outage regions are different and have limited overlaps. To some extent, these two kinds of restoration actions are related yet independent. In general, they can be coordinated in such a straightforward manner to improve grid survivability: For outage regions to be sufficiently restored by conventional restoration strategies, adopt those strategies; for islanded outage regions without power sources, send MEGs and form MGs once MEGs arrive; for outage regions insufficiently restored by conventional restoration strategies, adopt conventional strategies first, and then switch to coordinated restoration strategies once MEGs arrive. The real-time allocation problem introduced later in this chapter can be used to generate coordinated strategies.

3.3 Preallocation Optimization of Mobile Emergency Generators

Starting from this section, MEGs are used as an example to introduce an optimization methodology for preallocating emergency resources to improve grid survivability.

3.3.1 A Two-Stage Stochastic Optimization Model

The preallocation of MEGs and other emergency resources can be optimized by stochastic optimization models similar to the following one:

$$\min \sum_{n \in \mathcal{N}} u_n \sum_{i \in \mathcal{B}} w_i p_i \left[\left(1 - \sum_{k \in \{\mathcal{F}, \mathcal{G}\}} \gamma_{ikn} \right) T_{in} + \sum_{k \in \mathcal{G}} \sum_{s \in \mathcal{S}} \sum_{m \in \mathcal{M}} \gamma_{ikn} y_{smkn} t_{skn} \right] \quad (3.1)$$

$$\tau_{smikn} \leq \gamma_{ikn}, \tau_{smikn} \leq y_{smkn}, \tau_{smikn} \geq \gamma_{ikn} + y_{smkn} - 1, \forall s, \forall m, \forall i, \forall k, \forall n \quad (3.2)$$

$$\sum_{m \in \mathcal{M}} x_{sm} \leq X_s, \forall s \quad (3.3)$$

$$\sum_{s \in S} x_{sm} = 1, \forall m \tag{3.4}$$

$$\sum_{k \in \mathcal{G}} y_{smkn} \leq x_{sm}, \forall s, \forall m, \forall n \tag{3.5}$$

$$\sum_{s \in S} \sum_{m \in \mathcal{M}} y_{smkn} \leq 1, \forall k, \forall n \tag{3.6}$$

$$z_{kn} = \sum_{s \in S} \sum_{m \in \mathcal{M}} y_{smkn}, \forall k \in \mathcal{G}, \forall n \tag{3.7}$$

$$z_{kn} = 1, \forall k \in \mathcal{F}, \forall n \tag{3.8}$$

$$\sum_{k \in \{\mathcal{F}, \mathcal{G}\}} v_{ikn} \leq 1, \forall i, \forall n \tag{3.9}$$

$$v_{ikn} \leq z_{kn}, \forall k, \forall i, \forall n \tag{3.10}$$

$$v_{kkn} \geq z_{kn}, \forall k, \forall n \tag{3.11}$$

$$v_{ikn} \leq v_{jkn}, j = \theta_k(i), \forall k, \forall i, \forall n \tag{3.12}$$

$$c_{ijn} = \sum_{k \in \{\mathcal{F}, \mathcal{G}\}} v_{hkn}, h = \zeta_k(i, j), \forall(i, j), \forall n \tag{3.13}$$

$$\gamma_{ikn} = v_{ikn} l_{in}, \forall k, \forall i, \forall n \tag{3.14}$$

$$\sum_{j \in \mathcal{N}_i^k} P_{jn}^k = P_{in}^k - \gamma_{ikn} p_i, \forall k, \forall i, \forall n \tag{3.15}$$

$$\sum_{j \in \mathcal{N}_i^k} Q_{jn}^k = Q_{in}^k - \gamma_{ikn} q_i, \forall k, \forall i, \forall n \tag{3.16}$$

$$0 \leq P_{in}^k \leq v_{ikn} \Pi, \forall k, \forall i \neq k, \forall n \tag{3.17}$$

$$0 \leq Q_{in}^k \leq v_{ikn} \Pi, \forall k, \forall i \neq k, \forall n \tag{3.18}$$

$$0 \leq P_{kn}^k \leq \sum_{s \in S} \sum_{m \in \mathcal{M}} y_{smkn} P_m^{\max}, \forall k \in \mathcal{G}, \forall n \tag{3.19}$$

$$0 \leq Q_{kn}^k \leq \sum_{s \in S} \sum_{m \in \mathcal{M}} y_{smkn} Q_m^{\max}, \forall k \in \mathcal{G}, \forall n \tag{3.20}$$

$$(P_{in}^k)^2 + (Q_{in}^k)^2 \leq (S_{ij}^{\max})^2, j = \theta_k(i), \forall k, \forall i \neq k, \forall n \tag{3.21}$$

$$V_{kn}^k = z_{kn} V_R, \forall k, \forall n \tag{3.22}$$

$$V_{in}^k = V_{jn}^k - \frac{r_{ij} \cdot P_{in}^k + x_{ij} \cdot Q_{in}^k}{V_R} - \delta_{in}^k, j = \theta_k(i), \forall k, \forall i \neq k, \forall n \tag{3.23}$$

$$0 \leq V_{in}^k \leq v_{ikn} V_R, \forall k, \forall i, \forall n \tag{3.24}$$

$$0 \leq \delta_{in}^k \leq (1 - v_{ikn}) V_R, \forall k, \forall i, \forall n \tag{3.25}$$

$$\sum_{k \in \{\mathcal{F},\mathcal{G}\}} v_{ikn}(1 - \varepsilon) V_0 \leq \sum_{k \in \{\mathcal{F},\mathcal{G}\}} V_{in}^k \leq \sum_{k \in \{\mathcal{F},\mathcal{G}\}} v_{ikn}(1 + \varepsilon) V_0, \forall i, \forall n \tag{3.26}$$

$$c_{ijn} = 0, \forall n, \forall (i,j) \in \mathcal{LO}_n \tag{3.27}$$

The objective function (3.1) is minimizing the expected outage duration, considering the loads' sizes and priorities. It thus maximizes the loads' survivability. Particularly, all the terms in the square brackets in whole, i.e. $[(1 - \sum_{k \in \{\mathcal{F},\mathcal{G}\}} \gamma_{ikn}) T_{in} + \sum_{k \in \mathcal{G}} \sum_{s \in \mathcal{S}} \sum_{m \in \mathcal{M}} \gamma_{ikn} y_{smkn} t_{skn}]$, represent the outage duration of a load. The second term in the parentheses, i.e. $\sum_{k \in \{\mathcal{F},\mathcal{G}\}} \gamma_{ikn}$, indicates if the load at node i is recovered by a feeder root node (with a substation) or a MEG in scenario n. First, if it equals to 0, this load is not recovered and will be in outage for the estimated recovery time T_{in}. In this case, $(1 - \sum_{k \in \{\mathcal{F},\mathcal{G}\}} \gamma_{ikn})$ equals to 1, and $\sum_{k \in \mathcal{G}} \sum_{s \in \mathcal{S}} \sum_{m \in \mathcal{M}} \gamma_{ikn} y_{smkn} t_{skn}$ equals to 0. Therefore, T_{in} takes effect in the objective (3.1) as this load's outage duration. Second, if $\sum_{k \in \{\mathcal{F},\mathcal{G}\}} \gamma_{ikn}$ equals to 1, this load is recovered by one of the two possible power sources. One possibility is that it attains power from a feeder root node. As its restoration can be implemented immediately or shortly, its outage duration is regarded as zero here. In this case, both $(1 - \sum_{k \in \{\mathcal{F},\mathcal{G}\}} \gamma_{ikn})$ and $\sum_{k \in \mathcal{G}} \sum_{s \in \mathcal{S}} \sum_{m \in \mathcal{M}} \gamma_{ikn} y_{smkn} t_{skn}$ equal to 0, making the terms in the square brackets in whole equal to 0. The other possibility is that a MEG recovers this load. The travel time of the MEG on RNs is regarded as its outage duration. In this case, $(1 - \sum_{k \in \{\mathcal{F},\mathcal{G}\}} \gamma_{ikn})$ still equals to 0. And assuming that this load is recovered by a MEG at node k_1, and this MEG is from staging location s_1, then the term $\sum_{k \in \mathcal{G}} \sum_{s \in \mathcal{S}} \sum_{m \in \mathcal{M}} \gamma_{ikn} y_{smkn} t_{skn}$ equals to $t_{s_1 k_2 n}$ under scenario n. By using the objective function (3.1), preallocation decisions with the optimal expected performance of future real-time allocation are obtained. Note that (3.1) includes quadratic terms of two binary variables, i.e. $\gamma_{ikn} y_{smkn}$. To linearize them, each of them is replaced by an auxiliary binary variable τ_{smikn} and constraints (3.2) are added to achieve equivalent transformation.

The rest constraints can be separated into five different groups as below:

- Preallocation constraints (3.3) and (3.4) to preallocate the MEGs and to avoid violating the capacity of staging locations: at each staging location, the number of preallocated MEGs is restricted by its capacity as in (3.3). Constraint (3.4) ensures that a MEG is preallocated to exactly one staging location.
- Real-time allocation constraints (3.5) and (3.6) to dispatch MEGs to selected grid nodes from staging locations: Constraint (3.5) ensures that in each scenario, MEG m is sent from its preallocated staging location s to one of the candidate

DS nodes. Constraint (3.6) declares that for each candidate node, at most one MEG, if any, is allocated.

- DS network reconfiguration constraints (3.7)–(3.14) to represent feasible topologies that meet the radial topology requirement and separate the system into MGs (each MG has one power source): A subgrid powered by a MEG at node k or a feeder root node k is referred to as MG k. Auxiliary variables z_{kn} in (3.7) and (3.8) indicate if node k has a power source in scenario n (i.e. $z_{kn} = 0$ if it is not connected with a MEG and it is not a feeder root node, $z_{kn} = 1$, otherwise). Constraint (3.9) assures that each DS node is included in at most one MG ($v_{ikn} = 1$ if under scenario n, MG k includes node i, $v_{ikn} = 0$, otherwise). Constraint (3.10) avoids constructing a MG without any power sources (if no power sources exist at node k, i.e. $z_{kn} = 0$, then the model will not form MG k, i.e. $v_{ikn} = 0$ for all i). Constraint (3.11) determines that any node with a power source is included in the MG powered by itself (i.e. if $z_{kn} = 1$, then $v_{kkn} = 1$). Constraint (3.12) ensures that a node can be included in MG k only if its parent node is also included in MG k, due to a spanning tree's connectivity features (in a radial distribution network, each MG can be regarded as a subtree network, and the root node is its power source node). Equation (3.13) means that a DS branch should be closed if its child node is included in a MG (if both nodes i and j are included in MG k, i.e. $v_{ikn} = v_{jkn} = 1$, then the branch (i,j) is also included in MG k, i.e. $c_{ijn} = 1$ indicating a closed state of the branch; considering constraint (3.12), this relationship is equivalent to that the child node of branch (i,j) is included in MG k). The DS radial topology is ensured by constraints (3.9)–(3.13). Constraint (3.14), which can be linearized similarly as in (3.2), ensures that MG k picks up the load at node i only if node i is included in MG k and its load switch is closed. Constraints in this group and the next group define the feasible set of related variables that realize the strategy of forming MGs with MEGs. Note that besides isolated outage areas, regions that still have access to the main grid power are also formulated for reasons aforementioned.
- Distribution grid operational constraints (3.15)–(3.26), to constrain MEG power outputs, voltages, and line flows: Based on the DistFlow model [20, 21], Equations (3.15) and (3.16) express each node's real and reactive power balance conditions. Constraints (3.17) and (3.18) assure that if node i is not included in MG k, its in-flow power regarding MG k is zero (i.e. $P_{in}^k = Q_{in}^k = 0$ if $v_{ikn} = 0$). Constraints (3.19) and (3.20) restrict the real and reactive power injection at a DS node according to the connected MEG's capacity (if node k is connected with the m-th MEG, its power injection P_{kn}^k and Q_{kn}^k have to be smaller than its capacity P_m^{max} and Q_m^{max}). Constraint (3.21) is the apparent power capacity constraint of each line. It can be linearized using techniques such as that in [22]. Equation (3.22) sets the reference voltage value $V_R = (1 + \varepsilon)V_0$ as the voltage

at nodes with power sources. Equation (3.23) expresses the relationship of voltages between connected nodes regarding MG k. Constraint (3.24) assures that if a node is not included in MG k, its voltage value regarding MG k should be zero (i.e. $V_{in}^k = 0$ if $v_{ikn} = 0$). Constraint (3.25) specifies the range of voltage slack variables which ensure the satisfaction of (3.23). Constraint (3.26) sets the voltage limit if a node is included in one of the MGs.

- Distribution grid damage scenario constraint (3.27) to indicate scenario conditions: with $\mathcal{LO}_n$ defining the set of damaged branches in scenario n, constraint (3.27) restricts them to be inoperable, i.e. fixed in the open state.

Therefore, the preallocation model is a scenario-based, two-stage stochastic optimization problem. Its preallocation decisions in the first stage are assessed by a number of real-time allocation problems in the second stage associated with the considered scenarios of DS damages and RN damages or congestion. The preallocation decisions therefore use emergency resources to improve the expected survivability of the grid.

Given the optimal preallocation decisions, i.e. determined x_{sm}, the actual real-time allocation of MEGs and other emergency resources can be implemented after the natural disaster strikes and the assessment of damages to DSs and RNs is completed. With only one scenario n indicating the natural disaster's results, real-time allocation is optimized using the model as follows:

$$\min \sum_{i \in B} w_i p_i \left[\left(1 - \sum_{k \in \{F, G\}} \gamma_{ikn} \right) T_{in} + \sum_{k \in G} \sum_{s \in S} \sum_{m \in M} \tau_{smikn} t_{skn} \right] \tag{3.28}$$

$$\text{s.t. } (3.2), (3.5) - (3.27)$$

Other relevant constraints can also be incorporated, e.g. the tolerable interruption time of a critical load at node i:

$$\sum_{k \in G} \gamma_{ikn} \geq 1 \tag{3.29}$$

$$\sum_{k \in G} \sum_{s \in S} \sum_{m \in M} \gamma_{ikn} y_{smkn} t_{skn} \leq \bar{t}_i \tag{3.30}$$

If the MEG installation time is known, it can be simply added to t_{skn} in (3.1), (3.28), and (3.30), and we can add a requirement on MEGs' capacity utilization rate (CUR), i.e. R:

$$\sum_{i \in B} \gamma_{ikn} p_i \geq R \sum_{s \in S} \sum_{m \in M} y_{smkn} P_m^{\max}, \forall k \tag{3.31}$$

where the CUR of a MEG is defined as the sum of real power demand it serves, divided by its real power capacity.

3.3.2 Availability of Mobile Emergency Generators

The capacities of available MEGs are set as parameters in the above formulations. Dispatchers, in some cases, are allowed to or have to choose a budgeted number of emergency resources, e.g. MEGs of different types. The above formulations can accommodate such cases by listing a sufficient number of MEGs of each type and letting the optimizer to select. A budget constraint restricting the number of selected MEGs should also be included. By doing this, the MEG capacities are cooptimized with other decisions. Note that this might increase the computational burden, which can be relieved by experience-based heuristics. For instance, a good-quality initial solution can be set at first by experienced dispatchers.

3.3.3 Connection of Mobile Emergency Generators

To operate a MG including two or more generators, their coordination needs to be handled [23, 24]. This can be difficult not only for MEGs temporarily installed but also for generators preinstalled, due to communication obstructions and the system's postdisaster-degraded state. Therefore, though having two or more MEGs in a MG might perform better in matching power sources and loads, enhancing the voltage security and improving the grid survivability, here we choose to have only one power source in each MG [4]. For some types of emergency generators that have required functional modules, operating two or three MEGs in a parallel manner can be implemented by well-trained staff [12]. The premise is often that the MEGs are connected to the same distribution grid node. This special case can be accommodated by the above formulations using minor modifications. Particularly, constraints (3.6) and (3.7) should be modified as below:

$$\sum_{s \in S} \sum_{m \in M} y_{smkn} \leq 3, \forall k, \forall n \tag{3.32}$$

$$\frac{\sum_{s \in S} \sum_{m \in M} y_{smkn}}{3} \leq z_{kn} \leq \sum_{s \in S} \sum_{m \in M} y_{smkn}, \forall k \in \mathcal{G}, \forall n \tag{3.33}$$

which allow at most three MEGs connected to a DS node.

In terms of selecting candidate nodes for MEG connection, the following issues should be considered:

- Site requirements. The locations of candidate nodes should have appropriate space for installing and operating MEGs and/or other emergency resources, and not be threatened by risks such as flooding [12].

- Access requirements. Their locations need to be reachable for fuel trucks and truck-mounted MEGs via the RNs [12].
- Facility requirements. It is preferred to select nodes with connection panels to interface with MEGs' plug-terminated cables [11]. A location with an underground fuel tank is also preferred [11, 12].
- Other considerations such as the noise.

If needed, the number of candidate nodes can be further reduced based on dispatchers' preferences and other factors including the distance.

3.3.4   Coordination of Multiple Flexibility in Microgrids

Other than MEGs, there can be other power or flexibility resources in the formed microgrids. Besides MEGs, feeder root nodes representing substation power are also modeled in the current formulations. Other flexibility resources, such as DGs and the measures of partial restoration/curtailment of loads, can also be included to further improve grid survivability. Particularly, DGs can be formulated in a manner similar to that of feeder root nodes, so that DGs recover critical loads by MG formation similar to that in [4]. By some easy modifications of the model such as adding a continuous variable for each load, partial restoration/curtailment of loads can be formulated, too. Since this strategy is not always practically feasible, it is omitted here. More importantly, potentially it can result in over-positive CUR results; therefore, over-evaluating the presented method's performance in utilizing emergency resources. The focus of this chapter is to design an approach for better using MEGs and other emergency resources in response to natural disasters.

3.4   Solution Algorithms

Figs. 3.1 and 3.2 have shown the implementation timeline of dispatching MEGs and other emergency resources to enhance power grid survivability. Figure 3.3, which is self-explanatory, indicates the relationships among task modules and data sets that are needed to conduct the MEGs' preallocation and real-time allocation. Based on the framework shown in Figure 3.3, the proposed methodology can be easily adjusted to address the dispatch of other emergency resources. For example, the utility repair truck-scheduling problem can be resolved similarly using a framework with proactive preallocation and timely real-time response. The detailed grid operation model is also needed to optimize the electric service restoration and recovery process. Three selected issues relevant to implementing the proposed dispatch method are discussed as follows:

Algorithm 3.1 Scenario Decomposition Algorithm

1: Set *lower_bound* $= -\infty$, *upper_bound* $= +\infty$, $\mathbf{x}^* = \emptyset$, $\mathbf{S} = \emptyset$

2: Set $\lambda = \mathbf{0}$

3: Solve $\alpha_n = \min\{f_n(\mathbf{x}_n, \mathbf{y}_n) + \lambda^{\mathrm{T}}\mathbf{A}_n\mathbf{x}_n : \mathbf{x}_n \in \Lambda \setminus \mathbf{S}, \mathbf{y}_n \in \mathbf{\Omega}_n\}$ and attain $\mathbf{x}_n^*$ for all n

4: If meeting consensus criteria, update λ and go to *Step 3*; otherwise, go to *Step 5*

5: Update $\mathbf{S}' = \bigcup_n \{\mathbf{x}_n^*\}$, $\mathbf{S} = \mathbf{S} \bigcup \mathbf{S}'$, *lower_bound* $= \sum_n u_n \alpha_n - \lambda^{\mathrm{T}}\mathbf{h}$

6: For all $\mathbf{x}^0 \in \mathbf{S}'$ do

7: Solve $\beta_n = \min\{f_n(\mathbf{x}^0, \mathbf{y}_n) : \mathbf{y}_n \in \mathbf{\Omega}_n\}$ for all n

8: If *upper_bound* $\geq \sum_n u_n \beta_n$, update $\mathbf{x}^* = \mathbf{x}^0$, *upper_bound* $= \sum_n u_n \beta_n$

9: If (*upper_bound* $-$ *lower_bound*)/*lower_bound* $\geq$ *convergence_tolerance*, go to *Step 2*; otherwise, terminate

---

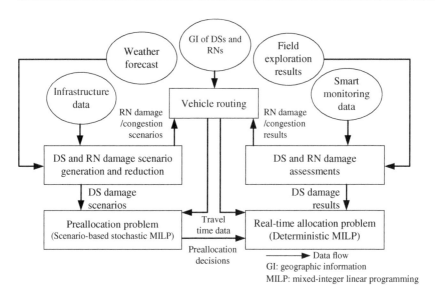

Figure 3.3   Relationships among task modules and data sets.

3.4.1   Scenario Generation and Reduction

On the one hand, environmental data and power grid data can be used in statistical data fitting models to estimate damages and outages due to natural disasters. Model performance can be evaluated by measurements of fitting goodness. For instance, reference [25] assesses models such as Bayesian Additive Regression Trees in predicting the number of damaged distribution grid poles. On the other hand, physical mechanisms of damages and outages are considered in simulation-based models to make predictions. For instance, reference [26] studies the mechanism of overhead lines damaged by localized high-intensity wind. References [16, 27], and [28] provide more detailed reviews. In general, it is an active, challenging, and critical research topic of using statistical or simulation-based models for generating scenarios of natural disaster damages. A potentially important topic is the use of graph-theoretic indicators. In this chapter, mature tools or extensive expert experience for scenario generation are assumed available, and we do not explore this topic as it is not the focus. Particularly, in Section 3.5, scenarios are generated following [29] and [30], i.e. comparing vulnerable components' failure probabilities with a uniformly distributed random variable in the interval $(0, 1)$. Three rules are used to reduce scenarios: First, scenarios that MEGs can recover more critical loads are prioritized, as it is naturally of little interest for MEG dispatch to consider scenarios that MEGs are not quite effective in improving grid survivability. Second, scenarios that have the same real-time allocation solution are aggregated, since they tend to have similar impact on preallocation decisions. Third, scenarios with higher probabilities are prioritized. Note that a lognormal distribution is assumed for the MEG travel time on each RN edge.

3.4.2 Dijkstra's Shortest-Path Algorithm

As abovementioned, the optimal dispatch of emergency resources, which have to travel on RNs to allocated locations, can be influenced by the traffic issue. Therefore, a VR module finding the fastest or shortest path from an origin to a destination is adopted. VR can be solved by the Floyd–Warshall algorithm [31], and the Dijkstra's algorithm [32], etc. Since only paths from staging locations to candidate nodes are of interest, the Dijkstra's algorithm is applied here. The VR module derives the travel time from staging locations to candidate nodes for each scenario. Then the data are used in both preallocation and real-time allocation problems. Therefore, the traffic issue is accounted for when optimizing emergency resource dispatch to improve grid survivability, e.g. by reducing the outage time of critical loads to be recovered by MEGs. One might also consider assigning must-pass locations for MEGs and other emergency resources to use their exploration role in assessing DS and RN damages.

3.4.3 Scenario Decomposition Algorithm

The scenario-based, two-stage stochastic preallocation optimization model has a block-diagonal structure. It can be recast in a compact form as follows:

$$\min \left\{ \sum_n f_n(\mathbf{x}, \mathbf{y}_n) : \mathbf{x} \in \Lambda, \mathbf{y}_n \in \Omega_n \right\} \tag{3.34}$$

where $\mathbf{x}, \mathbf{y}_n, f_n(\mathbf{x}, \mathbf{y}_n), \Lambda$, and Ω_n are first-stage preallocation variables, second-stage real-time allocation variables in scenario n, second-stage objective function for scenario n, the feasible set of $\mathbf{x}$ defined by constraints (3.3)–(3.4), and the feasible set of $\mathbf{y}_n$ defined by constraints (3.5)–(3.27), respectively. Generally, the scenario decomposition (SD) algorithm is to transform problem (3.34) into problem (3.35):

$$\min \left\{ \sum_n f_n(\mathbf{x}_n, \mathbf{y}_n) : \mathbf{x}_n \in \Lambda, \mathbf{y}_n \in \Omega_n, \sum_n A_n \mathbf{x}_n = \mathbf{h} \right\} \tag{3.35}$$

where the last constraint is the nonanticipativity constraint requiring $\mathbf{x}_1 = \mathbf{x}_2 = \cdots = \mathbf{x}_n$. Therefore, the preallocation model has a block-diagonal structure to be taken advantage of. Procedures of the SD algorithm are presented in Algorithm 3.1.

The SD algorithm is actually a modified progressive hedging algorithm [33, 34] that can be grouped as an augmented Lagrangian relaxation algorithm. Particularly, $f_n(\mathbf{x}_n, \mathbf{y}_n) + \lambda^{\mathrm{T}} A_n \mathbf{x}_n$, i.e. the objective function in Step 3 of Algorithm 3.1, can be rewritten in a more detailed form as follows:

$$\min_{\mathbf{x}_n, \mathbf{y}_n} \sum_l \left[\mu(l) + \lambda(l) - \rho \overline{\mathbf{x}}_n(l) + 0.5\rho \right] \mathbf{x}_n(l) + \vartheta(l) \mathbf{y}_n(l) \tag{3.36}$$

where $\overline{\mathbf{x}}_n$ denotes the weighted average of $\mathbf{x}_n$ over all scenarios $n \in \mathcal{N}$ in the previous iteration; μ and ϑ are cost-coefficient vectors, i.e. problem parameters; $(\bullet)(l)$ denotes the l-th element of the vector $(\bullet)$; ρ is an algorithm parameter set as 10 000 in this chapter. Note that, other than fixed ρ strategies, there exist variable ρ strategies. Interested readers can refer to [35–37], which discuss both of them. And λ is updated by

$$\lambda(l) \leftarrow \lambda(l) + \rho \left[\mathbf{x}_n(l) - \overline{\mathbf{x}}_n(l) \right] \tag{3.37}$$

which is a commonly used updating rule. In Step 4 of Algorithm 3.1, the consensus criterion is set as below:

$$td = \sum_{l,n} \frac{|\mathbf{x}_n(l) - \overline{\mathbf{x}}_n(l)|}{|\mathcal{N}|} \le td \tag{3.38}$$

where td is defined as the average per scenario total deviation of $\mathbf{x}_n$ from the average $\overline{\mathbf{x}}_n$; td denotes a threshold set as 0.5; and $|\mathcal{N}|$ is the number of scenarios. The consensus criteria can also be set based on other metrics such as the normalized average per-scenario deviation from the average [35] and the overall cost discrepancy [36]. And the postdisaster real-time allocation problem can be solved directly by a solver such as Gurobi with sufficient efficiency.

3.5 Case Studies

3.5.1 Test System Introduction

The test system used for illustration is shown in Figs. 3.4 and 3.5. A district with four DSs in four subdistricts is considered, since emergency resources dispatch usually involves a region with multiple DSs. There are 51 intersections

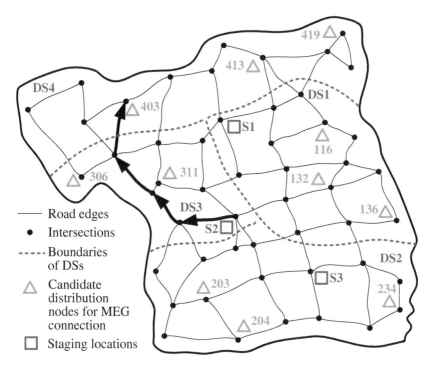

Figure 3.4 Geographic information for RNs, staging locations, DSs, and candidate distribution nodes for connecting MEGs.

and 82 edges in the RNs shown in Figure 3.4. The distribution grids in Figure 3.5 have totally 114 nodes. We randomly generate the loads' sizes and priority weights. Several nodes in DSs are picked as candidate nodes for MEG connection. It is assumed that the utility in this area has three staging locations with each being able to accommodate two MEGs and has emergency resources listed in Table 3.1. We consider a hurricane, with high wind being a major influential factor. Using branches' failure probabilities from fragility curves, DS damage scenarios are generated. The travel time on each RN edge in different scenarios is assumed a lognormal distribution.

3.5.2 Demonstration of the Proposed Dispatch Method

This subsection demonstrates how the proposed emergency resources dispatch approach, i.e. the two-stage method of preallocation and real-time allocation, is implemented step by step to dispatch MEGs for improving grid survivability:

- Before the hurricane, with appropriate scenario generation and reduction, the emergency resources preallocation problem is solved. It finds the

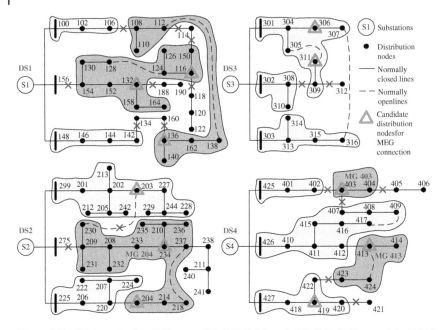

Figure 3.5 DSs' topologies (DS1: a modified IEEE 34-node DS [38]; DS2: a modified IEEE 37-node DS [39]; DS3: a modified 16-node DS [40]; DS4: a modified 27-node DS [41]), and a demonstration case.

Table 3.1 Available MEGs' capacities.

m	P_m^{max} (kW)	Q_m^{max} (kVar)
1	400	300
2	800	600
3	1000	800
4	1600	1200
5	2000	1600

preallocation of MEGs with the best-expected future real-time allocation performance in enhancing the survivability of critical loads. Twenty scenarios are considered. Decisions of preallocation for the studied case are listed in the second column of Table 3.2.

- After the hurricane, an evaluation on RN status and DS damages is conducted. Then the emergency resources real-time allocation problem is solved for the resulting scenario. It is assumed that branches with a cross in Figure 3.5 are damaged. The third column of Table 3.2 lists the real-time allocation decisions.

Table 3.2 Demonstration of the proposed MEG dispatch approach.

Staging location	Preallocation decisions (x_{sm})		Real-time allocation decisions (y_{smkn})
S1	MEG 3	$\longrightarrow$	Allocated to node 413
	MEG 4	$\longrightarrow$	Allocated to node 116
S2	MEG 1	$\longrightarrow$	Allocated to node 403
	MEG 2	$\longrightarrow$	Allocated to node 136
S3	MEG5	$\longrightarrow$	Allocated to node 204

- Emergency resources, i.e. MEGs in this study, are then sent to allocated DS nodes from staging locations. Their travel paths on RNs are decided in the former step that uses the VR module to compute needed parameters (i.e. the travel time t_{skn}) for the real-time allocation problem. For instance, MEG 1 at staging location S2 is allocated to node 403. The arrowed bold edges in Figure 3.4 indicate its travel path obtained by the VR module, and it uses around 21 minutes of traveling time. A new path can also be generated, if the RN status is updated.

- After MEGs arrive at allocated locations, they are connected to distribution grids and recover critical loads by forming MGs. Therefore, the survivability of critical loads is improved. In Step 2, when the emergency resources real-time allocation problem is solved, the formation of MGs is already determined. Figure 3.5 depicts the MG formation results for the assumed resulting scenario. Each region more deeply shadowed is a MG powered by a MEG. MG 413 and MG 204 marked in Figure 3.5 explain why regions still having main grid power should be included in the formulation. Though loads in MG 413 and MG 204 can reach a feeder root node via undamaged branches, they cannot be fully recovered due to operational constraints, e.g. the limit on branch flow. Thus, MEGs are connected for better recovery. Note that there are still some unserved loads. Further conventional repair and restoration efforts are needed to attain complete electric service recovery.

3.5.3 Capacity Utilization Rate

After a natural disaster strikes, the number of emergency resources typically will be less than enough to meet the demands. In this study, MEGs are expected to be fully used for supplying a maximum amount of critical loads, so that the survivability enhancement is maximized. Therefore, the matching between MEGs and critical loads is important, as mentioned above. In this subsection, the capacity utilization of MEGs is investigated. The evaluation process is to do simulations

Table 3.3 Statistics of capacity utilization rates in simulations.

MEG No. m	Using objective (3.28)		Using objective (3.40)	
	CUR	Times when CUR $\geq$ 85%	CUR	Times when CUR $\geq$ 85%
1	73.03%	221	74.58%	228
2	64.13%	97	80.87%	252
3	75.93%	242	87.28%	323
4	63.65%	149	71.36%	221
5	60.41%	126	72.14%	179
Average	67.43%	167	77.25%	240.6

by solving real-time allocation problems for different scenarios based on fixed preallocation decisions, and then summarize the simulation results. We use the CUR as an evaluation index here. If MEG m is connected to node k in scenario n, its CUR will be

$$CUR = \left(\sum_{i \in B} \gamma_{ikn} p_i \bigg/ P_m^{\max} \right) \times 100\% \tag{3.39}$$

Based on preallocation decisions discussed in Section 3.5.2, we conducted 500 real-time allocation simulations. Statistics for the CUR of each MEG with different capacities are listed in the second column of Table 3.3. With the average MEG CUR being 67.43%, the real-time allocation's performance in matching critical loads with emergency resources to improve grid survivability is acceptable. In fact, using (3.28) as the real-time allocation problem's objective, partial MEG capacities are sacrificed to attain prompt recovery of high-priority critical loads. Here, the CUR of a modified real-time allocation problem is further investigated. That is, the real-time allocation problem's objective function (3.28) is replaced with the following one that drops considerations of loads' priorities and emergency resources' travel time on RNs:

$$\max \sum_{i \in B} \sum_{k \in \{\mathcal{F}, \mathcal{G}\}} \gamma_{ikn} p_i \tag{3.40}$$

Again, 500 simulations of real-time emergency resources allocation were implemented. The fourth column of Table 3.3 lists the statistics for the CUR of MEGs. As indicated by the results, the MEG capacity utilization performance is now even better. System survivability is also improved in terms of the total amount of surviving loads. Therefore, one can change settings of the objective function based on one's preferences. Actually, whether choosing (3.28) or (3.40) as the objective, in general, the proposed real-time allocation model is able to appropriately match loads with MEGs as long as the distribution grid topology allows. As the third and fifth columns of Table 3.3 indicate, in a majority of the simulated cases, different types of MEGs reach a CUR level above 85%. However, note that in some other simulated cases, DS damages due to the hurricane may force some MEGs to recover small amounts of demands in some small islanded outage regions, e.g. the MG 403 shown in Figure 3.5.

In fact, it is reasonable to reserve 25~35% of MEGs' capacities, so as to accommodate the load fluctuations and help recover adjacent loads after quickly repairing some grid components. For instance, in Figure 3.5, after quickly repairing branch (404,405), the MEG at node 403 can recover the loads at nodes 405 and 406. Electric service survivability is thus enhanced, as the restoration of these loads will not have to wait for the postdisaster repairing and recovery of feeders and main grids, which can take much longer.

Results in this subsection demonstrate the proposed method's capability of matching emergency resources and loads; therefore, attaining better utilization of MEGs and greater improvement of system survivability. First, the strategy of forming MGs with MEGs can avoid insufficient utilization of MEGs' capacities. Second, MEG allocation and load pickup in forming MGs are cooptimized by our models. Various conditions and constraints, which are major challenges for the current practice of MEG dispatch, are appropriately included.

3.5.4 Importance of Considering Traffic Issue and Preallocation

As summarized in Table 3.4, real-time allocation of emergency resources, i.e. MEGs in this study, is simulated 100 times for each setting.

First, we show that to reduce the traffic issue's impact on postdisaster real-time allocation and to better improve grid survivability, it is necessary to conduct proactive preallocation of emergency resources. As shown by the third row of Table 3.4, with optimal preallocation, MEGs' time spent on RNs in real-time allocation is short. After the posthurricane damage and outage assessments of RNs and DSs

Table 3.4 Statistics of load restoration and MEG travel time in simulations.

	Average amount of loads recovered by MEGs	Average travel time to allocated locations
Nonoptimal preallocation	3866.8 kW	28.26 minutes
Optimal preallocation using objective (3.28)	3910.9 kW	21.58 minutes
Optimal preallocation using objective (3.41)	4082.5 kW	35.05 minutes

are partly or fully completed, the optimal real-time allocation plan can be attained. MEGs are then sent to allocated nodes and will arrive soon to recover critical loads. If preallocation is not conducted, of course, one can still use the same real-time allocation plan to recover the same critical loads. Nevertheless, MEGs might take longer time traveling on RNs to allocated nodes from other distant areas. If some administrative processes are further involved, the loads' outage time can be hours or even days longer, before they are restored by MEGs.

Besides, rather than a random preallocation, the optimal preallocation of emergency resources is desired for grid survivability improvement. The second row of Table 3.4 includes simulation statistics of real-time allocation when random and nonoptimal preallocation decisions are used. Compared to real-time allocation simulations based on the optimal preallocation, the amount of restored load is slightly less, and the travel time of MEGs on RNs is 30.95% longer.

Second, we show the importance of explicitly considering the traffic issue in real-time allocation of emergency resources. We experimentally alter the objective of real-time allocation to be the following one:

$$\min \sum_{i \in B} w_i p_i \left(1 - \sum_{k \in \{F, G\}} \gamma_{ikn} \right) T_{in} \tag{3.41}$$

That is, we drop the traffic issue consideration. As indicated by the last row of Table 3.4, based on the optimal preallocation, although the amount of restored load is slightly higher, this change leads to a longer outage duration of loads to be restored by MEGs. That is, the survivability enhancement is smaller.

Simulations validate the importance of accounting for the traffic issue and adopting preallocation. Solving the VR problem by approaches including the Dijkstra's algorithm, and then including relevant parameters in both preallocation and real-time allocation problems are effective in decreasing the outage time of critical loads to be recovered by MEGs. The effectiveness can be better in practice. In short, first, if proactive preallocation of emergency resources is not implemented, some critical loads' outage time can be much longer. Second, optimal preallocation does result in better real-time allocation performance than nonoptimal preallocation. Third, to attain the intended prompt restoration and better survivability enhancement, the traffic issue needs to be explicitly considered in real-time allocation.

3.5.5 Computational Efficiency

In this work, a computer with an Intel i5-4278U processor and 8 GB of memory and the Gurobi 6.0.4 solver with the default setting are used to solve involved mixed-integer linear programming problems. Algorithm 3.1's convergence tolerance is set as 0.5%.

Table 3.5 Comparing the computation time of the preallocation problem (minutes).

Solution method	Scenario number				
	10	20	30	40	50
Gurobi 6.0.4	8.38	20.24	54.71	86.62	134.12
SD algorithm	4.61	10.04	35.73	62.16	101.22

We consider 20 scenarios in the preallocation problem. The SD algorithm takes 8.2 iterations and 10.04 minutes on average to solve the problem. As preallocation is implemented as much as days in advance, the computational efficiency is acceptable. Table 3.5 further compares the SD algorithm's solution time with the computation time directly applying Gurobi with the *MIPGap* set as 0.5% for fair comparison.

In general, applying the SD algorithm decreases the computation time. Nevertheless, with the number of scenarios increasing, the improvement on computational efficiency becomes less significant. In this regard, Algorithm 3.1 can be implemented in a distributed parallel computing framework [34, 35] to achieve faster computation. Note that the convergence tolerance can be set differently by decision-makers based on situations including the computation environment, computation framework (parallel or serial), and scenario number. It is set as 0.5% here, so that the preallocation problems in this section are solved in an appropriate length of time. Much longer time is needed to further decrease the optimality gap. The determination of scenario number is also a critical issue. While including more scenarios can enhance the solution quality, less scenarios are desired to attain improved computational efficiency. The scenario number can be decided by decision-makers dependently based on the situations. First, if the solutions are not consistent, one can consider including more scenarios. Second, if including more scenarios is not an option, another choice is to run simulations for evaluating solutions.

Regarding the real-time allocation problem, it is mostly solved promptly in half a second, also indicating an acceptable computational efficiency.

3.6 Summary and Conclusions

Natural disasters often result in sustained DS component damages, causing prolonged outages for many consumers. Emergency resources, including MEGs, are valuable flexibility for fast electric service recovery enhancing power grid survivability. This chapter proposes to dispatch MEGs as DGs to recover critical loads by forming MGs. This method is implemented by a two-stage dispatch approach with preallocation and real-time allocation. Its effectiveness is validated by illustrative cases. Critical loads' outage time can be reduced by explicitly considering the traffic issue and proactively conducting preallocation. Real-time allocation achieves rational matching between loads and emergency resources. In summary, the proposed method for emergency resources dispatch is effective in decreasing the outage time and outage scale of critical loads after natural disasters and improves the survivability of critical loads. Compared with the current practice, emergency resources can be better utilized for resilient response to natural disasters.

Some other topics on utilizing emergency resources are worth studying. First, robust optimization based preallocation and real-time allocation approaches against uncertainties can be developed to attain fast response. In Chapter 9, a robust preallocation method will be presented. As for real-time allocation, currently, we are using perfect information for it. That is, damage assessments are assumed to be completed quickly for informing the system status. However, complete DS and RN damage assessments can be very time-consuming in some cases. Thus, real-time allocation of emergency resources with imperfect information needs to be studied. Second, multiperiod dynamic dispatch of emergency resources should be co-optimized, e.g. together with the sequence of repairing damaged grid components. After measures including conventional grid restoration and resilient MEG response, the grid stays in the postrestoration state for some time and will begin the infrastructure recovery process [42]. Damaged components will be repaired and the grid will be restored and recovered dynamically. It is important to study how MEGs should dynamically respond in this infrastructure recovery process to mitigate load outage. Chapters 9 and 10 will discuss relevant topics. Third, combined with the problem of optimal repair sequence, the proposed emergency resources dispatch methodology can be modified or reduced for scheduling utility repair trucks. Chapter 10 will present relevant models.

Nomenclature

Indices and Sets

h, i, j, k	indices of grid nodes
m	index of MEGs
n	index of scenarios
s	index of staging locations for preallocating MEGs
$\mathcal{B}$	set of all grid nodes, $h, i, j, k \in \mathcal{B}$

$\mathcal{F}, \mathcal{G}$	set of feeder root nodes, and set of candidate nodes for connecting MEGs, $k \in \{\mathcal{F}, \mathcal{G}\}$
$\mathcal{M}$	set of MEGs, $m \in \mathcal{M}$
$\mathcal{N}$	set of scenarios, $n \in \mathcal{N}$
S	set of staging locations for preallocating MEGs, $s \in S$
$\mathcal{L}$	set of distribution grid lines, $(i,j) \in \mathcal{L}$

Parameters

X_s	maximum number of MEGs preallocated to staging location s
t_{skn}	travel time of a MEG from staging location s to node k of the DS in scenario n
T_{in}	predicted restoration time for the load at node i in scenario n by conventional restoration
u_n	weight or probability of scenario n
p_i, q_i	real and reactive power demand at node i
ω_i	priority weight of power demand at node i
r_{ij}, x_{ij}	resistance and reactance of line (i,j)
V_0, ε	rated voltage and voltage deviation tolerance
$P_m^{\max}, Q_m^{\max}$	maximum real and reactive power output of MEG m
$S_{ij}^{\max}$	apparent power capacity of branch (i,j)
$\theta_k(i)$	parent node of node i regarding node k
S_i^k	set of child nodes of node i regarding node k
$\zeta_k(i,j)$	child node of line (i,j) regarding node k
Π	a sufficiently big number

Variables

x_{sm}	binary, 1 if MEG m is preallocated to staging location s, 0 otherwise
y_{smkn}	binary, 1 if MEG m is real-time allocated (sent) from staging location s to node k in scenario n, 0 otherwise
z_{kn}	binary, 1 if node k is a feeder root node or a MEG is connected to it in scenario n, 0 otherwise
v_{ikn}	binary, 1 if node i is included in the MG powered by the power source at node k in scenario n, 0 otherwise
c_{ijn}	binary, 1 if line (i,j) is closed under scenario n, 0 otherwise
l_{in}	binary, 1 if the switch of the load at node i is closed under scenario n, 0 otherwise
γ_{ikn}	binary, 1 if the load at node i is picked up by the power source at node k under scenario n, 0 otherwise

P_{in}^k, Q_{in}^k real and reactive in-flow power of node i regarding the MG powered by the power source at node k under scenario n

V_{in}^k, δ_{in}^k voltage and auxiliary voltage slack variable of node i regarding the MG powered by the power source at node k under scenario n

References

1 The North American Electric Reliability Corporation, *Hurricane Sandy Event Analysis Report*, 2014.

2 R. J. Campbell, *Weather-related power outages and electric system resiliency*. Congressional Research Service, Washington, DC, USA, 2012.

3 U.S. Electric Power Research Institute, *Enhancing Distribution Resiliency: Opportunities for Applying Innovative Technologies*, 2013.

4 C. Chen, J. Wang, F. Qiu, and D. Zhao, "Resilient distribution system by microgrids formation after natural disasters," *IEEE Transactions on Smart Grid*, vol. 7, no. 2, pp. 958–966, Mar. 2016.

5 U.S. Federal Emergency Management Agency, *Mitigation Assessment Team Report: Hurricane Sandy in New Jersey and New York*, 2013.

6 The GridWise Alliance, *Improving electric grid reliability and resilience: Lessons learned from Superstorm Sandy and other extreme events*, 2013.

7 D. Barrett, *Few big FEMA generators humming*, 2012.

8 C. Yin, J. Wang, and T.-C. Ma, "System design of mobile emergency power plant based on fuel cells," in *2011 International Conference on Electric Information and Control Engineering*, pp. 5419–5422, 2011.

9 C. Y. Jeong, J. G. Cho, J. W. Baek, and D. W. Yoo, "A new integrated controller based 100 kVA mobile engine generator for single/three phase distribution line backup," in *15th Annual IEEE Applied Power Electronics Conference and Exposition*, vol. 2, pp. 902–907, 2000.

10 F. T. Dai, "Risks of network protection for mobile generator applications," in *2008 IET 9th International Conference on Developments in Power System Protection (DPSP 2008)*, pp. 681–686, 2008.

11 C. R. Nightingale, "The design of mobile engine driven generating sets and their role in the British telecommunications network," in *INTELEC '83 - Fifth International Telecommunications Energy Conference*, pp. 144–150, 1983.

12 S. Iwai, T. Kono, M. Hashiwaki, and Y. Kawagoe, "Use of mobile engine generators as source of back-up power," in *INTELEC 2009 - 31st International Telecommunications Energy Conference*, pp. 1–6, 2009.

13 L. Zhou, M. Fan, and Z. Zhang, "A study on the optimal allocation of emergency power supplies in urban electric network," in *CIRED 2009 - 20th International Conference and Exhibition on Electricity Distribution - Part 1*, pp. 1–4, 2009.

14 J. Shang, X. Sheng, J. Zhang, and W. Zhao, "The optimized allocation of mobile emergency generator based on the loads importance," in *2009 Asia-Pacific Power and Energy Engineering Conference*, pp. 1–4, 2009.

15 P. S. Georgilakis and N. D. Hatziargyriou, "Optimal distributed generation placement in power distribution networks: Models, methods, and future research," *IEEE Transactions on Power Systems*, vol. 28, no. 3, pp. 3420–3428, 2013.

16 S. D. Whipple, *Predictive storm modelling and optimizing crew response to improve storm response operations*. MIT, Cambridge, MA, USA, 2014.

17 C.-C. Liu, S. J. Lee, and S. S. Venkata, "An expert system operational aid for restoration and loss reduction of distribution systems," *IEEE Transactions on Power Systems*, vol. 3, no. 2, pp. 619–626, 1988.

18 C. P. Nguyen and A. J. Flueck, "Agent based restoration with distributed energy storage support in smart grids," *IEEE Transactions on Smart Grid*, vol. 3, no. 2, pp. 1029–1038, 2012.

19 S. Khushalani, J. M. Solanki, and N. N. Schulz, "Optimized restoration of unbalanced distribution systems," *IEEE Transactions on Power Systems*, vol. 22, no. 2, pp. 624–630, 2007.

20 M. E. Baran and F. F. Wu, "Network reconfiguration in distribution systems for loss reduction and load balancing," *IEEE Transactions on Power Delivery*, vol. 4, no. 2, pp. 1401–1407, 1989.

21 H. Yeh, D. F. Gayme, and S. H. Low, "Adaptive VAR control for distribution circuits with photovoltaic generators," *IEEE Transactions on Power Systems*, vol. 27, no. 3, pp. 1656–1663, 2012.

22 X. Chen, W. Wu, and B. Zhang, "Robust restoration method for active distribution networks," *IEEE Transactions on Power Systems*, vol. 31, no. 5, pp. 4005–4015, 2016.

23 H. Qi, X. Wang, L. M. Tolbert, F. Li, F. Z. Peng, P. Ning, and M. Amin, "A resilient real-time system design for a secure and reconfigurable power grid," *IEEE Transactions on Smart Grid*, vol. 2, no. 4, pp. 770–781, 2011.

24 F. Katiraei and M. R. Iravani, "Power management strategies for a microgrid with multiple distributed generation units," *IEEE Transactions on Power Systems*, vol. 21, no. 4, pp. 1821–1831, 2006.

25 S. Guikema, S. Quiring, and S.-R. Han, "Prestorm estimation of Hurricane damage to electric power distribution systems.," *Risk Analysis*, vol. 30, no. 12, pp. 1744–1752, 2010.

26 G. McClure, S. Langlois, and J. Rogier, "Understanding how overhead lines respond to localized high intensity wind storms," in *Structures Congress 2008: Crossing Borders*, pp. 1–10, 2008.

27 S.-R. Han, *Estimating Hurricane outage and damage risk in power distribution systems*. Texas A&M University, College Station, TX, USA, 2008.

28 Y. Wang, C. Chen, J. Wang, and R. Baldick, "Research on resilience of power systems under natural disasters-a review," *IEEE Transactions on Power Systems*, vol. 31, no. 2, pp. 1604–1613, 2016.

29 M. Ouyang and L. Dueñas-Osorio, "Time-dependent resilience assessment and improvement of urban infrastructure systems," *Chaos*, vol. 22, no. 3, p. 033122, 2012.

30 J. Winkler, L. Dueñas-Osorio, R. Stein, and D. Subramanian, "Performance assessment of topologically diverse power systems subjected to hurricane events," *Reliability Engineering & System Safety*, vol. 95, no. 4, pp. 323–336, 2010.

31 T. H. Cormen, C. E. Leiserson, R. L. Rivest, and C. Stein, *Introduction to algorithms*. MIT press, 2009.

32 E. W. Dijkstra, "A note on two problems in connexion with graphs," *Numerische Mathematik*, vol. 1, no. 1, pp. 269–271, 1959.

33 R. T. Rockafellar and R. J.-B. Wets, "Scenarios and policy aggregation in optimization under uncertainty," *Mathematics of Operations Research*, vol. 16, no. 1, pp. 119–147, 1991.

34 S. Ahmed, "A scenario decomposition algorithm for 01 stochastic programs," *Operations Research Letters*, vol. 41, no. 6, pp. 565–569, 2013.

35 J.-P. Watson, D. L. Woodruff, and D. R. Strip, "Progressive hedging innovations for a stochastic spare parts support enterprise problem," tech. rep., Sandia National Lab., Albuquerque, NM, USA, 2007.

36 J.-P. Watson and D. L. Woodruff, "Progressive hedging innovations for a class of stochastic mixed-integer resource allocation problems," *Computational Management Science*, vol. 8, no. 4, pp. 355–370, 2011.

37 T. G. Crainic, X. Fu, M. Gendreau, W. Rei, and S. W. Wallace, "Progressive hedging-based metaheuristics for stochastic network design,"*Networks*, vol. 58, no. 2, pp. 114–124, 2011.

38 IEEE PES Power System Analysis, Computing and Economics Committee, *IEEE 34 Node Test Feeder*, Feb. 2014.

39 IEEE PES Power System Analysis, Computing and Economics Committee, *IEEE 37 Node Test Feeder*, Feb. 2014.

40 S. Civanlar, J. J. Grainger, H. Yin, and S. S. H. Lee, "Distribution feeder reconfiguration for loss reduction," *IEEE Transactions on Power Delivery*, vol. 3, no. 3, pp. 1217–1223, 1988.

41 D. Das, "A fuzzy multiobjective approach for network reconfiguration of distribution systems," *IEEE Transactions on Power Delivery*, vol. 21, no. 1, pp. 202–209, 2006.

42 M. Panteli and P. Mancarella, "The grid: Stronger, bigger, smarter?: Presenting a conceptual framework of power system resilience," *IEEE Power and Energy Magazine*, vol. 13, no. 3, pp. 58–66, 2015.

4

Grid Automation Enabling Prompt Restoration

Grid automation, especially the installation of remote-controlled switches (RCSs), plays a critical role in fast electric service recovery, which is an essential requirement for reliable and resilient power grids against natural disasters. However, the cost of RCSs and other automated devices and the vast footprint of distribution grids limits the widespread utilization of grid automation. This chapter presents a new approach to grid automation planning that allocates RCSs to improve the restoration performance and optimize resilience and reliability benefits with appropriate cost. Particularly, the optimal number and locations of to-be-upgraded switches can be determined by the approach for different objectives: (i) To maximize the reduction in customer interruption cost (CIC); (ii) To maximize the reduction in system average interruption duration index (SAIDI); or (iii) To maximize loads that can be recovered by the grid automation system with upgraded RCSs. This chapter shows that actually these grid automation optimization problems can be formulated as mixed-integer convex programming (MICP) models. This chapter also introduces a novel method for equivalently transforming and efficiently solving each of these optimization models. Globally optimal grid automation plans, therefore, can be obtained in a reasonable length of time. The IEEE 33-node and 123-node test systems are adopted for demonstrating the effectiveness of the proposed approach in enhancing grid resilience against outages caused by natural disasters, etc.

4.1 Smart Grid and Automation Systems

Reliable and resilient electricity supply is the prerequisite for many social and economic activities and an essential requirement for the smart grid. As reported in [1], around 70% of electric service interruptions are due to contingencies in distribution systems (DSs) caused by extreme weather events, etc. Achieving high reliability and resilience for DSs is critically important.

Power Grid Resilience against Natural Disasters: Preparedness, Response, and Recovery, First Edition.
Shunbo Lei, Chong Wang, and Yunhe Hou.

Reducing the restoration time is an effective measure for improving smart grid reliability and resilience [2]. After an extreme event causes an outage, the fault is located and isolated by the DS operator. Then, by DS reconfiguration, service restoration is implemented to recover the maximum amount of loads weighted by their priorities. For nonautomated DSs, field crews are sent to perform the involved switching actions. For partially or fully automated distribution grids installed with RCSs that can be controlled from the operation center, the reconfiguration can be conducted much more quickly. Thus, it is expected to improve the grid automation level by installing RCSs rather than manual switches in DSs in order to enable fast service restoration.

Having more RCSs in smart and automated distribution grids can naturally enhance the restoration performance with respect to the amount of restored loads and restoration time. Nevertheless, the cost of equipping RCSs and other necessary automated accessories and the resulting reliability and resilience benefits needs to be balanced considering the limited investment budget and vast extent of DSs. The optimal allocation of RCSs therefore becomes a combinatorial optimization problem seeking the optimal number and locations of RCSs. References [3–21], etc., have extensively investigated the RCS allocation problem. For its complexity, this problem is mostly solved by heuristics algorithms. For instance, reference [3] uses an immune algorithm to obtain the optimal placement of branch switches. Reference [4] proposes an ant colony system-based method for the problem of switch relocation. In [12], the genetic algorithm is used to solve an economics-based sectionalizer allocation problem. Other than heuristic methods, algorithms including two-stage decomposition [19], analytic hierarchical process [13], and direct and bisection search [5] are also used. Moreover, in [7], the globally optimal placement of automated sectionalizing switches is found by a mixed-integer linear programming (MILP) method. Reference [8] extends the MILP model for considering both earth faults and short-circuit faults.

Actually, a holistic smart grid automation optimization problem might have to take into account the exhaustive constraints of DS operation, potential coordination between RCSs and field crews, and the roles of RCSs in different outage recovery stages, etc. Nevertheless, such a holistic model is intractable in construction and solution. Prioritized objectives, in practice, have been proposed and investigated based on different settings or considerations of the problem. For instance, reference [20] proposes a RCS placement scheme of upgrading a near-minimum number of manual switches to RCSs, so that after fault isolation in any single-fault outage scenarios, the maximum loads can be restored using RCSs. This chapter concentrates on enhancing the service restoration performance. The number and locations of switches to be upgraded are optimized in order to appropriately enable fast restoration after natural disasters or other outage events.

Reliability and resilience of the DS thus can be improved. The characteristics of the approach in this chapter are summarized as below:

- The RCS allocation problem for enhancing the performance of smart grid restoration is studied and compared from different aspects. Specifically, this chapter proposes and studies grid automation optimization formulations for three different reliability and resilience objectives, i.e. to maximize the reduction in CIC, to maximize the reduction in SAIDI, and to maximize the loads that can be recovered by upgraded RCSs. Particularly, the last objective is attained by minimizing the number of RCSs, provided that the maximum loads can be recovered using RCSs [20].

- The proposed smart distribution grid automation planning method for RCS allocation optimization is based on mixed-integer programming (MIP) and a novel approach for transforming and solving the models. This chapter first shows that RCS allocation models oriented by the aforementioned different objectives can be formulated as MICP problems. Since the formulated MICP models can have quite large scales, this chapter then proposes the concept of *practical candidate restoration strategies*. A novel method is designed based on that for transforming and solving the large-scale MICP problems. Note that equivalent transformations are achieved, and globally optimal solutions can be obtained efficiently to improve DS resilience.

4.2 Distribution System Automation and Restoration

DS operators can utilize grid automation technologies including RCSs in service restoration to enhance reliability and resilience, in topology reconfiguration to reduce the losses, etc. Nevertheless, it is an excessively expensive distribution grid automation plan to upgrade all switches to RCSs. The RCS allocation problem finds a solution that cost-effectively balances the automated devices' installation cost, reliability and resilience improvement, and other objectives. In the literature, this problem is mostly restoration-oriented. Some references consider other objectives. For example, in [22], RCSs are placed to appropriately automate dynamic network reconfiguration for reducing DS power losses. This chapter focuses on the optimal allocation of RCSs for enhancing the restoration performance against scenarios such as natural disaster-induced power outages.

Automatic reclosers can enhance distribution grid reliability and resilience, too. Specifically, they are more effective in restoration for scenarios with temporary faults, but less effective for scenarios with permanent faults caused by natural disasters, etc. Distributed generations (DGs) also challenge the coordination between automatic reclosers and other protection devices. For example, miscoordination between reclosers and fuses can be caused by DGs [23, 24]. Fault voltage and current fed or sustained by DGs can prevent the extinction of fault arcs, resulting in failed recloses that raise stresses on grid components and develop a permanent fault from a temporary one. DGs' reactive power injection might make the protection miscoordination worsen [25]. Besides, out-of-phase reclosing may happen, which may damage grid devices seriously by large mechanical torques, overcurrent, or overvoltage. Automatic reclosing, in general, sometimes can be a problematic strategy [26]. The involved protection coordination currently becomes more complex and sometimes intractable. This chapter considers permanent faults. Coordination between automatic reclosers and RCSs is beyond the scope. This topic is worth studying yet also challenging by itself.

Distribution grid restoration is a process of multiple stages. As Figure 4.1 shows, in Stage 1 (right after the occurrence of the fault caused by a natural disaster or other events), automatic reclosers can respond to maintain some customers' power supply. In Stage 2 (after the fault is located), RCSs may be used by operators to implement preisolation and prerestoration. In Stage 3 (after the fault is precisely isolated by field crews), RCSs may be used for further recovering more consumers. If the maximum amount of loads is not recovered, manual restoration using manual switches is necessary in Stage 4. In Stage 5, field crews repair damaged elements and turn the grid back to its normal state. In some cases or for some systems, several stages of the above overall recovery process may be unnecessary or ineffective. Utilities do not have to conduct all recovery stages showed in Figure 4.1. In general, existing works on allocating switches do not

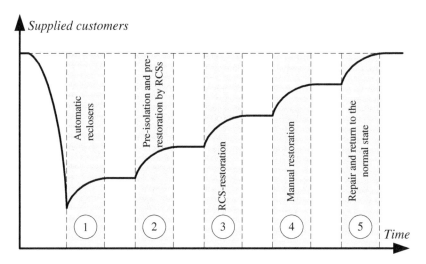

Figure 4.1 A schematic curve of supplied customers associated with a permanent fault event.

consider the whole recovery process for its complexity. This chapter follows [13, 17, 19, 20, 27] and concentrates on allocating RCSs to enable reliability- and resilience-enhancing restoration. References [21, 28, 29], etc., can be referred to for specific studies on the allocation of automatic reclosers.

4.3 Prompt Restoration with Remote-Controlled Switches

Optimal reliability- and resilience-oriented allocation of automated RCSs is to efficiently assist fast service recovery of distribution grids. The efficiency of RCSs in enhancing DS reliability and resilience essentially relies on the recovery strategy applied to respond to outage scenarios caused by natural disasters, etc. It is a critical component for an approach to reliability- and resilience-oriented allocation of RCSs, as it sets the pattern of utilizing grid automation technologies and therefore affects the allocation of RCSs. The restoration strategy of closing normally open branches can be effective. Still, distribution grid operational constraints might prohibit it from recovering the maximum number of loads. Such cases may also require load-switching actions. Note that they are both essentially network reconfiguration strategies. Network reconfiguration is considered in this chapter as the restoration strategy. With notations listed at the end of this chapter, a MIP-based restoration model is formulated as below:

$$\max \sum_{i \in \Phi_k} \omega_i \cdot p_i - \delta_1 \cdot \sum_{(i,j) \in \mathcal{L}} |\alpha_{ij} - \varphi_{ij}| \tag{4.1}$$

$$\text{s.t. } |\alpha_{ij} - \varphi_{ij}| \leq z_{ij}, \forall (i,j) \in \mathcal{L} \tag{4.2}$$

$$\sum_{(i,j) \in \mathcal{L}} \alpha_{ij} = |\mathcal{N}| - |\mathcal{N}_s| \tag{4.3}$$

$$p_i + \sum_{j \in \Omega_i} P_{ij} = 0, \forall i \in \mathcal{N} \tag{4.4}$$

$$q_i + \sum_{j \in \Omega_i} Q_{ij} = 0, \forall i \in \mathcal{N} \tag{4.5}$$

$$p_i = \bar{p}_i, \forall i \in \mathcal{N} \backslash \Phi_k \tag{4.6}$$

$$\varepsilon \leq p_i \leq \bar{p}_i, \forall i \in \Phi_k \tag{4.7}$$

$$q_i = \frac{\bar{q}_i}{\bar{p}_i} \cdot p_i, \forall i \in \mathcal{N} \tag{4.8}$$

$$v_i - v_j \leq (1 - \alpha_{ij}) \cdot M + 2 \cdot (r_{ij} \cdot P_{ij} + x_{ij} \cdot Q_{ij}), \forall (i,j) \in \mathcal{L} \tag{4.9}$$

$$v_i - v_j \geq (\alpha_{ij} - 1) \cdot M + 2 \cdot (r_{ij} \cdot P_{ij} + x_{ij} \cdot Q_{ij}), \forall (i,j) \in \mathcal{L} \tag{4.10}$$

$$\underline{V}_i^2 \leq v_i \leq \overline{V}_i^2, \forall i \in \mathcal{N} \tag{4.11}$$

$$P_{ij}^2 + Q_{ij}^2 \leq \alpha_{ij} \cdot \overline{S}_{ij}^2, \forall (i,j) \in \mathcal{L} \tag{4.12}$$

The objective function (4.1) represents a common reliability and resilience indication maximizing the priority-weighted sum of restored loads and, meanwhile, minimizing the number of switching actions. The notation $|\cdot|$ denotes the absolute value. A small value is used for δ_1, so that maximizing restored loads is still the dominating objective. Constraint (4.2) is the switch-type-dependent operation constraint that only permits the grid automation technologies enabling RCSs to be used in restoration. Prompt restoration therefore is targeted. Equation (4.3) is a necessary condition to assure the distribution grid to operate in radial topologies. Equations (4.4) and (4.5) require all nodes to meet the real and reactive power balance conditions, respectively. Equation (4.6) stops the recovery strategy from curtailing loads in intact regions to pick up loads in outage regions. Equation (4.6) can be dropped, if the system allows curtailing loads of low priorities to restore loads of high priorities. Constraint (4.7) represents the given range of the recovered load for each outage node. To get rid of transfer nodes, a small positive number ε and inequality constraints $\varepsilon \leq p_i$ are added [30]. Constraints (4.3)–(4.7) together sufficiently assure a radial topology of the system without isolated islands. Equation (4.8) assumes that loads have fixed power factors during recovery. Constraints (4.9) and (4.10) express the DistFlow model-based power flow equation [31]. The quadratic terms, which are much smaller, are neglected [32]. A sufficiently large positive number M is adopted for relaxing these constraints for disconnected lines. Constraint (4.11) limits the voltage magnitude. Constraint (4.12) is nonlinear but convex. It restricts each branch's apparent power by its capacity and also enforces disconnected branches' power flow to be zero.

Since the formulation (4.1)–(4.12) obtains restoration strategies using only RCSs, it is called the *RCS-restoration model*. It targets for prompt restoration strategies to improve DS resilience with grid automation technologies. It is a MICP formulation based on [30] and [32]. It can be further transformed into a MILP model, if constraint (4.12) is linearized by techniques such as that in [33]. Other constraints may also be added to better indicate natural disaster scenarios and other conditions, e.g. the limit on the number of switching actions. This chapter considers single-fault outage scenarios. Henceforth, the formulation (4.1)–(4.12) with a fixed k, i.e. the remote-controlled switches-restoration model (RCS-RM) for the k-th scenario with a single fault, is also referred to as *RCS-RM-k*. Let z be the vector of z_{ij}. If other parameters are given, both the optimal solution(s) and the feasible set of RCS-RM-k are actually determined by z. That is, both of them are functions of z.

Restoration strategies can be generated by many other algorithms, including multiagent system [34], expert system [35], and heuristic search [36]. Also, network reconfiguration decisions can be generated by many other methods, including particle swarm optimization [37], spanning tree search [38], stimulated annealing [39], and genetic algorithm [40]. A MICP approach is adopted in this chapter for resilience-enhancing restoration against outage events including natural disasters, as its global optimum can be promptly obtained by off-the-shelf solvers such as Gorubi. Moreover, with appropriately formulated cuts added, it can produce all practical candidate restoration strategies (PCRSs), which are needed in constructing the transformed models for optimizing grid automation and RCS allocation. Here, a PCRS is a restoration strategy which maximizes the amount of restored loads with given switching actions and does not have redundant switching actions. Section 4.5 will provide more details.

4.4 Remote-Controlled Switch Allocation Models

Existing works extensively use two objectives in formulating reliability- and resilience-oriented grid automation and RCS allocation optimization problems: (1) To decrease the CIC [3–17]; and (2) To reduce the SAIDI [12–18]. Another worth-mentioning publication is [20]. The authors upgrade the minimum number of switches to be RCSs, provided that for any single-fault outage scenarios, the selected RCSs can restore the maximum loads. This section formulates different RCS allocation optimization models for the three objectives aforementioned, with an aim to improve grid reliability and resilience against natural disasters. Notations are listed at the end of this chapter.

4.4.1 Minimizing Customer Interruption Cost

CIC is a commonly adopted metric for assessing DS reliability and resilience. Many previous works on grid automation allocate switches to minimize the sum of CIC plus the installation, maintenance, and operation cost of switches. A modified objective function is used here:

$$\max\ RCIC - (IC + M\&OC) \tag{4.13}$$

In (4.13), it maximizes the *reduction in CIC (RCIC)* minus RCSs' *installation cost (IC)* and *maintenance and operation cost (M&OC)*. The formulations of *RCIC*, *IC*, and *M&OC* are as below:

$$RCIC = \sum_{k \in \mathcal{N}_f} \sum_{i \in \Phi_k} \lambda_k \cdot p_{ik} \cdot [f_i(t_{ik}^{MS}) - f_i(t_{ik}^{RCS})] \tag{4.14}$$

$$IC + M\&OC = \sum_{(i,j) \in \mathcal{L}} z_{ij} \cdot (C_{RCS}^{INV} + C_{RCS}^{M\&O}) \tag{4.15}$$

The time of operating manual switches may include time of waiting for available field crews and time of traveling to the sites, etc., while with grid automation technologies, the RCSs can be operated from the control center. Therefore, t_{ik}^{MS} is naturally larger than t_{ik}^{RCS}. Under natural disaster scenarios, t_{ik}^{MS} can be so much larger than t_{ik}^{RCS}. Since the interruption cost function $f_i(t)$ is normally increasing with respect to the outage duration t, $f_i(t_{ik}^{MS})$ is larger than $f_i(t_{ik}^{RCS})$, too. Suppose that p_{ik} is the load at node i that can be restored using RCSs under the k-th single-fault outage scenario, the system CIC is reduced by $\lambda_k \cdot p_{ik} \cdot [f_i(t_{ik}^{MS}) - f_i(t_{ik}^{RCS})]$. As in (4.14), we sum over all outage scenarios to calculate *RCIC* of the system to assess system reliability and resilience improvement. And we only have to sum over all outage loads for each scenario, as the *CIC* of unaffected loads is already zero. Note that it is acceptable to assign a typical or average value for t_{ik}^{MS} and t_{ik}^{RCS}, respectively [3, 8, 14]. Regarding *IC* and *M&OC*, they are represented by (4.15) that sums over RCSs' candidate locations.

This model includes two groups of variables, i.e. z_{ij} and p_{ik}. First, the grid automation variables, i.e. RCS allocation variables z_{ij}, have to be binary:

$$z_{ij} \in \{0, 1\}, \forall (i, j) \in \mathcal{L} \tag{4.16}$$

Second, the real power loads restored using RCSs in a given outage scenario k, i.e. $p_k = \{p_{1k}, p_{2k}, ..., p_{Nk}\}$, are determined by the number and locations of RCSs, i.e. z. Therefore, the following implicit constraint is introduced:

$$p_k = \{p_{1k}, p_{2k}, ..., p_{Nk}\} \in \Psi_k(z), \forall k \in \mathcal{N}_f \tag{4.17}$$

In (4.17), the feasible set of tuple p_k, i.e. $\Psi_k(z)$, is a function of z. Let $F_k(z)$ be RCS-RM-k's feasible set. As aforementioned, $F_k(z)$ is also a function of z. Actually, $\Psi_k(z)$ can be attained by projecting $F_k(z)$ with respect to p_k [41, 42]. That is, for each point in $\Psi_k(z)$, at least one point in $F_k(z)$ has the same value for p_k; for each point in $F_k(z)$, $\Psi_k(z)$ also has a point with the same value for p_k. Thus, the implicit constraint (4.17) can be explicitly represented by the constraints of RCS-RM-k for all k. The CIC-oriented RCS allocation model therefore can be formulated as follows:

> obj.: (4.13)

> s.t. (4.14) − −(4.16); (4.2) − −(4.12) for all k

which is a MICP model (or a MILP model if constraint (4.12) is linearized) for grid automation to enhance resilience against natural disasters. Note that no restriction is imposed on $f_i(t)$. In the proposed model, it is considered as a common nonlinear or piecewise linear customer damage function (CDF), which is different for each customer sector. For a mixture of customer sectors, a composite CDF can be constructed [43]. In some cases, a linear CDF may also be acceptable [9]. Whichever form of the CDF is used, with given t_{ik}^{MS} and t_{ik}^{RCS}, the values of $f_i(t_{ik}^{MS})$ and $f_i(t_{ik}^{RCS})$

are determined and serve as CIC parameters in the above model [7, 8]. That is, the above model remains a MICP problem, and its computational complexity generally does not change.

4.4.2 Minimizing System Average Interruption Duration Index

SAIDI is another widely used reliability index for distribution grids. Some works also use it or its variants to evaluate grid resilience against natural disasters. Many works adopt the objective of minimizing SAIDI for allocating RCSs in grid automation. Here, this objective is adjusted to be

$$\max \delta_2 \cdot RSAIDI - (IC + M\&OC) \tag{4.18}$$

In (4.18), since the two objectives' units are different, a weighting coefficient δ_2 is added. IC and $M\&OC$ have the same definitions as in equation (4.13). $RSAIDI$ denotes the *reduction in SAIDI* that is modeled as follows:

$$RSAIDI = \frac{\sum_{k \in \mathcal{N}_f} \sum_{i \in \Phi_k} \lambda_k \cdot n_{ik} \cdot (t_{ik}^{MS} - t_{ik}^{RCS})}{\sum_{i \in \mathcal{N}} n_i^0} \tag{4.19}$$

Suppose that node i has n_{ik} customers that can be recovered by RCSs in the k-th single-fault scenario, SAIDI of the system is decreased by $\lambda_k \cdot n_{ik} \cdot (t_{ik}^{MS} - t_{ik}^{RCS})/ \sum_{i \in \mathcal{N}} n_i^0$. The system's $RSAIDI$ is therefore computed by equation (4.19) that sums over all outage nodes under each considered outage scenario.

Constraint (4.16) is again needed in this grid allocation model for SAIDI-oriented RCS allocation, as variables z_{ij} always are binary. Regarding the other group of variables n_{ik}, they need to meet an implicit constraint that is similar to (4.17):

$$n_k = \{n_{1k}, n_{2k}, ..., n_{Nk}\} \in \Lambda_k(z), \forall k \in \mathcal{N}_f \tag{4.20}$$

where the tuple n_k's feasible set is denoted as $\Lambda_k(z)$. It is, again, a function of z. Similar to the CIC-oriented RCS allocation model, the implicit constraint (4.20) can be explicitly formulated using the constraints of an adapted version of RCS-RM-k for all k. By doing so, a MICP model, again, is obtained. Since the space is limited and the needed modifications are straightforward, the adjusted RCS-RM-k model is not elaborated on here.

If the electricity market is deregulated and performance-based regulations are used, a penalty-based scheme is often adopted, and the privatized and liberalized utilities might prefer a grid automation model minimizing the penalty associated with the violation of a reliability or resilience index threshold. In this regard, the above formulation can be adjusted accordingly to be

$$\min \rho \cdot SAIDI^{\text{vio}} + (IC + M\&OC) \tag{4.21}$$

$$\text{s.t. } SAIDI^{\text{vio}} \geq SAIDI^0 - RSAIDI - SAIDI^{\text{tgt}} \tag{4.22}$$

$$SAIDI^{\text{vio}} \geq 0 \tag{4.23}$$

$$(4.15) - -(4.16); (4.19) - -(4.20)$$

where ρ is the penalty coefficient; $SAIDI^{\text{vio}}$ denotes how much $SAIDI^{\text{tgt}}$, i.e. the targeted SAIDI, is violated; $SAIDI^0$ is the original SAIDI level before grid automation. As can be seen, only slight modifications are needed. The transformation and solution methods proposed in Section 4.5 will still be applicable. Normally, with information and communications technologies, etc., utilities are capable of attaining the reliability and resilience index target to avoid penalties. In many cases, they are also motivated to maximize the reliability and resilience metric to get the contracted bonus or reward, since regulators hope to not only safeguard but also enhance customers' electric service [44, 45] against events including natural disasters. Besides, in some cases, the regulation is conducted ex post rather than ex ante [46]. In those cases, the reliability and resilience index threshold is not pregiven. Here, this chapter focuses on the unadjusted SAIDI-oriented grid automation model for RCS allocation.

4.4.3 Maximizing System Restoration Capability

In [20], a novel grid automation optimization method for allocating RCSs to improve distribution grid recovery capability is introduced. A near-minimum number of switches are upgraded to RCSs so that in each scenario under consideration, the upgraded RCSs are capable of restoring an amount of demands equal to that using all switches. DS reliability against single-fault scenarios and resilience against natural disasters thus are enhanced. This chapter develops different formulation and solution approaches for this problem, so as to obtain the global optimum. The objective is minimizing the number of upgraded RCSs:

$$\min \sum_{(i,j) \in \mathcal{L}} z_{ij} \tag{4.24}$$

In (4.24), by summing binary variables z_{ij} over all candidate locations of RCSs, the number of RCSs is computed.

For the k-th outage scenario, let restorable load RL_k and RCS-restorable load RL_k^{RCS} be the maximum amount of loads that can be recovered using all switches and only RCSs, respectively. The following constraint requires that for each considered outage scenario, the restorable load and the RCS-restorable load are the same:

$$RL_k^{\text{RCS}} = RL_k, \forall k \in \mathcal{N}_{\text{f}} \tag{4.25}$$

RL_k is a parameter. Its value can be computed by solving RCS-RM-k with all z_{ij} fixed as 1. RL_k^{RCS} is a variable represented as follows:

$$RL_k^{\text{RCS}} \leq \sum_{i \in \Phi_k} p_{ik}, \forall k \in \mathcal{N}_{\text{f}} \tag{4.26}$$

In this model, variables p_{ik} also need to meet the implicit constraint (4.17). As mentioned above, it can be explicitly formulated by the constraints of RCS-RM-k for all k. Therefore, the restorable load-oriented (RL-oriented) grid automation model for RCS allocation is expressed as below:

> obj.: (4.24)

> s.t. (4.16); (4.25) $- -$(4.26); (4.2) $- -$(4.12) for all k

As one of the three proposed grid automation models for RCS allocation, the above formulation is a MICP problem of a large scale. It is to be efficiently transformed and solved, by being decomposed into MILP problems of small scales.

Note that there is not a perfect DS reliability and resilience regulation model. For instance, penalty-based schemes have quite limited effects in providing economic signals for resilience and reliability investments such as grid automation [45]. Resilience and reliability metrics, such as SAIDI and CIC, can have high correlations. As there are larger needs for quality regulation and cost-efficiency incentives, performance-based regulation might be preferred, which helps improve the cost efficiency of utilities and enhance the electric service resilience and reliability for customers [44]. Nevertheless, if these are less such needs, cost-based regulation might be a better approach [46].

4.5 Solution Method

The MICP-based grid automation optimization models presented in Section 4.4 can have quite large scales. They may be computationally insolvable even with state-of-the-art solvers. This section equivalently transforms them into MILP models that have much less integer variables and are easier to solve. Each of the RCS allocation models can use the proposed transformation and solution method. The method is presented based on the CIC-oriented model. The other two models can be similarly transformed and solved.

4.5.1 Practical Candidate Restoration Strategies

In this subsection, all discussions are with respect to RCS-RM-k without constraint (4.2), which is referred to as RM-k. The number of feasible restoration strategies of RM-k can be infinite. However, only a limited number of them are actually of interest. Here, we set $\omega_i = f_i(t_{ik}^{MS}) - f_i(t_{ik}^{RCS})$, and define a *practical candidate restoration strategy (PCRS)* as a feasible restoration strategy satisfying the following two conditions:

- A larger amount of priority-weighted loads cannot be restored by the same set of switching actions. That is, a PCRS maximizes $\sum_{i \in \Phi_k} \omega_i \cdot p_i$ with a given α

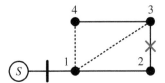

Figure 4.2 A sample four-node system $(\bar{p}_i = 1$ p.u. for $i = 1, 2, 3, 4)$.

(vector of α_{ij}). If two feasible restoration strategies have the same value for α, the strategy with a smaller value of $\sum_{i \in \Phi_k} \omega_i \cdot p_i$ is not a PCRS.

- A PCRS does not include switching actions that are redundant. That is, the same amount of loads weighted by priorities cannot be restored by any proper subset of a PCRS' set of switching actions.

The first condition is generally straightforward. It states that, with switching actions given, other variables are to maximize the priority-weighted sum of restored loads. The second condition is explained in more details using a sample four-node system in Figure 4.2.

Assume that a natural disaster causes a fault forcing line $(2, 3)$ to be inoperable and disconnected. Then nodes 3 and 4 are without electricity supply. Assume that normally open lines $(1, 4)$ and $(1, 3)$ have sufficiently large capacities. Further assume that other constraints, such as voltage limits, will not be violated. Three restoration strategies can be selected by the operator:

- Strategy 1: Close line $(1, 3)$ to recover 2 p.u. loads;
- Strategy 2: Close line $(1, 4)$ to recover 2 p.u. loads;
- Strategy 3: Open line $(3, 4)$ and close lines $(1, 4)$ and $(1, 3)$ to recover 2 p.u. loads.

Since Strategies 1 and 2 do not involve redundant switching actions and they fully recover electricity supply for outage nodes, they are PCRSs. Though Strategy 3 achieves full restoration, too, it involves redundant switching actions. Specifically, a subset of its switching actions, e.g. only closing line $(1, 3)$, can achieve the same effect. Therefore, it is not a PCRS.

If the capacities of lines $(1, 4)$ and $(1, 3)$ are both restricted by 1.5 p.u., the set of PCRSs is different. Strategies 1 and 2 are still PCRSs. Nevertheless, due to the new restriction, each of them can recover only 1.5 p.u. loads. In contrast, Strategy 3 can restore electric service to all consumers. Therefore, it is a PCRS, too, since its extra-switching actions are no longer redundant.

The number of PCRSs is finite and limited. It is straightforward that RM-k's optimal solution is a PCRS. In fact, for each considered outage scenario, the set of all PCRSs can be obtained by recursively solving RM-k and adding appropriately formulated cuts. The approach is as below:

Algorithm 4.1 Finding PCRSs for a Fixed k.

1: Initialize $itr = 1, R_k = \emptyset$
2: Solve RM-k with cuts (4.27) added:

   obj.: (4.1)

   s.t. (4.3) $- -$(4.12)

   $$\sum_{(i,j)\in\Theta_k^\sigma} |\alpha_{ij} - \varphi_{ij}| \leq |\Theta_k^\sigma| - 1, \forall \sigma \in R_k \qquad (4.27)$$

3: If the above optimization problem becomes infeasible, return results and terminate; otherwise, go to *Step 4*
4: Obtain the optimal solution; record p_k^σ (the vector of p_{ik}^σ) and Θ_k^σ for $\sigma = itr$; update $R_k = R_k \bigcup \{itr\}$, $itr = itr + 1$; and go to *Step 2*

In Algorithm 4.1, itr is the iteration number; σ is the index for PCRSs; R_k is the set of PCRSs for the k-th outage scenario; p_{ik}^σ is the restored real power demand at node i using the σ-th PCRS for the k-th outage scenario; Θ_k^σ and $|\Theta_k^\sigma|$ are the set and number of switches used in the σ-th PCRS for the k-th outage scenario. Other notations are defined hereinbefore or in the Nomenclature at the end of this chapter.

In Step 2, cuts (4.27) are added to RM-k, so as to preclude the PCRSs that are already obtained and included in R_k. Thus, we can recursively solve RM-k with cuts (4.27) to obtain a new PCRS. After all PCRSs are obtained and included in R_k, the optimization problem in Step 2 becomes infeasible as cuts (4.27) preclude all feasible solutions. In this case, Algorithm 4.1 can be terminated. In Step 4, p_k^σ is also recorded to be used in Section 4.5.2. Actually, the sequentially generated PCRSs will have nonincreasing values of $\sum_{i\in\Phi_k} \omega_i \cdot p_i$, i.e. the sum of restored priority-weighted loads. According to their preferences, distribution grid operators can choose to only generate and use a specified number of PCRSs that are the most effective in enhancing resilience against natural disasters, etc.

Note that the formulation of (4.27) is nontrivial. In this chapter, cuts (4.27) are formulated to exclude the combination of switching actions of each obtained PCRS. In fact, for a given σ, the cut (4.27) not only precludes the σ-th PCRS but also precludes any recovery strategy with switching actions being a proper superset of Θ_k^σ. This characteristic of (4.27) ensures that each optimal solution attained in Step 4 is a PCRS. In [47, 48], similar cuts are formulated based on the logic expression "exclusive or." It can be used here to formulate cuts excluding each obtained PCRS's network topology. Nevertheless, their method is not used here, because their cuts only preclude the obtained PCRSs. Using them instead of (4.27) will result in solutions that are not PCRSs in Step 4 of the algorithm.

PCRSs for each considered outage scenario can be obtained by running Algorithm 4.1 for each k. Note that here since constraint (4.2) is discarded, the switch type is not taken into account. That is, it is assumed that all switches are usable and available in generating PCRSs.

Here, the sample four-node system in Figure 4.2 is used for illustrating how Algorithm 4.1 generates PCRSs. Again, assume that lines $(1, 3)$ and $(1, 4)$ have sufficiently large capacities. First, the restoration model, i.e. equations (4.1) and (4.3)–(4.12), is constructed for the outage scenario shown in Figure 4.2. Solving it obtains the first PCRS, i.e. Strategy 1 mentioned above. Second, the cut (4.27) is formulated according to the first PCRS. For this PCRS, $|\Theta_k^\sigma| = 1$ and $\Theta_k^\sigma = \{(1, 3)\}$. And line $(1, 3)$ is normally open. That is, $\varphi_{13} = 0$. Therefore, the cut (4.27) is formulated as follows:

$$|\alpha_{13} - 0| \leq 1 - 1 \quad \Rightarrow \quad |\alpha_{13}| \leq 0 \tag{4.28}$$

Third, the restoration model, with this cut added, is solved again. Note that with the cut (4.28), not only the first PCRS is precluded but also any recovery strategy containing the first PCRS's switching action becomes infeasible. For example, Strategy 3 mentioned above, which includes the switching action of closing line $(1, 3)$, is infeasible now. Solving the model obtains the second PCRS, i.e. Strategy 2 mentioned above. As these two PCRSs have the same number of switching actions and the same effect, the sequence to obtain them can be reversed. Fourth, a new cut is formulated according to the second PCRS:

$$|\alpha_{14} - 0| \leq 1 - 1 \quad \Rightarrow \quad |\alpha_{14}| \leq 0 \tag{4.29}$$

The above cut is also added to the restoration model that is solved again. The solver reports that the model is infeasible, since we have obtained all PCRSs, and all other recovery strategies which are not PCRSs are eliminated by the cuts (4.28) and (4.29). Therefore, Algorithm 4.1 terminates. In general, before the restoration model becomes infeasible, Algorithm 4.1 repeats Steps 2–4 to sequentially generate PCRSs.

Note that if the capacities of lines $(1, 4)$ and $(1, 3)$ are both restricted by 1.5 p.u., Strategy 3 mentioned above will be the first generated PCRS. For this PCRS, $|\Theta_k^\sigma| = 3$, $\Theta_k^\sigma = \{(3, 4), (1, 3), (1, 4)\}$, and $\varphi_{34} = 1$, $\varphi_{13} = \varphi_{14} = 0$. Its associated cut is as follows:

$$|\alpha_{34} - 1| + |\alpha_{13}| + |\alpha_{14}| \leq 2 \tag{4.30}$$

Neither Strategy 1 nor Strategy 2 violates this cut. In this regard, 3 PCRSs will be obtained.

4.5.2 Model Transformation

In Section 4.4.1, the CIC-oriented grid automation optimization model for allocating RCSs is modeled as a large-scale MICP. Particularly, the purpose of including constraints (4.3)–(4.12) for all k is to define the set of feasible recovery strategies for each outage scenario. Besides, constraint (4.2) restricts these restoration strategies' availability, which is determined by RCSs' number and locations that impact grid reliability and resilience against natural disasters.

Normally, the number of feasible recovery strategies defined by (4.3)–(4.12) is infinite. Nevertheless, owing to PCRSs' features, in fact, it is equivalent to only consider PCRSs in this problem. That is, constraints (4.3)–(4.12) can be replaced by a limited number of PCRSs. Using this idea, we rewrite the CIC-oriented grid automation model as below:

obj.: (4.13)

s.t. (4.15) $--$ (4.16)

$$RCIC = \sum_{k \in \mathcal{N}_f} \sum_{\sigma \in R_k} \sum_{i \in \Phi_k} \lambda_k \cdot \gamma_k^\sigma \cdot p_{ik}^\sigma \cdot [f_i(t_{ik}^{MS}) - f_i(t_{ik}^{RCS})] \tag{4.31}$$

$$\gamma_k^\sigma \in \{0, 1\}, \forall \sigma \in R_k, \forall k \in \mathcal{N}_f \tag{4.32}$$

$$\gamma_k^\sigma \leq \prod_{(i,j) \in \Theta_k^\sigma} z_{ij}, \forall \sigma \in R_k, \forall k \in \mathcal{N}_f \tag{4.33}$$

$$\sum_{\sigma \in R_k} \gamma_k^\sigma \leq 1, \forall k \in \mathcal{N}_f \tag{4.34}$$

where γ_k^σ is a binary variable indicating if the σ-th PCRS for the k-th outage scenario is available; the symbol $\prod$ is an operation obtaining the product over a set's elements. Other notations are defined hereinbefore or in the nomenclature at the end of this chapter.

Equation (4.31) is the new expression of $RCIC$. For a given outage scenario k, if the σ-th PCRS is available, i.e. $\gamma_k^\sigma = 1$, $RCIC$ can be reduced by $\sum_{i \in \Phi_k} \lambda_k \cdot p_{ik}^\sigma \cdot [f_i(t_{ik}^{MS}) - f_i(t_{ik}^{RCS})]$. Summing it over all PCRSs and all considered outage scenarios attains the system $RCIC$, indicating the reliability and resilience improvement. Constraint (4.33) restricts that a PCRS is available only if all used switches are automated to be RCSs. Constraint (4.34) ensures that at most one PCRS is activated for each scenario so that the system $RCIC$ is not over-estimated. As we set $\omega_i = f_i(t_{ik}^{MS}) - f_i(t_{ik}^{RCS})$ when generating PCRSs, the above model is equivalent to the CIC-oriented RCS allocation model in Section 4.4.1.

4.5.3 Linearization and Simplification Techniques

Because constraint (4.33) includes the product of multiple integer variables, the resilience-enhancing grid automation optimization model in Section 4.5.2 is a nonlinear integer programming problem. It can be linearized by replacing (4.33) with the following equivalent constraint:

$$\gamma_k^\sigma \le z_{ij}, \forall (i,j) \in \Theta_k^\sigma, \forall \sigma \in R_k, \forall k \in \mathcal{N}_f \tag{4.35}$$

Thus, the model becomes an integer linear programming problem. We can further simplify it by treating binary variables γ_k^σ as continuous variables constrained by

$$0 \le \gamma_k^\sigma \le 1, \forall \sigma \in R_k, \forall k \in \mathcal{N}_f \tag{4.36}$$

For each outage scenario k, the optimal solution will naturally set $\gamma_k^\sigma = 1$ for the PCRS maximizing the system $RCIC$ (i.e. maximizing the reliability and resilience improvement) and $\gamma_k^\sigma = 0$ for the other PCRSs. Therefore, it is still an equivalent model. In short, the original CIC-oriented grid automation optimization model for RCS allocation, a large-scale MICP with $(|\mathcal{N}_f| + 1) \times |\mathcal{L}|$ integer variables, is transformed and becomes

obj.: (4.13)

s.t. (4.15) $--$(4.16); (4.31); (4.34) $--$(4.36)

which is a MILP problem and has only $|\mathcal{L}|$ integer variables.

4.5.4 Overall Solution Process

Before adopting the transformed model introduced above, all PCRSs for each outage scenario have to be generated. Algorithm 4.2 describes the overall process to solve the original model.

Algorithm 4.2 Solving the CIC-oriented RCS Allocation Model.

1: Initialize $k = 1$
2: Call Algorithm 4.1 to obtain R_k, and Θ_k^σ and p_k^σ for $\sigma \in R_k$
3: If $k = |\mathcal{N}_f|$, go to *Step 4*; otherwise, update $k = k + 1$ and go to *Step 2*
4: Formulate and solve the transformed model; return z and terminate

---

In general, the original model is decomposed into multiple small-scale MILP problems that can be efficiently solved by state-of-the-art solvers. Compared with directly solving the original large-scale MICP, Algorithm 4.2 reduces the worst-case computational complexity from $\mathcal{O}(2^{(|\mathcal{N}_f|+1)\cdot|\mathcal{L}|} \cdot \rho(|\mathcal{N}_f|, |\mathcal{N}|, |\mathcal{L}|))$ to $\mathcal{O}(2^{|\mathcal{L}|} \cdot \rho(|\mathcal{N}_f|, |\mathcal{N}|, |\mathcal{L}|))$ approximately [49]. Here, $\rho(|\mathcal{N}_f|, |\mathcal{N}|, |\mathcal{L}|)$ is a polynomial indicating the computational complexity of solving a linear programming

(LP) problem, which depends on the applied LP algorithm, and $|\mathcal{N}_f|$, $|\mathcal{N}'|$, and $|\mathcal{L}|$ here.

Similarly, the SAIDI-oriented grid automation optimization model for allocating RCSs can be transformed equivalently into

obj.: (4.18)

s.t. $(4.15) - -(4.16); (4.34) - -(4.36)$

$$RSAIDI = \frac{\sum_{k\in\mathcal{N}_f}\sum_{\sigma\in R_k}\sum_{i\in\Phi_k}\lambda_k \cdot \gamma_k^\sigma \cdot n_{ik}^\sigma \cdot (t_{ik}^{MS} - t_{ik}^{RCS})}{\sum_{i\in\mathcal{N}}n_i^0} \qquad (4.37)$$

where n_{ik}^σ is the number of restored customers at node i using the σ-th PCRS for the k-th outage scenario. Other notations are defined hereinbefore or in the nomenclature at the end of this chapter. Equation (4.37) is the new representation of $RSAIDI$. Note that for this model, the definition of PCRSs has to be slightly adjusted. With fixed k and α, a PCRS of this model maximizes $\sum_{i\in\Phi_k}(t_{ik}^{MS} - t_{ik}^{RCS}) \cdot n_{ik}$ with $n_{ik} = n_{ik}^\sigma$, rather than maximizing $\sum_{i\in\Phi_k}[f_i(t_{ik}^{MS}) - f_i(t_{ik}^{RCS})] \cdot p_{ik}$ with $p_{ik} = p_{ik}^\sigma$.

The RL-oriented grid automation optimization model for RCS allocation can be equivalently transformed into

obj.: (4.24)

s.t. (4.16); (4.35)

$$\sum_{\sigma\in R_k}\gamma_k^\sigma \geq 1, \forall k \in \mathcal{N}_f \qquad (4.38)$$

Here, the definition of PCRSs has to be adjusted, too. In this model, PCRSs for a given outage scenario k are the feasible recovery strategies recovering RL_k loads and having no redundant switching actions. Particularly, constraint (4.38) assures that RCSs are upgraded so as to have at least one available PCRS for each outage scenario considered.

Therefore, with minor modifications, Algorithms 4.1 and 4.2 can be used to similarly transform and solve these two models. In general, first, PCRSs for each outage scenario are attained. Then, they are included in the transformed RCS allocation models in Section 4.5.2 or this section. Furthermore, using the techniques presented in Section 4.5.3, the transformed models are linearized and simplified. This approach reduces the number of integer variables in optimization problems and also reduces the computation time.

4.6 Case Studies

This section demonstrates the proposed models and algorithms on two test systems. A computer with 8 GB of memory and an Intel i5-4278U processor is used. Gurobi 6.0.4 with the default setting solves the involved MIP problems.

4.6.1 Illustration on a Small Test System

Figure 4.3 shows the first test system. Line capacities are set according to the rules in [50], and the minimum and maximum voltage values are set to be 0.95 p.u. and 1.05 p.u., respectively. The average failure rate is set as 0.18 failures/year-km [14]. The lifetime and investment cost of a RCS are set as 15 years and $9071 USD, respectively [3]. The annual maintenance and operation cost are set to be 2% of the annualized investment cost. The operating time of remote-controlled and manual switches are assumed as 5 and 60 minutes, respectively. The loads are classified randomly into different types. The interruption cost for a commercial, industrial, and residential customer is set as $31.98/kWh, $26.16/kWh, and $4.59/kWh, respectively, [8]. Other data can be found in [31].

We conduct illustrative cases for each of the three grid automation optimization models for RCS allocation. First, their results are described separately. Then, a brief comparison is provided.

4.6.1.1 Results of the CIC-oriented Model

Using this model, the optimal resilience-enhancing solution is to automate the switches to RCSs on five lines, i.e. (30, 31), (25, 29), (18, 33), (12, 22), and (8, 9). The cost associated with RCSs is $3084.14/year. As a return, CIC of the system is decreased by $31142.77/year. This result implies that several RCSs can efficiently decrease CIC of a distribution grid. When running Algorithm 4.2, 424 PCRSs in total are obtained within one minute. The number of PCRSs for each single-fault outage scenario is listed in Table 4.1. Though for each scenario, theoretically the number of combinations of switching actions can be up to $2^{|\mathcal{L}|}$,

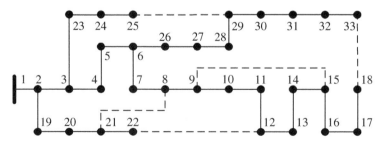

Figure 4.3 IEEE 33-node test system's topology.

Table 4.1 Number of PCRSs for each single-fault outage scenario.

Fault location	3	4	5	6	7	8	9	10	11
Number of PCRSs	2	83	74	11	9	8	4	5	5
Fault location	12	13	14	15	16	17	19	20	21
Number of PCRSs	4	2	2	1	1	1	10	2	1
Fault location	23	24	26	27	28	29	30	31	32
Number of PCRSs	15	1	36	33	31	14	39	29	1

Table 4.2 CIC parameters' impact on the optimal RCS allocation (IEEE 33-Node Test System).

CIC parameters	Number of RCSs	RCS cost ($/yr)	RCIC ($/yr)	Optimal RCS locations
25%	4	2467.31	7257.55	$(8, 9), (12, 22), (18, 33),$ $(25, 29)$
50%	5	3084.14	15571.38	$(8, 9), (12, 22), (18, 33),$ $(25, 29), (30, 31)$
100%	5	3084.14	31142.77	$(8, 9), (12, 22), (18, 33),$ $(25, 29), (30, 31)$
200%	5	3084.14	62285.53	$(8, 9), (12, 22), (18, 33),$ $(25, 29), (30, 31)$
400%	7	4317.80	124727.83	$(6, 7), (8, 21), (11, 12),$ $(12, 22), (18, 33), (25, 29),$ $(31, 32)$

the number of PCRSs is very small. Based on the obtained PCRSs, the transformed RCS allocation model is formulated, and it is solved in 0.5 seconds.

Sensitivity analysis is also conducted regarding CIC parameters, for studying their impact on this RCS allocation model's solution. Table 4.2 shows the results. As indicated, in this case, halving or doubling CIC parameters does not change the optimal number and locations of RCSs. In other words, this model's solution for this test system is robust to CIC parameters' changes and has robust performance in enhancing resilience with grid automation.

4.6.1.2 Results of the SAIDI-oriented Model

Essentially, this grid automation optimization model is a problem of multiple objectives. The optimal reliability- and resilience-improving solution is impacted

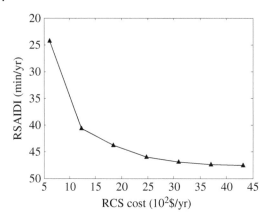

Figure 4.4 Pareto-optimal solutions of the SAIDI-oriented RCS allocation model (IEEE 33-node test system).

Table 4.3 RCS switching time's impact on the optimal RCS allocation (IEEE 33-Node Test System).

Switching time of RCSs	Number of RCSs	RCS cost ($/yr)	RSAIDI (min /yr)	Optimal RCS locations
10 secs	5	3084.14	51.03	$(9, 10), (12, 22), (18, 33),$ $(25, 29), (30, 31)$
1 min	5	3084.14	50.32	$(9, 10), (12, 22), (18, 33),$ $(25, 29), (30, 31)$
5 mins	4	2467.31	45.96	$(6, 7), (8, 21), (18, 33), (25, 29)$
10 mins	3	1850.48	39.86	$(8, 21), (18, 33), (25, 29)$

by the value of δ_2. This model is solved for different values of δ_2 to obtain different Pareto-optimal solutions that are shown in Figure 4.4. It is shown that, with the number of RCSs rising, upgrading one more switch to RCS results in a reducing marginal benefit with respect to the reduction in SAIDI. For each solution, in total 483 PCRSs are obtained in one minute and eight seconds. Then the transformed model is formulated and solved in 0.5 seconds.

Sensitivity analyses are conducted based on the base case here, which is the solution with $\delta_2 = 500$. First, the switching time of RCSs is changed to explore its impact on the solution. Table 4.3 shows the results. Compared to the base case, when the RCS switching time is reduced to 1 minutes or 10 seconds, one more switch is upgraded to be a RCS. In contrast, if the switching time of RCSs is increased to 10 minutes, the number of RCSs reduces by 1. These are naturally true, as the advantages of RCSs in enhancing grid reliability and resilience can be significantly diminished by longer switching time. Actually, in Table 4.3, the values of RCS switching time and *RSAIDI* are strongly correlated. The impact

Table 4.4 Failure rates' impact on the optimal RCS allocation (IEEE 33-Node Test System).

Failure rate changes	Number of RCSs	RCS cost ($/yr)	RSAIDI (min/yr)	Optimal RCS locations
25%	2	1233.66	10.19	$(12, 22), (25, 29)$
50%	4	2467.31	22.98	$(6, 7), (8, 21), (18, 33), (25, 29)$
100%	4	2467.31	45.96	$(6, 7), (8, 21), (18, 33), (25, 29)$
200%	6	3700.97	94.85	$(8, 21), (11, 12), (12, 22),$ $(18, 33),$ $(25, 29), (27, 28)$
400%	6	3700.97	189.71	$(8, 21), (11, 12), (12, 22),$ $(18, 33),$ $(25, 29), (27, 28)$

Table 4.5 RL-oriented optimal RCS allocation (IEEE 33-Node Test System).

Number of RCSs	Optimal RCS locations
8	$(9, 10), (9, 15), (12, 22), (14, 15), (18, 33), (25, 29), (28, 29), (30, 31)$

of failure rate changes on this model's optimal solution is further investigated. Table 4.4 lists the results. Generally, with increasing failure rates, the number of RCSs also increases.

4.6.1.3 Results of the RL-oriented Model

Table 4.5 reports the optimal reliability- and resilience-enhancing solution of this grid automation model. For transforming the original model, 296 PCRSs are obtained in around 45 seconds. It then takes less than one second to solve the transformed model. In total, eight switches are upgraded to be RCSs, costing $4934.62/yr. With the upgraded RCSs, the maximum loads can be quickly recovered after fault location and isolation in each considered outage scenario, i.e. $RL_k^{RCS} = RL_k$ for all k. Upgrading more switches to be RCSs does not increase the value of RL_k^{RCS} for any considered scenario here. That is, the amount of loads that can be recovered by the upgraded RCSs is maximized.

4.6.1.4 Comparisons

Table 4.6 compares the base-case solutions of the third grid automation optimization models for RCS allocation. The allocation decisions, for clarity, are again listed in column 2. As shown, both the number and locations of RCSs are very different.

Table 4.6 Comparing different RCS allocation models (IEEE 33-Node Test System).

Allocation model	Allocation result	RCIC minus RCS cost ($/yr)	RSAIDI (min/yr)	Restoration capability
CIC oriented	$(8,9),(12,22),(18,33),$ $(25,29),(30,31)$	28058.63	46.89	24/27
SAIDI oriented	$(6,7),(8,21),(18,33),(25,29)$	27351.96	45.96	18/27
RL -oriented	$(9,10),(9,15),(12,22),$ $(14,15),(18,33),(25,29),$ $(28,29),(30,31)$	26249.58	47.14	27/27

First, the CIC-oriented RCS allocation plan has the largest value of *RCIC* minus the cost of RCSs. In other words, this model is the most effective in choosing RCSs to decrease CIC of a distribution grid. Second, the SAIDI-oriented model, with the number of RCSs given, provides a grid automation scheme that decreases SAIDI the most. Applying this model, four RCSs are chosen to attain a *RSAIDI* of 45.96 min/year. It is slightly less than 46.89 min/year and 47.14 min/year, which are achieved by five RCSs selected by the CIC-oriented model and eight RCSs selected by the RL-oriented model, respectively. In fact, the SAIDI-oriented model can decrease SAIDI by 46.92 min/year and 47.43 min/year with five and six RCSs, respectively. Third, only the RL-oriented model ensures that the selected RCSs can recover the maximum loads after fault location and isolation for all 27 considered scenarios, i.e. $RL_k^{RCS} = RL_k$ for all k. Nevertheless, eight switches have to be automated. The four RCSs selected by the SAIDI-oriented model just ensure $RL_k^{RCS} = RL_k$ for 18 considered outage scenarios. The five RCSs selected by the CIC-oriented model have $RL_k^{RCS} = RL_k$ for 24 scenarios, which can be very acceptable.

In general, different objectives can result in different optimal reliability- and resilience-enhancing grid automation schemes. Distribution grid planners may apply or develop an appropriate model if they have specific objectives, or compare different models' results to finalize decisions if they want to balance different reliability and resilience objectives. In this case, specifically, the RCS allocation scheme by the CIC-oriented grid automation model is the most balanced.

4.6.2 Results on a Large Test System

Figure 4.5 depicts a larger test system used for demonstrating the proposed approach' scalability. Reference [51] includes the system data. Other parameters are set the same as those in the first smaller test system.

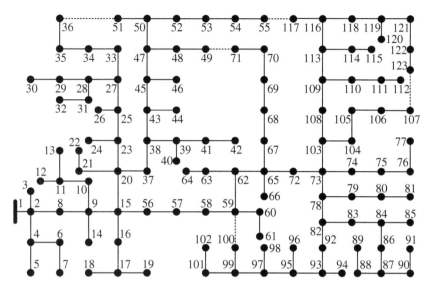

Figure 4.5 IEEE 123-node test system' topology.

Table 4.7 CIC parameters' impact on the optimal RCS allocation (IEEE 123-Node Test System).

CIC parameters	Number of RCSs	RCS cost ($/yr)	RCIC ($/yr)	Optimal RCS locations
25%	5	3084.14	7031.90	(36, 51), (49, 71), (55, 117), (59, 100), (73, 103)
50%	7	4317.80	15930.03	(36, 51), (49, 71), (50, 51), (55, 117), (59, 100), (73, 103), (107, 123)
100%	9	5551.45	33133.82	(36, 51), (47, 50), (49, 71), (55, 117), (59, 100), (72, 73), (73, 78), (78, 82), (107, 123)
200%	11	6785.11	66907.37	(36, 51), (47, 50), (49, 71), (55, 117), (59, 100), (72, 73), (73, 103), (78, 82), (107, 123), (121, 122), (122, 123)
400%	12	7401.94	136162.63	(36, 51), (47, 50), (49, 71), (50, 51), (55, 117), (59, 100), (65, 72), (72, 73), (73, 103), (78, 82), (107, 123), (116, 118)

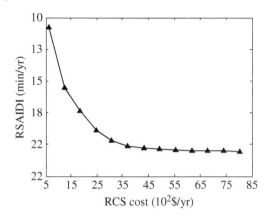

Figure 4.6 Pareto-optimal solutions of the SAIDI-oriented RCS allocation model (IEEE 123-node test system).

Results of the CIC-oriented grid automation optimization model are listed in Table 4.7. Sensitivity analysis is conducted with respect to CIC parameters. In general, if CIC parameters increase, the number of RCSs also increases to improve reliability and resilience. For the base case, we use 18 minutes to generate 1156 PCRSs. Then the transformed model is formulated and solved in 0.5 seconds.

Regarding the SAIDI-oriented model, the value of δ_2 is changed to attain different Pareto-optimal solutions that are shown in Figure 4.6. With the number of RCSs increasing, the marginal cost of further decreasing the SAIDI to enhance reliability and resilience also increases. This model involves 1023 PCRSs that are generated in 15 minutes. It takes 0.6 seconds to solve the transformed model.

Finally, the RL-oriented allocation scheme is listed in Table 4.8. In total, 12 switches are upgraded to be RCSs. This model involves only 321 PCRSs that are generated in five minutes. It takes 0.7 seconds to solve the transformed model.

Table 4.9 compares the computation time and the number of integer variables for both test systems. In Table 4.9, "Direct solution" means directly applying an applicable solver, e.g. Gurobi 6.0.4, to solve the original large-scale grid automation models. "Proposed method" stands for the transformation and solution approach presented in Section 4.5. And "N/A" means that within the maximum time limit (set as two hours, i.e. 7200 seconds), the problem is not solved. As shown in Table 4.9, the proposed approach reduces the number of integer

Table 4.8 Optimal RL-oriented RCS allocation result (IEEE 123-node test system).

Number of RCSs	Optimal RCS locations
12	$(36, 51), (47, 50), (49, 71), (50, 51),$ $(55, 117), (59, 100), (73, 78), (73, 103),$ $(99, 100), (107, 123), (108, 109), (116, 118)$

Table 4.9 Computational efficiency comparisons.

IEEE 33-node test system				
Allocation model	**Number of integer variables**		**Computation time (s)**	
	Direct solution	**Proposed method**	**Direct solution**	**Proposed method**
CIC-oriented	1036	37	858.7	60.2
SAIDI-oriented	1036	37	924.5	68.3
RL-oriented	1036	37	803.2	45.8

IEEE 123-node test system				
Allocation model	**Number of integer variables**		**Computation time (s)**	
	Direct solution	**Proposed method**	**Direct solution**	**Proposed method**
CIC-oriented	6223	127	N/A	1055.6
SAIDI-oriented	6223	127	N/A	879.2
RL-oriented	6223	127	N/A	310.1

variables in involved optimization problems and thereby also decreases the computation time.

4.7 Impacts of Remote-Controlled Switch Malfunction

Switch malfunction is a critical issue in distribution grid automation and postdisaster recovery. Failures of the protection system due to supervisory control malfunction, etc., as indicated in [52, 53], have critical influences on bulk power grid reliability and resilience. This is a more challenging issue for distribution grids with less sophisticated equipment [54]. Reference [55] studies RCS malfunction's impacts on distribution grid reliability and shows that as RCS malfunction paralyzes fast service recovery, the efficacy of RCSs declines. It is also indicated that this issue may impact the optimal grid automation plan, including RCS allocation. In other words, assuming full reliability for grid automation systems leads to overestimation of their merits; therefore, misguiding the most resilient and optimal grid automation plan.

However, in the literature, most papers on reliability- and resilience-enhancing grid automation optimization do not take into account the switch malfunction issue. Only [54], to the best of our knowledge, has considered this issue in allocating sectionalizing switches. In [54], the malfunction probability and interruption cost associated with switch malfunction are modeled and considered in the objective function to be minimized. Reference [54] uses the restoration strategy of closing a normally open switch, while this chapter uses a MIP-based network reconfiguration model to obtain postfault or postdisaster recovery strategies that may involve load switching, etc. Moreover, for each outage scenario, multiple candidate restoration strategies are considered in this chapter. Because of these differences, the approach of considering switch malfunction in [54] is not applicable in this chapter.

To extend the proposed grid automation optimization models in this chapter to incorporate the RCS malfunction issue for better resilience improvement against natural disasters, potential methods include the following: (i) For the RL-oriented model, change the right-hand side number of constraint (4.38) to be larger than 1. By doing this, more than 1 PCRSs, if any, are prepared for each considered outage scenario, thus decreasing the probability of RCS malfunction paralyzing prompt service restoration. (ii) For the CIC-oriented and SAIDI-oriented models, multiply the functioning probability of each PCRS with related variables and parameters, and/or add another cost due to RCS malfunction to the objective function. This cost should be modeled based on analyses of historical data, etc. (iii) Integrate the proposed method with simulation-based optimization [56].

Still, to consider malfunction issues in grid automation optimization (especially against natural disasters), many challenges need to be resolved. For example, the selection of RCS malfunction scenarios has to be investigated. Note that there exist different types of RCS malfunctions, i.e. isolation capability malfunction and remote capability malfunction, which have different probabilities. The number of concurrent RCS malfunctions is also uncertain. Other challenges include the much heavier computational burden, etc., which makes the consideration of the RCS malfunction issue quite complex.

Note that, an optimal solution of a grid automation method that does not take into account the issue of switch malfunction, as in most of the relevant works, should still be acceptable. By appropriately coordinating field crews and the control center, RCS malfunction's impacts on system resilience can be reduced. Besides, the optimal allocation of RCSs, compared with manual switches, can be less sensitive to the malfunction issue [54].

4.8 Consideration of Distributed Generations

With demand response, energy storage systems, and DGs, etc., distribution grids are becoming more active. However, it is a challenging task to holistically consider these components in grid automation optimization. Moreover, renewable DGs' uncertain and intermittent natures make it somewhat more challenging to consider them in resilience-enhancing grid automation optimization.

From the viewpoint of distribution grid operation, using DGs in postdisaster recovery should be needed and beneficial. For example, reference [57] uses DG island restoration to promptly restore electricity supply to critical loads. Nevertheless, many challenging issues have to be resolved, so as to pragmatize the application of DGs in distribution grid recovery. To name a few, first, reliable supports from a DG forecasting system and the distribution SCADA system, etc., are required. Second, voltage limit violation and/or branch overload can be caused by DG uncertainties, resulting in further load shedding and recovery results that are worse than expected or even infeasible [50]. Third, in many cases, DGs are uncontrollable. Forth, some DGs do not have the black-start capability. Fifth, to avoid out-of-phase synchronization that can lead to large mechanical torques, overcurrent, and overvoltage, etc., DGs may have to be disconnected before closing switches. Distribution grid restoration with DGs is quite complex due to such issues. In fact, because of the increased number and severity of faults, DGs may even result in longer restoration time [58].

From the viewpoint of distribution grid planning, it is challenging to consider DGs in grid automation optimization. The reasons include the aforementioned challenges of using DGs in grid restoration. Above all, the availability of DGs in considered outage scenarios (such as those caused by natural disasters) is far from ensured and difficult to quantify. It is not easy to prepare reliable and resilient DG-involving recovery strategies for considered outage scenarios when planning a distribution grid. Besides, considering that the extent of distribution grids is vast and the cost of active network devices/technologies is high, many distribution grids remain passive.

Still, if DGs' penetration in a distribution grid achieves a critically high level, or the distribution grid is equipped with microturbine-based DGs, the placement of RCSs and other grid automation devices indeed has to consider DGs. In this case, our approach, with slight adjustments, is still applicable. Particularly, the restoration model has to be modified to obtain restoration strategies using DGs. Nevertheless, the challenges aforementioned still have to be resolved. For instance, robust optimization or stochastic optimization [59], etc., may be adopted to deal with DG uncertainties.

4.9 Summary and Conclusions

This chapter presents a new approach to reliability- and resilience-oriented grid automation optimization for enhancing the restoration performance against natural disasters. The number and locations of switches to be upgraded are optimized to enable fast service recovery of distribution grids. We propose MICP-based grid automation models for three different objectives, i.e. maximizing the RCIC, maximizing the RSAIDI, and maximizing the RCS-restorable load. We also propose a novel transformation and solution approach applicable to each of these models. Based on two test systems, illustrative cases demonstrate the proposed models and algorithms' effectiveness.

As mentioned above, RCSs and other grid automation devices can also be placed to enable dynamic reconfiguration of distribution grids in order to decrease power losses and improve distributed renewable generation hosting capacity, etc. Moreover, the proposed transformation and solution approach may be useful for stochastic MILP problems of a certain type. It is worth investigating these problems in the future.

Nomenclature of RCS-Restoration Models

Indices and Sets

i, j	indices of nodes
$\mathcal{N}$	set of all nodes
$\mathcal{N}_s$	set of substation nodes
$\mathcal{L}$	set of all lines
Ω_i	set of nodes connected with node i
Φ_k	set of outage nodes because of a fault in zone k

Parameters

$\overline{p}_i, \overline{q}_i$	real and reactive power loads at node i
$\overline{V}_i, \underline{V}_i$	maximum and minimum voltage values of node i
r_{ij}, x_{ij}	resistance and reactance of branch (i, j)
$\overline{S}_{ij}$	apparent power capacity of branch (i, j)
φ_{ij}	binary, $\varphi_{ij} = 1$ or 0 if line (i, j) is normally closed or open, respectively
z_{ij}	binary, $z_{ij} = 1$ or 0 if line (i, j) is installed with a remote-controlled or manual switch, respectively
ω_i	priority weight of the load at node i
δ_1	a coefficient adjusting relative weights of two objectives
M	a sufficiently large positive number
ε	a small positive number

Variables

α_{ij}	binary, $\alpha_{ij} = 1$ or 0 if line (i, j) is closed or open, respectively
P_{ij}, Q_{ij}	real and reactive power flow on line (i, j)
p_i, q_i	restored real and reactive power demand at node i
v_i	squared voltage magnitude of node i

Nomenclature of RCS Allocation Models

Indices and Sets

k	index of zones (possible fault locations)
$\mathcal{N}_f$	set of zones (possible fault locations)

Parameters

λ_k	total failure rate of distribution components in zone k
$t_{ik}^{MS}, t_{ik}^{RCS}$	restoration time of the load at node i in the k-th outage scenario using manual or remote-controlled switches
C_{RCS}^{INV}	investment cost of a RCS
$C_{RCS}^{M\&O}$	maintenance and operation cost of a RCS
ρ	a penalty coefficient
δ_2	a coefficient adjusting relative weights of two objectives
n_i^0	number of customers at node i
RL_k	restorable load in the k-th outage scenario

Variables

z_{ij}	binary, $z_{ij} = 1$ or 0 if line (i, j) is installed with a remote-controlled or manual switch, respectively
n_{ik}, p_{ik}	restored number of customers or restored real power demand, respectively, at node i using RCSs in the k-th outage scenario
RL_k^{RCS}	RCS-restorable load in the k-th outage scenario

Functions

$f_i(t)$	customer damage function expressing the interruption cost of the load at node i, which depends on the outage duration t
$\Psi_k(z)$	feasible set of the tuple $p_k = \{p_{1k}, p_{2k}, ..., p_{Nk}\}$, which depends on $z = \{z_{ij} : \forall(i,j) \in \mathcal{L}\}$
$\Lambda_k(z)$	feasible set of the tuple $n_k = \{n_{1k}, n_{2k}, ..., n_{Nk}\}$, which depends on $z = \{z_{ij} : \forall(i,j) \in \mathcal{L}\}$

References

1 R. Billington and R. N. Allan, "Reliability evaluation of power systems," 1984.

2 A. Chowdhury and D. Koval, *Power distribution system reliability: Practical methods and applications*. John Wiley & Sons, 2011.

3 C.-S. Chen, C.-H. Lin, H.-J. Chuang, C.-S. Li, M.-Y. Huang, and C.-W. Huang, "Optimal placement of line switches for distribution automation systems

using immune algorithm," *IEEE Transactions on Power Systems*, vol. 21, no. 3, pp. 1209–1217, 2006.

4 J.-H. Teng and Y.-H. Liu, "A novel ACS-based optimum switch relocation method," *IEEE Transactions on Power Systems*, vol. 18, no. 1, pp. 113–120, 2003.

5 P. Wang and R. Billinton, "Demand-side optimal selection of switching devices in radial distribution system planning," *IEE Proceedings-Generation, Transmission and Distribution*, vol. 145, no. 4, pp. 409–414, 1998.

6 A. Moradi and M. Fotuhi-Firuzabad, "Optimal switch placement in distribution systems using trinary particle swarm optimization algorithm," *IEEE Transactions on Power Delivery*, vol. 23, no. 1, pp. 271–279, 2007.

7 A. Abiri-Jahromi, M. Fotuhi-Firuzabad, M. Parvania, and M. Mosleh, "Optimized sectionalizing switch placement strategy in distribution systems," *IEEE Transactions on Power Delivery*, vol. 27, no. 1, pp. 362–370, 2011.

8 O. K. Siirto, A. Safdarian, M. Lehtonen, and M. Fotuhi-Firuzabad, "Optimal distribution network automation considering earth fault events," *IEEE Transactions on Smart Grid*, vol. 6, no. 2, pp. 1010–1018, 2015.

9 G. Celli and F. Pilo, "Optimal sectionalizing switches allocation in distribution networks," *IEEE Transactions on Power Delivery*, vol. 14, no. 3, pp. 1167–1172, 1999.

10 H. Falaghi, M.-R. Haghifam, and C. Singh, "Ant colony optimization-based method for placement of sectionalizing switches in distribution networks using a fuzzy multiobjective approach," *IEEE Transactions on Power Delivery*, vol. 24, no. 1, pp. 268–276, 2008.

11 J.-H. Teng and C.-N. Lu, "Feeder-switch relocation for customer interruption cost minimization," *IEEE Transactions on Power Delivery*, vol. 17, no. 1, pp. 254–259, 2002.

12 G. Levitin, S. Mazal-Tov, and D. Elmakis, "Optimal sectionalizer allocation in electric distribution systems by genetic algorithm," *Electric Power Systems Research*, vol. 31, no. 2, pp. 97–102, 1994.

13 D. P. Bernardon, M. Sperandio, V. J. Garcia, L. N. Canha, A. da Rosa Abaide, and E. F. B. Daza, "AHP decision-making algorithm to allocate remotely controlled switches in distribution networks," *IEEE Transactions on Power Delivery*, vol. 26, no. 3, pp. 1884–1892, 2011.

14 L. S. de Assis, J. F. V. González, F. L. Usberti, C. Lyra, C. Cavellucci, and F. J. Von Zuben, "Switch allocation problems in power distribution systems," *IEEE Transactions on Power Systems*, vol. 30, no. 1, pp. 246–253, 2014.

15 R. Billinton and S. Jonnavithula, "Optimal switching device placement in radial distribution systems," *IEEE Transactions on Power Delivery*, vol. 11, no. 3, pp. 1646–1651, 1996.

16 W. Tippachon and D. Rerkpreedapong, "Multiobjective optimal placement of switches and protective devices in electric power distribution systems using ant colony optimization," *Electric Power Systems Research*, vol. 79, no. 7, pp. 1171–1178, 2009.

17 D. Bernardon, M. Sperandio, V. Garcia, J. Russi, L. Canha, A. Abaide, and E. Daza, "Methodology for allocation of remotely controlled switches in distribution networks based on a fuzzy multi-criteria decision making algorithm," *Electric Power Systems Research*, vol. 81, no. 2, pp. 414–420, 2011.

18 A. S. Bouhouras, G. T. Andreou, D. P. Labridis, and A. G. Bakirtzis, "Selective automation upgrade in distribution networks towards a smarter grid," *IEEE Transactions on Smart Grid*, vol. 1, no. 3, pp. 278–285, 2010.

19 P. Carvalho, L. Ferreira, and A. C. Da Silva, "A decomposition approach to optimal remote controlled switch allocation in distribution systems," *IEEE Transactions on Power Delivery*, vol. 20, no. 2, pp. 1031–1036, 2005.

20 Y. Xu, C.-C. Liu, K. P. Schneider, and D. T. Ton, "Placement of remote-controlled switches to enhance distribution system restoration capability," *IEEE Transactions on Power Systems*, vol. 31, no. 2, pp. 1139–1150, 2015.

21 I.-h. Lim, T. Sidhu, M. Choi, S. Lee, and B. Ha, "An optimal composition and placement of automatic switches in DAS," *IEEE Transactions on Power Delivery*, vol. 28, no. 3, pp. 1474–1482, 2013.

22 Z. Ghofrani-Jahromi, M. Kazemi, and M. Ehsan, "Distribution switches upgrade for loss reduction and reliability improvement," *IEEE Transactions on Power Delivery*, vol. 30, no. 2, pp. 684–692, 2014.

23 P. H. Shah and B. R. Bhalja, "New adaptive digital relaying scheme to tackle recloser-fuse miscoordination during distributed generation interconnections," *IET Generation, Transmission and Distribution*, vol. 8, no. 4, pp. 682–688, 2014.

24 A. Fazanehrafat, S. Javadian, S. Bathaee, and M.-R. Haghifamt, "Maintaining the recloser-fuse coordination in distribution systems in presence of DG by determining DG's size," in *2008 IET 9th International Conference on Developments in Power System Protection (DPSP 2008)*, pp. 132–137, 2008.

25 H. Yazdanpanahi, Y. W. Li, and W. Xu, "A new control strategy to mitigate the impact of inverter-based DGs on protection system," *IEEE Transactions on Smart Grid*, vol. 3, no. 3, pp. 1427–1436, 2012.

26 L. Kumpulainen and K. Kauhaniemi, "Distributed generation and reclosing coordination," in *Nordic Distribution and Asset Management Conference*, 2004.

27 L. S. de Assis, J. F. V. González, F. L. Usberti, C. Lyra, and F. Von Zuben, "Optimal allocation of remote controlled switches in radial distribution systems," in *2012 IEEE Power and Energy Society General Meeting*, pp. 1–8, 2012.

28 Q. Qin and N. E. Wu, "Recloser allocation and placement for rural distribution systems," in *2015 IEEE Power & Energy Society General Meeting*, pp. 1–5, 2015.

29 M. Hajinia, S. M. Mahaei, and J. Jafarzadeh, "Simultaneous placement of recloser and fuse based on society development algorithm," in *2015 20th Conference on Electrical Power Distribution Networks Conference (EPDC)*, pp. 48–52, 2015.

30 M. Lavorato, J. F. Franco, M. J. Rider, and R. Romero, "Imposing radiality constraints in distribution system optimization problems," *IEEE Transactions on Power Systems*, vol. 27, no. 1, pp. 172–180, 2011.

31 M. Baran and F. Wu, "Network reconfiguration in distribution systems for loss reduction and load balancing," *IEEE Transactions on Power Delivery*, vol. 4, no. 2, pp. 1401–1407, 1989.

32 J. A. Taylor and F. S. Hover, "Convex models of distribution system reconfiguration," *IEEE Transactions on Power Systems*, vol. 27, no. 3, pp. 1407–1413, 2012.

33 X. Chen, W. Wu, and B. Zhang, "Robust restoration method for active distribution networks," *IEEE Transactions on Power Systems*, vol. 31, no. 5, pp. 4005–4015, 2015.

34 Y. Liu, Y. Hou, S. Lei, and D. Wang, "A distribution network restoration decision support algorithm based on multi-agent system," in *2016 IEEE PES Asia-Pacific Power and Energy Engineering Conference (APPEEC)*, pp. 33–37, 2016.

35 C.-C. Liu, S. J. Lee, and S. Venkata, "An expert system operational aid for restoration and loss reduction of distribution systems," *IEEE Transactions on Power Systems*, vol. 3, no. 2, pp. 619–626, 1988.

36 A. Morelato and A. Monticelli, "Heuristic search approach to distribution system restoration," *IEEE Transactions on Power Delivery*, vol. 4, no. 4, pp. 2235–2241, 1989.

37 A. Y. Abdelaziz, F. Mohammed, S. Mekhamer, and M. Badr, "Distribution systems reconfiguration using a modified particle swarm optimization algorithm," *Electric Power Systems Research*, vol. 79, no. 11, pp. 1521–1530, 2009.

38 J. Li, X.-Y. Ma, C.-C. Liu, and K. P. Schneider, "Distribution system restoration with microgrids using spanning tree search," *IEEE Transactions on Power Systems*, vol. 29, no. 6, pp. 3021–3029, 2014.

39 H.-C. Chang and C.-C. Kuo, "Network reconfiguration in distribution systems using simulated annealing," *Electric Power Systems Research*, vol. 29, no. 3, pp. 227–238, 1994.

40 K. Nara, A. Shiose, M. Kitagawa, and T. Ishihara, "Implementation of genetic algorithm for distribution systems loss minimum re-configuration," *IEEE Transactions on Power Systems*, vol. 7, no. 3, pp. 1044–1051, 1992.

41 D. Bertsimas and J. N. Tsitsiklis, *Introduction to linear optimization*. Athena Scientific, Belmont, MA, USA, 1997.

42 G. B. Dantzig, "Fourier-Motzkin elimination and its dual," tech. rep., Department of Operations Research, Stanford University, CA, USA, 1972.

43 G. Wacker and R. Billinton, "Customer cost of electric service interruptions," *Proceedings of the IEEE*, vol. 77, no. 6, pp. 919–930, 1989.

44 R. Billinton and Z. Pan, "Incorporating reliability index probability distributions in financial risk assessment with performance based regulation," *Electric Power Components and Systems*, vol. 33, no. 6, pp. 685–697, 2005.

45 A. L. da Silva, L. C. Nascimento, A. Guimar aes, and J. Mello, "Reliability indices applied to performance-based mechanisms in electric power distribution systems," *International Journal of Systems Assurance Engineering and Management*, vol. 1, no. 2, pp. 105–112, 2010.

46 T. Solver and L. Soder, "Comparison of incentives for distribution system reliability in performance-based regulations," in *2004 IEEE International Conference on Electric Utility Deregulation, Restructuring and Power Technologies*, vol. 2, pp. 485–490, 2004.

47 T. Ding, C. Li, C. Yan, F. Li, and Z. Bie, "A bilevel optimization model for risk assessment and contingency ranking in transmission system reliability evaluation," *IEEE Transactions on Power Systems*, vol. 32, no. 5, pp. 3803–3813, 2016.

48 S. Ahmed, "A scenario decomposition algorithm for 0–1 stochastic programs," *Operations Research Letters*, vol. 41, no. 6, pp. 565–569, 2013.

49 M. Conforti, G. Cornuéjols, and G. Zambelli, *Integer programming*. Springer, 2014.

50 K. Chen, W. Wu, B. Zhang, and H. Sun, "Robust restoration decision-making model for distribution networks based on information gap decision theory," *IEEE Transactions on Smart Grid*, vol. 6, no. 2, pp. 587–597, 2014.

51 IEEE PES Power System Analysis, Computing and Economics Committee, *IEEE 123 Node Test Feeder*, Feb. 2014.

52 K. Jiang and C. Singh, "New models and concepts for power system reliability evaluation including protection system failures," *IEEE Transactions on Power Systems*, vol. 26, no. 4, pp. 1845–1855, 2011.

53 M. Eliassi, H. Seifi, and M.-R. Haghifam, "Incorporation of protection system failures into bulk power system reliability assessment by Bayesian networks," *IET Generation, Transmission and Distribution*, vol. 9, no. 11, pp. 1226–1234, 2015.

54 M. Farajollahi, M. Fotuhi-Firuzabad, and A. Safdarian, "Optimal placement of sectionalizing switch considering switch malfunction probability," *IEEE Transactions on Smart Grid*, vol. 10, no. 1, pp. 403–413, 2017.

55 A. Safdarian, M. Farajollahi, and M. Fotuhi-Firuzabad, "Impacts of remote control switch malfunction on distribution system reliability," *IEEE Transactions on Power Systems*, vol. 32, no. 2, pp. 1572–1573, 2016.

56 A. Gosavi, "Parametric optimization techniques and reinforcement learning," 2003.

57 X. Huang, Y. Yang, and G. A. Taylor, "Service restoration of distribution systems under distributed generation scenarios," *CSEE Journal of Power and Energy Systems*, vol. 2, no. 3, pp. 43–50, 2016.

58 P. Hallberg, J. J. Alba Rios, C. Bergerland, A. Blanquet, M. Cailliau, C. Clifford, P. De Wit, E. Diskin, H. Feuk, B. Gouverneur, M. Kay, P. Lawson, M. Lombardi, P. Mandatova, R. Otter, P. Simon Caballero, J. Tello Guijarro, W. Tenschert, D. Trebolle, S. Wanzek, "Active distribution system management: A key tool for the smooth integration of distributed generation," *Eurelectric TF Active System Management*, 2013, 8 pages.

59 A. Ben-Tal, L. El Ghaoui, and A. Nemirovski, *Robust optimization*. Princeton University Press, 2009.

Part III

Response as a Natural Disaster Unfolds

5

Security Region-Based Operational Point Analysis for Resilience Enhancement

The increasing extreme weather events pose unprecedented challenges on power system operation in consideration of their sequential and uncertain influences on power systems.

This chapter proposes a concept with regard to a sequential steady-state security region (SSSR) and SSSR-based resilience enhancements for transmission systems, considering uncertain varying topology changes, due to extreme weather events are investigated. SSSR is a polytope that describes an operational region. In consideration of uncertain varying topology changes with SSSR, the resilience enhancement problem is constructed as a bilevel programming optimization model, which is utilized for the system operators to find an optimal strategy against the most threatening weather-related scenario. Karush–Kuhn-Tucker (KKT) condition is used to transform the suboptimization model in the lower-level model into a group of equivalent constraints in the upper-level model.

5.1 Resilience-Oriented Operational Strategies

Extreme weather events with higher frequency and heavy intensity have great impacts on power system operation [1, 2]. The countermeasures against possible extreme weather events are usually not included in the conventional operational strategies. Many organizations, e.g. the North American Electric Reliability Corporation [3, 4], the House of Lords in the United Kingdom [5], the United States National Research Council (NRC) [6], and the United States Electric Power Research Institute [7], have already clarified the importance of system resilience against extreme weather events.

According to the framework of power system resilience, strategies in three stages, i.e. prior to the event, during the event, and after the event [4, 6], are needed. *Prior to the event*, preventive strategies and assessments are expected to be implemented to enhance the system capacity of keeping operation in the face

Power Grid Resilience against Natural Disasters: Preparedness, Response, and Recovery, First Edition.
Shunbo Lei, Chong Wang, and Yunhe Hou.
© 2023 John Wiley & Sons Ltd. Published 2023 by John Wiley & Sons Ltd.

of disasters. The historical data-based regression models are established in [8–10] to assess the possible outages caused by hurricanes. In addition, the impacts of wildfires and extreme floods on power systems are analyzed [11, 12]. With the assessments, some preventive strategies, e.g. network hardening, prehurricane restoration planning, microgrid construction, and mobile energy storage preallocation, are investigated to increase system capacities to stay standing in the face of extreme weather events. For example, a proactive resource allocation model in consideration of repair and restoration is investigated in [13] to mitigate the potential damages on the path of an upcoming hurricane. Amirioun et al. [14] present conservative schedules for microgrids in consideration of the minimum number of vulnerable lines in service on the path of an approaching hurricane. A trilevel optimal hardening strategy is proposed in [15] to enhance the distribution system resilience against extreme weather events. A model for resilient routing and scheduling of mobile power sources is established to enable rapid prerestoration for high survivability of critical loads. *During the event*, real-time strategies are needed to enhance the system capacity of managing the impacts as it unfolds. An operational enhancement approach, assessing the impact of severe weather events on power systems, is presented in [16] to avoid the potential cascading failures during extreme weather events. A microgrid sectionalization method, dividing an interconnected system into several microgrids, is investigated in [17] to improve distribution system resilience. An integrated resilience response framework, including topology switching, generator dispatch, and load shedding, is proposed in [18]. A resilience-constrained unit commitment (RCUC) model, with operational constraints, heterogeneity of power flow distribution, and lines forced outages, is constructed in [19]. A Markov-based operation strategy is proposed in [20] to enhance system resilience during an unfolding extreme event. System topologies on the trajectory of the extreme weather-related event are modeled as Markov states, with which a recursive optimization in consideration of operation constraints is established. *After the event*, restoration strategies are needed to recover the system to a normal operating condition as quickly as possible. Microgrid construction is one of effective methods to restore loads quickly after disasters. Chen et al. [21] proposed a distribution system operational approach by constructing multiple microgrids with distributed generators to restore critical loads from the power outage. In consideration of the scarcity of power generation resources, Gao et al. [22] employed the concept of continuous operating time (COT) to determine the availability of microgrids to evaluate the service time and in consequence to restore critical load by using a chance-constrained model. To avoid subsequent outages, Che and Shahidehpour [23] presented the formation of adaptive multimicrogrids with mobile emergency resources as part of the critical service restoration strategy. After disasters, appropriately dispatching repair crews and mobile power sources can accelerate the load restoration. Lei et al. [24] proposed a resilient scheme for disaster recovery logistics to cooptimize distribution system restoration in consideration of dispatching repair crews and mobile power sources by using a mixed integer linear programming.

Based on the literature review, most of research studies focused on system hardening, mobile emergency resource allocation/dispatch, microgrid construction, and repair crew dispatch to enhance power system resilience for distribution systems. For transmission systems, the effects of these strategies are limited due to different characteristics between distribution systems and transmission systems. In practice, a proper initial operating point before an extreme weather event impacts the transmission systems is critical in consideration of operational constraints. An inappropriate initial operating point may need to perform load shedding in the face of uncertain sequential system topologies on the trajectory of a weather-related event in consideration of the operational constraints. Therefore, this chapter focuses on the investigation of the operating point for the transmission system to enhance the system resilience against the extreme weather event. The contributions of this chapter are listed as follows: (i) The concept of a SSSR is proposed; (ii) Based on the SSSR, resilience enhancement in consideration of uncertain varying topology changes caused by the extreme weather events is implemented; (iii) Resilience enhancement is constructed as a bilevel programming optimization model, with which an optimal operating point against the threatening scenarios caused by the sequential weather event can be achieved.

5.2 Security Region during an Unfolding Disaster

5.2.1 Sequential Security Region

Different transmission lines could be in failure with a certain probability in sequential time periods on the trajectory of an unfolding weather-related event, and this can lead to different topologies in different time periods. Take the scenario in Figure (5.1) as an example, the set of failure components at t_1 and t_2 could be $\{b_{12}\}$ and $\{b_{12}, b_{14}\}$, respectively. This means that the line b_{12} is in failure at t_1 and t_2 and the line b_{14} is in failure at t_2. Because of uncertain influences of the unfolding event on the network, the failure components are also uncertain and the failure sets are also different. For all of these possible scenarios, it is expected that the operating points could be located in the feasible region. However, the system operational points over sequential time periods depends on operational constraints, e.g. thermal capacity limits of lines, generators' ramping rates, generators' capacity limits. An appropriate initial operating point in the

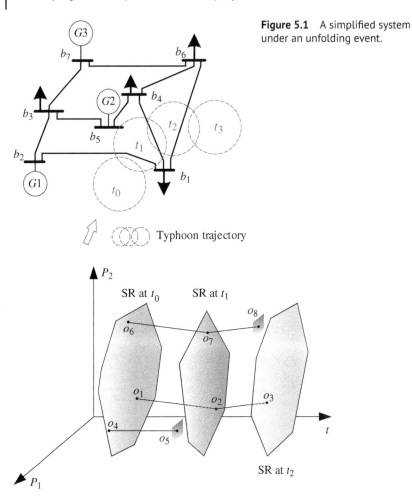

Figure 5.1 A simplified system under an unfolding event.

Typhoon trajectory

Figure 5.2 Security region (SR) over sequential time periods.

time period t_0 plays an important role in the subsequent operating points. The steady-state security region (SSR) is employed to illustrate the significance of the operating point.

The SSR is represented by a multidimensional space of which boundaries are the operational constraints. By utilizing the SSR, it is not difficult to check if an operational condition is steady-state secure. Take the scenario in Figure 5.2 as an example. The operating conditions o_1, o_4, and o_6 at time t_0 are steady-state secure because these operational points are within the SSR at t_0. However, sequential strategies should be needed when considering the impacts of sequential time periods.

For example, in Figure 5.2, o_1 can reach o_2, and then can reach o_3 constrained by the generators' ramp-rates at t_1, t_2, and t_3, respectively. However, o_4 can only reach the region that does not belong to SSR at t_2. Similarly, o_7 can only reach the red rectangular region at t_3, where the corresponding operating region is not within SSR at t_3. Based on the above analysis, it is concluded that the impacts of the sequential strategies over different time periods need be considered. Therefore, a SSSR in consideration of sequential time periods is proposed based on the existing SSR.

SSR can be modeled as (5.1).

$$\psi = \{y | a \cdot y \leq b\} \tag{5.1}$$

where y is the variable vector, where each element denotes power generation of each generator, and a and b are coefficient matrices, in which each element represents the operating constraints.

Based on (5.1), SSSR can be extended as (5.2).

$$\Psi = \{\mathbf{Y} | \mathbf{A} \cdot \mathbf{Y} \leq \mathbf{B}\} \tag{5.2a}$$

$$\mathbf{Y} = [\mathbf{y}_1^T, \ldots, \mathbf{y}_t^T \ldots, \mathbf{y}_N^T]^T \tag{5.2b}$$

$$\mathbf{A} = \begin{bmatrix} \mathbf{a}_1 & \mathbf{a}_{12} & & & \\ \mathbf{a}_{21} & \mathbf{a}_2 & & & \\ & & \ddots & & \\ & & & \mathbf{a}_{N-1} & \mathbf{a}_{N-1,N} \\ & & & \mathbf{a}_{N,N-1} & \mathbf{a}_N \end{bmatrix} \tag{5.2c}$$

$$\mathbf{B} = [\mathbf{b}_1^T, \ldots, \mathbf{b}_t^T \ldots, \mathbf{b}_N^T]^T \tag{5.2d}$$

where $\mathbf{A}$, $\mathbf{B}$, and $\mathbf{Y}$ are extended forms of a, b, and y, respectively. They include variables and coefficients related with operating constraints over sequential time periods. The elements $\mathbf{a}_{12}$ and $\mathbf{a}_{21}$ denote the coupling relations between the first time period and the second time period, and they depend on generators' ramp-rates.

5.2.2 Uncertain Varying System Topology Changes

SSSR can describe the sequential characteristics in consideration of sequential strategies. However, the uncertain influences of extreme weather-related events on the system should be also included.

Take the scenario in Figure 5.3 as an example, and the three-dimension figure representing the corresponding SSSR is projected into three two-dimension figures to make them read easily. The feasible regions at t_0, t_1, and t_2 are represented in Figure 5.3a–c, respectively. When determining the strategy at t_0, the system

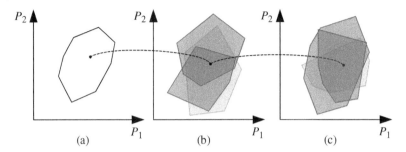

Figure 5.3 Sequential security regions illustrated by three two-dimension figures. (a) 2-dimension sequential security region at time t_0 (b) 2-dimension sequential security region at time t_1 (c) 2-dimension sequential security region at time t_2.

topology at the future time t_2 and t_3 cannot be known because the lines on the trajectory may be in failure due to extreme weather events, and this will cause different possible SSSRs.

When determining the strategies at t_0, it is expected to obtain a strategy that can reach a feasible region at t_1 and t_2 in consideration of different possible SSSRs. To have a mathematical model representing this region, a new set Ω, including all possible SSSRs caused by severe weather-related events, is defined as (5.3).

$$\Omega = \{\Psi_1, \Psi_2, \ldots, \Psi_K\} \tag{5.3}$$

where Ψ_1, Ψ_2, and Ψ_K are different SSSRs, representing different sequential system topologies $\Pi_1 = \{\pi_{11}, \ldots, \pi_{1N}\}$, $\Pi_2 = \{\pi_{21}, \ldots, \pi_{2N}\}$, and $\Pi_K = \{\pi_{K1}, \ldots, \pi_{KN}\}$. π_{KN} is a system topology scenario of the Kth SSSR at time N. The Monte Carlo method can be used to stochastically generate the sequential system topologies Π_1, Π_2, and Π_K based on failure probability due to the extreme weather events [25–27].

5.3 Operational Point Analysis Resilience Enhancement

This section first presents the mathematical formulation of SSSR in consideration of uncertain varying topology changes, and then the bilevel optimization model is established.

5.3.1 Sequential Security Region

For a given sequential system topology Π_k, we can list its corresponding SSSR as follows.

$$\sum_{g \in G_b} P_{gt} - P_{bt} + \sum_{b' \in B_b} P_{bb't} = 0 \quad \forall b, t \tag{5.4a}$$

$$B_{bb'}(\theta_{bt} - \theta_{b't}) = P_{bb't} \quad \forall (b, b') \in \Pi_K, t \tag{5.4b}$$

$$\underline{P}_{bb'} \leq P_{bb't} \leq \overline{P}_{bb'} \quad \forall (b, b') \in \Pi_K, t \tag{5.4c}$$

$$\underline{R}_g \leq P_{g(t+1)} - P_{gt} \leq \overline{R}_g \quad \forall g, t \tag{5.4d}$$

$$\underline{P}_g \leq P_{gt} \leq \overline{P}_g \quad \forall g, t \tag{5.4e}$$

$$\underline{\theta}_b \leq \theta_{bt} \leq \overline{\theta}_b \quad \forall b, t \tag{5.4f}$$

where (5.4a) denotes power balance at each bus, (5.4b) denotes the DC power flow through the line, (5.4c) is the line power limit, (5.4d) denotes the constraints of generators' ramp-rates, (5.4e) shows the constraints of generators' outputs, and (5.4f) presents the voltage constraints. An explicit function of voltage angles θ_b and θ'_b is used to represent $P_{bb't}$ to guarantee that topology changes can be easily embedded in the model. The generic form of (5.4) can be expressed as (5.5).

$$\exists \mathbf{Y} : \mathbf{A} \cdot \mathbf{Y} \leq \mathbf{B} \tag{5.5}$$

The model (5.5) indicates that there is a strategy satisfying the operational constraints with a given SSSR. To find this strategy, a new optimization model including positive slack vectors $\mathbf{s}^+$ and $\mathbf{s}^-$ can be constructed.

$$f(\mathbf{Y}) = \min_{\mathbf{Y}, \mathbf{s}^+, \mathbf{s}^-} \mathbf{1}^T \mathbf{s}^+ + \mathbf{1}^T \mathbf{s}^- \tag{5.6a}$$

$$s.t. \ \mathbf{A} \cdot \mathbf{Y} + \mathbf{1}^T \cdot \mathbf{s}^+ - \mathbf{1}^T \cdot \mathbf{s}^- \leq \mathbf{B} \tag{5.6b}$$

$$\mathbf{s}^+ \geq \mathbf{0}, \quad \mathbf{s}^- \geq \mathbf{0} \tag{5.6c}$$

where $\mathbf{1}$ and $\mathbf{0}$ are vectors that include 1 and 0 with proper dimensions, respectively. It is concluded that $\mathbf{Y} \neq \emptyset$ if we have $f(\mathbf{Y}) = 0$. When we have $f(\mathbf{Y}) > 0$, no strategy can satisfy the operating constraints, i.e. $\mathbf{Y} = \emptyset$. So we can conclude that the necessary and sufficient condition for $\mathbf{Y} \neq \emptyset$ is $f(\mathbf{Y}) = 0$.

5.3.2 Sequential Security Region with Uncertain Varying Topology Changes

An extreme weather event may lead to different system topologies on its trajectory. When considering possible topology changes, SSSR can be represented as follows:

$$\sum_{g \in G_b} P_{gt} - P_{bt} + \sum_{b' \in B_b} P_{bb'tk} = 0 \quad \forall b, t, k \tag{5.7a}$$

$$B_{bb'}(\theta_{btk} - \theta_{b'tk}) = P_{bb'tk} \quad \forall(b, b') \in \Pi_k, t, k \tag{5.7b}$$

$$\underline{P}_{bb'} \leq P_{bb'tk} \leq \overline{P}_{bb'} \quad \forall(b, b') \in \Pi_k, t, k \tag{5.7c}$$

$$\underline{R}_g \leq P_{g(t+1)} - P_{gt} \leq \overline{R}_g \quad \forall g, t \tag{5.7e}$$

$$\underline{P}_g \leq P_{gt} \leq \overline{P}_g \quad \forall g, t \tag{5.7e}$$

$$\underline{\theta}_b \leq \theta_{btk} \leq \overline{\theta}_b \quad \forall b, t, k \tag{5.7f}$$

where the possible topology changes are represented by the topology scenario notation k in (5.7a)–(5.7f). Even though (5.7a)–(5.7f) are directly related to the system topologies, it is not easy to optimize this model because the model is not an explicit model with regard to optimization variables. Therefore, integer variables, representing on-off states of lines, are introduced into the model to directly connect with varying system topologies.

$$\sum_{g \in G_b} P_{gt} - P_{bt} + \sum_{b' \in B_b} P_{bb'tk} = 0 \quad \forall b, t, k \tag{5.8a}$$

$$B_{bb'}(\theta_{btk} - \theta_{b'tk}) - P_{bb'tk} + (1 - u_{bb'tk})M \geq 0$$
$$\forall(b, b') \in l, l \in \mathcal{L}, t \tag{5.8b}$$

$$B_{bb'}(\theta_{btk} - \theta_{b'tk}) - P_{bb'tk} - (1 - u_{bb'tk})M \leq 0$$
$$\forall(b, b') \in l, l \in \mathcal{L}, t \tag{5.8c}$$

$$u_{bb'tk}\underline{P}_{bb'} \leq P_{bb'tk} \leq u_{bb'tk}\overline{P}_{bb'} \quad \forall(b, b') \in l, l \in \mathcal{L}, t \tag{5.8d}$$

$$B_{bb'}(\theta_{btk} - \theta_{b'tk}) = P_{bb'tk} \quad \forall(b, b') \in l, l \in \overline{\mathcal{L}}, t, k \tag{5.8e}$$

$$\underline{P}_{bb'} \leq P_{bb'tk} \leq \overline{P}_{bb'} \quad \forall(b, b') \in l, l \in \overline{\mathcal{L}}, t, k \tag{5.8f}$$

$$\underline{R}_g \leq P_{g(t+1)} - P_{gt} \leq \overline{R}_g \quad \forall g, t \tag{5.8g}$$

$$\underline{P}_g \leq P_{gt} \leq \overline{P}_g \quad \forall g, t \tag{5.8h}$$

$$\underline{\theta}_b \leq \theta_{btk} \leq \overline{\theta}_b \quad \forall b, t, k \tag{5.8i}$$

where (5.8b)–(5.8d) represent the constraints with regard to the lines that are impacted by the extreme weather event. $u_{bb'tk} \; \forall(b, b') \in l, l \in \mathcal{L}, t$ can cover all system topologies in all possible SSSRs. In practice, a large-scale system will suffer from the curse of dimensionality with regard to system topologies, but some system topologies caused by the extreme weather event occur with a very small probability. To avoid the curse of dimensionality, the uncertain system topologies with very small probabilities can be ignored. To this end, we can use Monte Carlo to generate system topology scenarios.

5.3.3 Mapping System Topology Changes

Additional constraints with regard to system topology scenarios as variables in the optimization are constructed. A simple example is used to example this. There are three line l_1, l_2, and l_3, which will lead to $2^3 = 8$ topology scenarios if we consider all varying topology scenarios. Define x_1, x_2, and x_3 as binary variables to represent the states of three lines. It is assumed that we only consider two topology scenarios, and they are $\{x_1, x_2, x_3\} \in \{\{1,1,1\}, \{0,1,1\}\}$. The numbers "0" and "1" mean the off-state and on-state of the lines, respectively.

Define z_1 and z_2 as binary variables, which denote the selection of scenarios. If the scenario $\{1,1,1\}$ is selected, we have $z_1 = 1$. If the scenario $\{0,1,1\}$ is selected, we have $z_2 = 1$. Based on this case, the additional constraints can be expressed as (5.9).

$$z_1 = x_1 x_2 x_3 \tag{5.9a}$$

$$z_2 = (1 - x_1)x_2 x_3 \tag{5.9b}$$

$$z_1 + z_2 = 1 \tag{5.9c}$$

$$z_1, z_2, x_1, x_2, x_3 \in \{0,1\} \tag{5.9d}$$

where (5.9a) and (5.9b) denote the relations between the line states and the scenario selection. (5.9c) constrains that only one scenario can be selected. The multilinear terms in (5.9a) can be equivalently rewritten as several linear constraints by introducing new binary variables $\alpha = x_1 x_2$ with the McCormick envelope method.

$$\alpha \geq x_1 + x_2 - 1, \quad \alpha \leq x_1, \quad \alpha \leq x_2 \tag{5.10a}$$

$$z_1 \geq \alpha + x_3 - 1, \quad z_1 \leq \alpha, \quad z_1 \leq x_3 \tag{5.10b}$$

where (5.10a) and (5.10b) are equivalent to (5.9a) because all variables in (5.9a), (5.10a), and (5.10b) are binary.

Similarly, the multilinear terms in (5.9b) can be also rewritten with new binary variables $\beta = (1 - x_1)x_2$ as follows:

$$\beta \geq -x_1 + x_2, \quad \beta \leq 1 - x_1, \quad \beta \leq x_2 \tag{5.11a}$$

$$z_2 \geq \beta + x_3 - 1, \quad z_2 \leq \beta, \quad z_2 \leq x_3 \tag{5.11b}$$

The above case can be extended to a generic form:

$$z_k = \prod_{(b,b')\in l, l \in L, t} (1 - 2r_{bb'tk})(1 - u_{bb'tk} - r_{bb'tk}) \quad \forall k \tag{5.12a}$$

$$\sum_k z_k = 1 \tag{5.12b}$$

where $u_{bb'tk}$ and z_k are binary variables. The multilinear functions in (5.12a) are equivalently represented by a group of linear constraints by utilizing a recursive McCormick envelope method. In (5.12a), the multilinear function with all binary variables is $u_1 u_2 \cdots u_S$, in which the subscript S is the number of binary variables for one Ψ_k. $u_1 u_2 \cdots u_S$ can be expressed by introducing binary variables $\zeta_2, \zeta_3, \ldots, \zeta_S$ as follows:

$$\zeta_2 = u_1 u_2$$
$$\zeta_3 = \zeta_2 u_3 \tag{5.13}$$
$$\cdots$$
$$\zeta_S = \zeta_{S-1} u_S$$

Based on the recursive McCormick envelope method, the equivalent constraints for (5.13) can be expressed as follows:

$$\zeta_2 \geq u_2 + u_1 - 1$$
$$\zeta_2 \leq u_1$$
$$\zeta_i \geq 0 \quad (i = 2, \ldots, S)$$
$$\zeta_i \leq u_i \quad (i = 2, \ldots, S) \tag{5.14}$$
$$\zeta_i \geq u_i + \zeta_{i-1} - 1 \quad (i = 3, \ldots, S)$$
$$\zeta_i \leq \zeta_{i-1} \quad (i = 3, \ldots, S)$$

5.3.4 Bilevel Optimization Model

With the above reformulation, the optimization model (5.8) can be expressed as a generic form as follows:

$$\forall \mathbf{U}, \exists \mathbf{Y} : \quad \mathbf{A} \cdot \mathbf{Y} + \mathbf{C} \cdot \mathbf{U} \leq \mathbf{B} \tag{5.15}$$

where $\mathbf{Y}$ is the vector of continuous variables that represent operational conditions and $\mathbf{U}$ denotes the vector of binary variables in the model. $\mathbf{A}$, $\mathbf{B}$, and $\mathbf{C}$ are the coefficient matrices corresponding to (5.8a)–(5.8i). The physical meaning of (5.8) is that we can find a strategy that satisfies the operational constraints for all possible SSSRs. The model can be expressed as follows:

$$F(\mathbf{Y}) = \min_{\mathbf{Y}, \mathbf{s}^+, \mathbf{s}^-} \mathbf{1}^T \cdot \mathbf{s}^+ + \mathbf{1}^T \cdot \mathbf{s}^- \tag{5.16a}$$

$$s.t. \quad \forall \mathbf{U}, \ \exists \mathbf{Y}, \mathbf{s}^+ \geq \mathbf{0}, \mathbf{s}^- \geq \mathbf{0} \tag{5.16b}$$

$$\mathbf{A} \cdot \mathbf{Y} + \mathbf{C} \cdot \mathbf{U} + \mathbf{s}^+ - \mathbf{s}^- \leq \mathbf{B} \tag{5.16c}$$

where the optimization model (5.16) is to obtain a strategy satisfying the worst-case scenario. Therefore, a linear max–min optimization model can be used to express (5.16).

$$F(\mathbf{Y}) = \max_{\mathbf{U}} \min_{\mathbf{Y}, \mathbf{s}^+, \mathbf{s}^-} \mathbf{1}^T \cdot \mathbf{s}^+ + \mathbf{1}^T \cdot \mathbf{s}^- \tag{5.17a}$$

$$s.t. \quad \mathbf{A} \cdot \mathbf{Y} + \mathbf{C} \cdot \mathbf{U} + \mathbf{s}^+ - \mathbf{s}^- \leq \mathbf{B} \tag{5.17b}$$

$$\mathbf{s}^+ \geq \mathbf{0}, \quad \mathbf{s}^- \geq \mathbf{0} \tag{5.17c}$$

where (5.17) can be further reformulated as a bilevel optimization model as follows:

$$F(\mathbf{Y}) = \max_{\mathbf{U}} \mathbf{1}^T \cdot (\mathbf{s}^+)^* + \mathbf{1}^T \cdot (\mathbf{s}^-)^* \tag{5.18a}$$

$$s.t. \quad \min_{\mathbf{Y}, \mathbf{s}^+, \mathbf{s}^-} \mathbf{1}^T \cdot \mathbf{s}^+ + \mathbf{1}^T \cdot \mathbf{s}^- \tag{5.18b}$$

$$s.t. \quad \mathbf{A} \cdot \mathbf{Y} + \mathbf{C} \cdot \mathbf{U} + \mathbf{s}^+ - \mathbf{s}^- \leq \mathbf{B} \tag{5.18c}$$

$$\mathbf{s}^+ \geq \mathbf{0} \tag{5.18d}$$

$$\mathbf{s}^- \geq \mathbf{0} \tag{5.18e}$$

where $(\mathbf{s}^+)^*$ and $(\mathbf{s}^-)^*$ are the optimal solutions of the suboptimization model (5.18b)–(5.18d).

5.3.5 Solution Process

Because the constraint of the main model includes a suboptimization model (5.18b)–(5.18e), it is not easy to solve it directly. A group of equality and inequality constraints, based on KKT conditions, can be used to represent the suboptimiza-tion model (5.18b)–(5.18e). First, the Lagrangian of the suboptimization model (5.18b)–(5.18e) can be expressed as follows:

$$\begin{aligned} L &= \mathbf{1}^T \mathbf{s}^+ + \mathbf{1}^T \mathbf{s}^- \\ &+ \alpha^T (\mathbf{AY} + \mathbf{CU} + \mathbf{s}^+ - \mathbf{s}^- - \mathbf{B}) \\ &- \beta^T \mathbf{s}^+ - \gamma^T \mathbf{s}^- \end{aligned} \tag{5.19}$$

Based on the Lagrangian, $\mathbf{Y}^*$, $(\mathbf{s}^+)^*$, and $(\mathbf{s}^-)^*$ of the suboptimization model (5.18b)–(5.18e) satisfy the KKT conditions. Since there are many existing studies on the KKT conditions based on the Lagrangian function, we will not list the detailed models of KKT conditions.

5.4 Case Studies

One test system is used to validate the proposed model. Figure 5.1 shows the typhoon trajectory and the system topology. The buses b_2, b_5, and b_7 are connected to the generators G_1, G_2, and G_3, respectively. Loads connected to the buses b_1, b_3, b_4, and b_6 are 0.4, 0.4, 0.6, and 0.6 (p.u.), respectively. Table 5.1 shows the line parameters. The lower/upper power limits of each generator are 0.2 and 2.5 (p.u.), respectively. It is assumed that there are four decision periods. At time t_1 and t_2, the typhoon impacts the system topology directly. Even though the typhoon does not impact the system directly at t_0, the strategy at t_0 has great impacts on the strategies in the following time periods, e.g. t_1 and t_2, in consideration of the operational constraints, the generators' ramping rates, and possible topology changes caused by the typhoon.

The possible topology changes at t_1 and t_2, i.e. the SSSR, should be considered when making decisions at t_0. The possible topology changes at t_1 and t_2 are listed in Table 5.2. When the generators' ramping rates are 0.15 and 0.35 (p.u.), the corresponding feasible regions are depicted as the regions in Figure 5.4a and b. We can find that the higher generators' ramping rates lead to a larger feasible dispatch region. For the operational point A at t_0, it corresponds to $P_{G_1} = 0.62$, $P_{G_2} = 0.70$, and $P_{G_3} = 0.68$, which will be located in the feasible region with the generators' ramping rates as 0.15 (p.u.). For the possible system topologies caused

Table 5.1 Line data.

No.	From bus	To bus	Line limit (p.u.)
1	1	2	0.65
2	1	4	0.65
3	1	6	0.65
4	2	3	0.9
5	3	5	0.9
6	4	5	0.8
7	4	6	0.8
8	6	7	0.9
9	3	7	0.9

Table 5.2 Component failure scenarios.

No.	Component failure	
	t_1	t_2
1 (1)	—	—
2 (2)	—	b_1-b_4
3 (3)	—	b_1-b_6
4 (11)	—	b_1-b_4, b_1-b_6
5 (4)	b_1-b_2	—
6 (5)	b_1-b_2	b_1-b_4
7 (6)	b_1-b_2	b_1-b_6
8 (12)	b_1-b_2	b_1-b_4, b_1-b_6
9 (7)	b_1-b_4	—
10 (8)	b_1-b_4	b_1-b_6
11 (9)	b_1-b_2, b_1-b_4	—
12 (10)	b_1-b_2, b_1-b_4	b_1-b_6

Table 5.3 Component failure scenarios.

No.	Line power flow	
	t_1	t_2
1	—	0.5238
2	0.2050	—
3	0.1950	0.1238
4	0.6050	0.2312
5	0.0900	0.0345
6	⬚0.8000⬚	0.5255
7	0.0100	0.0745
8	0.7850	0.5507
9	0.1150	0.1343

by the typhoon, the corresponding line power flows at t_1 and t_2 are presented in Figure 5.5a and b. It is observed that all scenarios are within the limits. For the operational point B with $P_{G_1} = 0.62$, $P_{G_2} = 0.90$, and $P_{G_3} = 0.48$ at t_0, it is beyond the feasible dispatch region with the generators' ramping rates as 0.15 (p.u.). Table 5.3 lists line power flows. Under the optimal strategy at t_1, the power flow of

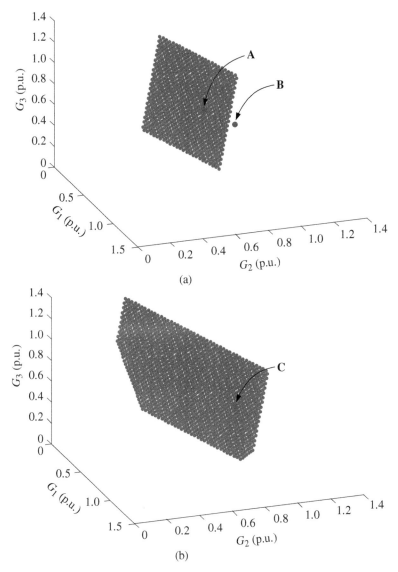

Figure 5.4 Feasible dispatch region in t_0 with ramping rates 0.15 p.u. (a) and 0.35 p.u. (b).

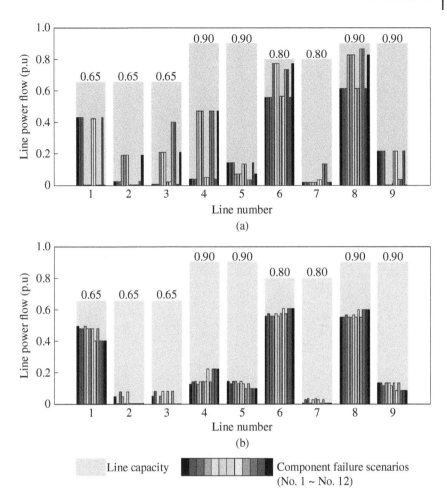

Figure 5.5 Power flows under different failure scenarios with ramping rate as 0.15 p.u. in t_1 (a) and t_2 (b), respectively.

the 6th line reaches the upper limit with the loss of load 0.015 (p.u.) at bus b_4. The point C has the same operational conditions as the point B except the generators' ramping rates. For the point C, the corresponding generators' ramping rates are 0.35 and in consequence, it is within the feasible region in consideration of possible

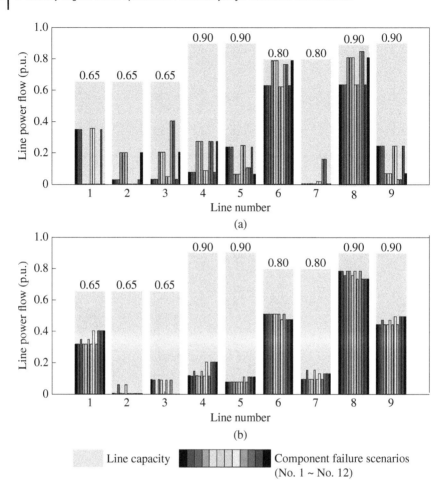

Figure 5.6 Power Flow with ramping rate 0.35 p.u. in t_1 (a) and t_2 (b), respectively.

system topologies caused by the typhoon. The corresponding line power flows at t_1 and t_2 are listed in Figure 5.6a and b, and all scenarios are within the limits.

Furthermore, the feasible sequential regions under different system topology changes are also impacted by system parameters. When having generators' ramping rates as 0.15 (p.u.), the feasible regions at t_0, with different capacities of the line b_4–b_5, i.e. 0.7, 0.8, 0.9, are presented in Figure 5.7a–c. Results show that a larger capacity limit can lead to a larger feasible dispatch region under the same adjustment ability of the system against potential varying topology changes. The system operators can dispatch the system to the feasible dispatch region to decrease the potential damages.

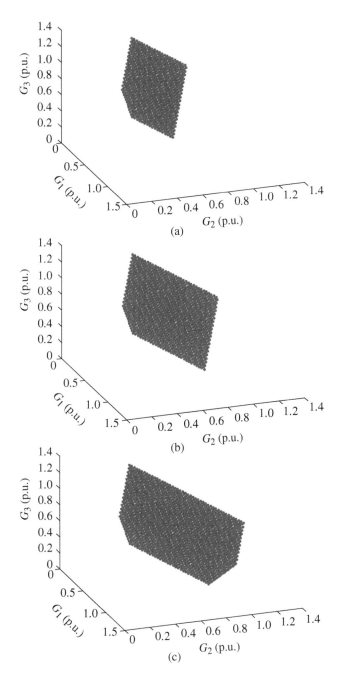

Figure 5.7 Feasible dispatch regions in t_0 with different line capacities 0.7 (a), 0.8 (b), 0.9 (c), respectively.

5.5   Summary and Conclusions

A SSSR is proposed in this chapter to contribute to construct resilient strategies in consideration of potential varying topology changes due to extreme weather-related events. The problem is to find a strategy that satisfies the operational constraints considering varying topology changes. To construct a mathematical model for this problem, a bilevel optimization model is established to search a strategy satisfying the worst-case scenario. To deal with the curse of dimensionality caused by possible system topologies for a large-scale system, the Monte Carlo method is utilized to produce the scenarios of system topology. A recursive McCormick envelope-based approach is employed to map the generated system topologies into the optimization variables. One system is used to validate the proposed model. The detailed results show that SSSR-based enhancement strategies are impacted by operating limits such as generators' ramping rates and line thermal capacity limits.

Nomenclature

Indices and Sets

b, b'	index of terminal buses of a line
g	index of generators
k	index of sequential system topology scenarios
l	index of lines
t	index of time periods
$\mathcal{G}_b$	set of units at power bus b
$\mathcal{B}_b$	set of power buses connected to power bus b
$\mathcal{L}$	set of lines that are in failure states for all sequential system topology scenarios
$\overline{\mathcal{L}}$	set of lines that are not in failure states for all sequential system topology scenarios
Π_k	set of system topologies from the first decision time period to the last decision time period for the kth sequential system topology scenario
Ω	set of all possible SSSRs corresponding to all sequential system topology scenarios

Parameters

Ψ_K	different sequential steady-state security region
$\mathbf{a}, \mathbf{b}$	coefficient matrices based on operational constraints
$\mathbf{A}, \mathbf{B}, \mathbf{C}$	coefficient matrices based on operational constraints
$B_{bb'}$	electrical susceptance
M	a large number
$\overline{P}_{bb'}$	upper active power capacities of line $b - b'$
P_{bt}	load at bus b in the time period t
$\underline{P}_g, \overline{P}_g$	lower and upper active power limits of generator g
$\underline{R}_g, \overline{R}_g$	ramp-up and ramp-down limits of generator u
$\underline{\theta}_b, \overline{\theta}_b$	limits of voltage angle at bus b

Variables

P_{gt}	active power of generator g at t and $t + 1$, respectively
$P_{bb'tk}$	active power through line $b - b'$ in the time period t under the sequential system topology scenario k
θ_{btk}	phase angle of bus b in the time period t under the sequential system topology scenario k
$u_{bb'tk}$	binary variable to indicate the state of line $b - b'$ in the time period t under the sequential system topology scenario k. '1' and '0' denote on-state and off-state, respectively
$\mathbf{U}$	binary variable vector representing different system topology
$\mathbf{Y}$	continuous variable vector representing system operating conditions
$\mathbf{Y}^*$	optimal value of the suboptimization model
$\mathbf{s}^+, \mathbf{s}^-$	vectors of positive slack variables
$(\mathbf{s}^+)^*$	optimal values of $\mathbf{s}^+$ for the suboptimization model
$(\mathbf{s}^-)^*$	optimal values of $\mathbf{s}^-$ for the suboptimization model
L	Lagrangian function
f, F	optimization objective functions
α, β, γ	Lagrange multiplier vectors
$\hat{\alpha}, \hat{\beta}, \hat{\gamma}$	Lagrange multiplier diagonal matrices corresponding to α, β, γ

References

1 Department of Energy, "Insurance as a risk management instrument for energy infrastructure security and resilience."

2 International Energy Agency, "RD&D needs for energy system climate preparedness and resilience."

3 North American Electric Reliability Corporation, "High-impact, low-frequency event risk to the North American bulk power system."

4 North American Electric Reliability Corporation, "Severe impact resilience: Considerations and recommendations."

5 House of Lords U.K., Science and Technology Select Committee, "The resilience of the electricity system."

6 National Academy of Sciences USA, *National Research Council: Disaster Resilience: A National Imperative*. National Academies Press, Washington, DC, USA, 2012.

7 Electric Power Research Institute, "Enhancing distribution resiliency: Opportunities for applying innovative technologies."

8 H. Liu, R. A. Davidson, J. R. Stedinger, D. V. Rosowsky, and D. V. Rosowsky, "Negative binomial regression of electric power outages in hurricanes," *Journal of Infrastructure Systems*, vol. 11, no. 4, pp. 258–267, 2015.

9 S. D. Guikema, R. Nateghi, S. M. Quiring, A. Staid, A. C. Reilly, and M. Gao, "Predicting hurricane power outages to support storm response planning," *IEEE Access*, vol. 2, pp. 1364–1373, 2014.

10 R. Nateghi, S. Guikema, and S. Quiring, "Forecasting hurricane-induced power outage durations," *Natural Hazards*, vol. 74, no. 3, pp. 1795–1811, 2014.

11 D. N. Trakas and N. D. Hatziargyriou, "Optimal distribution system operation for enhancing resilience against wildfires," *IEEE Transactions on Power Systems*, vol. 33, pp. 2260–2271, 2018.

12 M. H. Amirioun, F. Aminifar, and H. Lesani, "Towards proactive scheduling of microgrids against extreme floods," *IEEE Transactions on Smart Grid*, vol. 9, pp. 3900–3902, 2018.

13 A. Arab, A. Khodaei, S. K. Khator, K. Ding, V. A. Emesih, and Z. Han, "Stochastic pre-hurricane restoration planning for electric power systems infrastructure," *IEEE Transactions on Smart Grid*, vol. 6, pp. 1046–1054, 2015.

14 M. H. Amirioun, F. Aminifar, and H. Lesani, "Resilience-oriented proactive management of microgrids against windstorms," *IEEE Transactions on Power Systems*, vol. 33, pp. 4275–4284, 2018.

15 S. Ma, B. Chen, and Z. Wang, "Resilience enhancement strategy for distribution systems under extreme weather events," *EEE Transactions on Smart Grid*, vol. 9, pp. 1442–1451, 2018.

16 M. Panteli, D. N. Trakas, P. Mancarella, and N. D. Hatziargyriou, "Boosting the power grid resilience to extreme weather events using defensive islanding," *IEEE Transactions on Smart Grid*, vol. 7, pp. 2913–2922, 2016.

17 Z. Wang and J. Wang, "Self-healing resilient distribution systems based on sectionalization into microgrids," *IEEE Transactions on Power Systems*, vol. 30, pp. 3139–3149, 2015.

18 G. Huang, J. Wang, C. Chen, J. Qi, and C. Guo, "Integration of preventive and emergency responses for power grid resilience enhancement," *IEEE Transactions on Power Systems*, vol. 32, pp. 4451–4463, 2017.

19 Y. Wang, L. Huang, M. Shahidehpour, L. L. Lai, H. Yuan, and F. Y. Xu, "Resilience-constrained hourly unit commitment in electricity grids," *IEEE Transactions on Power Systems*, vol. 33, pp. 5604–5614, 2018.

20 C. Wang, Y. Hou, F. Qiu, S. Lei, and K. Liu, "Resilience enhancement with sequentially proactive operation strategies," *IEEE Transactions on Power Systems*, vol. 32, pp. 2847–2857, 2017.

21 C. Chen, J. Wang, F. Qiu, and D. Zhao, "Resilient distribution system by microgrids formation after natural disasters," *IEEE Transactions on Smart Grid*, vol. 7, pp. 958–966, 2016.

22 H. Gao, Y. Chen, Y. Xu, and C. Liu, "Resilience-oriented critical load restoration using microgrids in distribution systems," *IEEE Transactions on Smart Grid*, vol. 7, pp. 2837–2848, 2016.

23 L. Che and M. Shahidehpour, "Adaptive formation of microgrids with mobile emergency resources for critical service restoration in extreme conditions," *IEEE Transactions on Power Systems*, vol. 34, pp. 742–753, 2019.

24 S. Lei, C. Chen, Y. Li, and Y. Hou, "Resilient disaster recovery logistics of distribution systems: Co-optimize service restoration with repair crew and mobile power source dispatch," *IEEE Transactions on Smart Grid*, vol. 10, no. 6, pp. 6187–6202, 2019, early access.

25 M. P. Bhavaraju, R. Billinton, G. L. Landgren, M. F. McCoy, and N. D. Reppen, "Proposed terms for reporting and analyzing outages of electrical transmission and distribution facilities," *IEEE Transactions on Power Apparatus and Systems*, vol. PAS-104, pp. 337–348, 1985.

26 M. Ouyang and L. Dueñas-Osorio, "Multi-dimensional hurricane resilience assessment of electric power systems," *Structural Safety*, vol. 48, pp. 15–24, 2014.

27 C. D. Canham, M. J. Papaik, and E. F. Latty, "Interspecific variation in susceptibility to windthrow as a function of tree size and storm severity for northern temperate tree species," *Canadian Journal of Forest Research*, vol. 31, pp. 1–10, 2001.

6

Proactive Resilience Enhancement Strategy for Transmission Systems

Many of climate change-related weather events occur with increasing frequency and intensity, and cause many catastrophic outages, which remind the need to enhance the resilience of power systems. A proactive operation strategy is proposed in this chapter to enhance system resilience during an unfolding extreme event. A Markov process is used to represent the uncertain sequential transition of system states driven by the evolution of extreme events. Transition probabilities depend on failure rates caused by extreme events. For each state, a recursive value function with a current cost and a future cost is modeled in consideration of inter-temporal constraints and operation constraints. An optimal strategy is established by optimizing the recursive model, which is transformed into a mixed integer linear programming by using the linear scalarization method, with the probability of each state as the weight of each objective. The results demonstrate that the proposed proactive operation strategies can reduce the loss of load (LoL) due to the development of extreme events.

6.1 Proactive Strategy Against Extreme Weather Events

Extreme weather events, e.g. wind storms, typhoons, and hurricanes, are occurring with increasing intensity and causing complete or partial power outages. These outages suggest the vulnerability of current power systems. Since power systems are critical infrastructures for society and economic development, an outage might cause severe consequences. In the United States, weather-related outages cause estimated $25 billion economic losses each year. In China, Typhoon Rammasun, which struck the Guangdong province on July 2014, took several 220 kV transmission lines out of service. The severe consequences of such extreme weather events in power systems have brought power system resilience to the attention of organizations and governments in the world. A Policy Framework

Power Grid Resilience against Natural Disasters: Preparedness, Response, and Recovery, First Edition.
Shunbo Lei, Chong Wang, and Yunhe Hou.

for the Twenty-First Century Grid, which was released by the US government in June 2011, emphasized the significance of resilient grids in countering the effects of increasingly intense weather events. The US National Research Council (NRC) and the House of Lords in the United Kingdom have also emphasized the importance of a resilient energy infrastructure. The North American Electric Reliability Corporation (NERC) and the US Electric Power Research Institute (EPRI) have both recognized the functionalities of system resilience.

Based on the requirements of power system resilience, some conceptual frameworks have been proposed. To ensure resilience against extreme weather events, the strategies in the three stages of a severe event (i.e. prior to the event, during the event, and after the event) should be considered.

Prior to an extreme weather-related event, an accurate outage prediction contributes to manage preparedness and restoration efforts. To improve accuracy of predictions, a negative binomial regression model is proposed [1]. Since this model is based on data regarding outages caused by three hurricanes, i.e. Fran (1996), Bonnie (1998), and Floyd (1999), it is only suitable for a specific service area. To overcome this limit, a generic model for the full US coastline is proposed in [2]. To estimate power outage durations in the face of hurricanes, a statistical model is proposed in [3]. With outages and the duration predictions, some preventive strategies prior to an extreme weather-related event can be performed to increase power system resilience. Considering the stochastic and sequential characteristics of events, the events' potential impacts on the resilience of power systems are analyzed by using sequential Monte Carlo simulations. To minimize negative impacts, the response before a hurricane is modeled as a mixed-integer programming problem [4]. In addition, preparation of sufficient black start generating units and emergency generators also plays an important role in improving power system resilience before an extreme weather-related event [5, 6]. Furthermore, some strategies, e.g. maintenance planning [7] and wide-area controls in response to communication failures [8, 9], could also be performed to enhance power system resilience before a weather-related event.

During the event, hardening, which refers to physically changing power systems, is a measure to make systems less susceptible to weather-related events. Islanding schemes are important to improve power system resilience. In [10], a unified resilience evaluation and an enhancement method, including a novel defensive islanding algorithm, are proposed. The proposed islanding scheme can mitigate potential cascading effects during weather-related events. Considering nondispatchable and dispatchable distributed generators, Wang and Wang [11] proposes a novel comprehensive operation and self-healing scheme, which sectionalizes a distribution system into several micro-grids, to improve distribution system resilience.

After an extreme weather-related event, it is necessary for system operators to implement restoration strategies [12–14] to restore loads as quickly as possible. A generic conventional power system restoration can be divided into three stages, i.e. preparation, system restoration, and load restoration [15]. In the preparation stage, the system status, i.e. black start units, non-black start units, and critical loads, should be evaluated. In the system restoration stage, the main goal is to establish a strong bulk power network by restarting appropriate black start and non-black start units associated with appropriate transmission lines and some critical loads [16]. Many approaches, e.g. expert systems [17] and heuristic approaches [18], have been proposed to deal with load restoration.

Most research studies have focused on assessment/strategies prior to a weather-related event and restoration strategies after a weather-related event. However, strategies during an event are still in their infancy. In this work, we focus on operational strategies during an event to enhance power system resilience against extreme weather-related events. During extreme weather events, operating strategies should be established subject to both current system/equipment statuses and potential future statuses as the weather-related events unfolding. Due to the essentially sequential characteristics during an event unfolding, the operation strategies should be a sequence of actions associated with uncertainties caused by development of the event and faults of components. Therefore, a Markovian method for sequentially proactive generation redispatch is proposed. At each decision epoch, the system topology, which may change due to the failure of some components (such as transformers or transmission lines) due to extreme event, constitutes a Markov state. Transition probabilities between different states, i.e. different topologies, are determined by component failure rates and development of the event. In each state, a recursive value function that includes a current cost and a future cost is established subject to operation constraints (such as ramping rates of generators). The optimal strategy for each state is obtained by optimizing the proposed recursive model. The recursive model is transformed into a mixed integer linear programming by using the linear scalarization method, with the probability of each state used as the weight of each objective.

6.2 System States Caused by Unfolding Disasters

The influences of extreme weather events on system states are introduced in this chapter. Several component failure rate models and system states on the trajectory of extreme weather events are presented, respectively. Transition probabilities between different system states are modeled.

6.2.1 Component Failure Rate

The intensity of the hazardous forces of some extreme weather events, e.g. typhoons, windstorms, floods, lightning storms [19], will change both temporally and geographically as the trajectories of the weather events move passing a region. The component failures depend on hazardous forces and a generic fragility curve in Figure 6.1 can be employed to represent the component failure probabilities.

6.2.2 System States on Disasters' Trajectories

Different components, e.g. lines, transformers, loads, and generators, may be in failure over sequential time intervals because of sequential extreme weather events. For example Figure 6.2 shows two components A and B on the trajectory of a typhoon. Since the failure rates depend on wind speeds, the failure rates of each component at different sequential time intervals are usually different, as shown in Figure 6.3a.

The system topology may change because of component failures caused by extreme weather events over the decision periods. The system operators need to make decisions according to the real-time system topology at each decision epoch as well as possible topology scenarios in the subsequent time intervals.

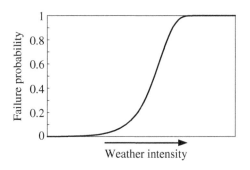

Figure 6.1 Generic fragility curve.

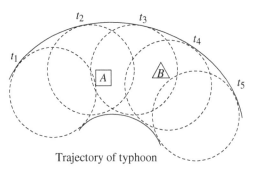

Figure 6.2 Two components on the trajectory of a typhoon.

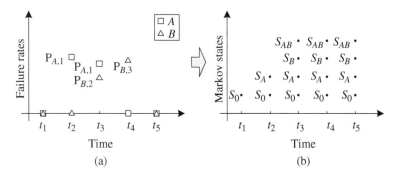

Figure 6.3 (a) Failure rates at different time intervals. (b) Markov states at different time intervals.

This chapter defines the system topology as a Markov state. Let $\Omega_{C,t}$ be the set of components that might be in failure at time t. Sets at different time intervals satisfy the following equations:

$$\Omega_{C,t} \subset \Omega_{C,t+1}, \quad t = \{1,2,\ldots\} \tag{6.1}$$

where (6.1) indicates that possible failed components at one interval should include the possibly failed components from the previous time intervals. For example the component B is impacted by the typhoon at t_3, but the component A and the component B are both included in $\Omega_{C,t}$ at t_3. The Markov states include S_0, S_A, S_B, and S_{AB}, as shown in Figure 6.3b. The subscript "0" denotes that no components are in failure.

Usually, the trajectory of a topology is uncertain, and the forecast of the trajectory cannot be entirely accurate. The uncertainty of a trajectory and their uncertain influences on component failures could be included in the component failure probabilities.

6.2.3 Transition Probabilities Between Different System States

Let $\Omega_{S,t}$ be the set of states at time t. Let $S_{i,t}$ and $S_{i',t+1}$ be states at time t and $t+1$, respectively. The transition probability from $S_{i,t}$ to $S_{i',t}$ can be expressed as follows:

$$\Pr\left(S_{i,t}, S_{i',t+1}\right) = \prod_{k \in \Omega_{C,t+1}} \Pr\left(s_{k,t}, s_{k,t+1}\right), \quad i \in \Omega_{S,t} \tag{6.2}$$

$$\Pr\left(s_{k,t}, s_{k,t+1}\right) = \begin{cases} 1, & s_{k,t} = 0, s_{k,t+1} = 0 \\ 0, & s_{k,t} = 0, s_{k,t+1} = 1 \\ 1 - \lambda_{k,t+1}, & s_{k,t} = 1, s_{k,t+1} = 1 \\ \lambda_{k,t+1}, & s_{k,t} = 1, s_{k,t+1} = 0 \end{cases} \tag{6.3}$$

where $\lambda_{k,t+1}$ is the failure probability of the component k at time $t + 1$, and Pr is the transition probability. $s_{k,t}$ is the status of the component k at time t. It can be a non-failure status or a failure status. "1" and "0" denote a normal operating status and a failure status. The uppercase S denotes a system state, including all statuses of electrical devices on the trajectory of a weather event.

6.3 Sequentially Proactive Operation Strategy

6.3.1 Sequential Decision Processes

Because it is uncertain whether a component on the trajectory of an extreme weather event is in failure or not, the best strategy is to make decisions based on real-time states of the system to optimize the optimization objective. When making state-based strategies, the system operators need to consider the current influences as well as the future influences caused by the decisions. The decision process is modeled as a Markov decision process. To simplify the model, it is assumed that decisions are made at discrete decision epochs.

In consideration of current and future influences caused by decisions, the value function for each state can be expressed as a recursive formula.

$$v_t(S_{i,t}, A_{a,t}) = R_t(S_{i,t}, A_{a,t}) + \sum_{i' \in \Omega_{i,t+1}^S} \left[\Pr(S_{i,t}, S_{i',t+1}) \cdot v_{t+1}(S_{i',t+1}, A_{a',t+1}) \right]$$

$$i \in \Omega_{S,t}, a \in \Omega_A, a' \in \Omega_A, t \in \Omega_T, t+1 \in \Omega_T$$

$$(6.4)$$

where $A_{a,t}$ denotes actions, i.e. generation redispatch in this chapter. $v_t(S_{i,t}, A_{a,t})$ is the expected cost of the state $S_{i,t}$ with the action $A_{a,t}$ from time t to the terminal time. $R(S_{i,t}, A_{a,t})$ is the immediate cost of the state $S_{i,t}$ with action $A_{a,t}$ at time t. The immediate cost is evaluated by the LoL. Based on (6.4), the optimal strategy with state $S_{i,t}$ at t can be obtained by using the following formula:

$$v_t^*(S_{i,t}) = \min \left\{ v_t(S_{i,t}, A_{a,t}), a \in \Omega_A \right\}, \quad i \in \Omega_{S,t}, \ t \in \Omega_T \tag{6.5}$$

where $v_t^*(S_{i,t})$ is the minimal expected cost of state $S_{i,t}$ at time t. The immediate cost $R_t(S_{i,t}, A_{a,t})$ is the cost of LoL, which is expressed as follows:

$$R_t(S_{i,t}, A_{a,t}) = \gamma_{L,t} \cdot \sum_{n \in \Omega_N} \left(\Delta L_{n,t,i} \cdot \Delta T \right) \tag{6.6}$$

6.3.2 Sequentially Proactive Operation Strategy Constraints

When dispatching the system under the state $S_{i,t}$ at time t, the operational constraints, e.g. power balance, voltage limits, generation limits, power flow limits, should be satisfied. Typically, the ramping rates between possible states should be satisfied.

1. Power balance. The power balance constraint in state $S_{i,t}$ at time t is expressed as follows:

$$\sum_{j \in \Omega_n^G} P_{j,t,i}^G - (L_{n,t,i} - \Delta L_{n,t,i}) + \sum_{n' \in \Omega_n^N} P_{n,n',t,i}^L = 0, \ \forall n \quad (6.7)$$

where (6.7) denotes power balance at each node in state $S_{i,t}$ at time t.

2. Ramping rates of generators. The ramping rates of each generator should be within the limits.

$$
\begin{aligned}
P_{j,t+1,i'}^G - P_{j,t,i}^G &\leq (2 - o_{j,t,i} - o_{j,t+1,i'}) \cdot P_j^{\min} \\
&+ (1 + o_{j,t,i} - o_{j,t+1,i'}) \cdot R_j^{UP} \quad i' \in \Omega_{i,t+1}^S, \ \forall j
\end{aligned} \quad (6.8)
$$

$$
\begin{aligned}
P_{j,t,i}^G - P_{j,t+1,i'}^G &\leq (2 - o_{j,t,i} - o_{j,t+1,i'}) \cdot P_j^{\min} \\
&+ (1 - o_{j,t,i} + o_{j,t+1,i'}) \cdot R_j^{DN} \quad i' \in \Omega_{i,t+1}^S, \ \forall j
\end{aligned} \quad (6.9)
$$

where the term $i' \in \Omega_{i,t+1}^S$ ensures that ramping rates should be satisfied between the state $S_{i,t}$ at t and its possible following states at $t + 1$. Considering potential online/offline statuses of generators, binary variables regarding generators' statuses are included in constraints (6.8) and (6.9).

3. Minimum up-time and down-time constraints of generators. Power generation of each generating unit should be within the limits.

$$-o_{j,t,i} + o_{j,t+1,i'} - o_{j,t',i''} \leq 0, \ 1 \leq t' - t \leq D_j^{ON}$$
$$i' \in \Omega_{i,t+1}^S, i'' \in \Omega_{i,t+}^S, \ \forall j \quad (6.10)$$

$$o_{j,t,i} - o_{j,t+1,i'} + o_{j,t',i''} \leq 1, \ 1 \leq t' - t \leq D_j^{OFF}$$
$$i' \in \Omega_{i,t+1}^S, i'' \in \Omega_{i,t+}^S, \ \forall j \quad (6.11)$$

where the terms $i' \in \Omega_{i,t+1}^S$ and $i'' \in \Omega_{i,t+}^S$ ensure that minimum up-time and down-time constraints should be satisfied between possible transition states.

4. Power flows of lines. The limits for power flows through online lines in state $S_{i,t}$ at time t should be satisfied.

$$B_{n,n'} \cdot (\theta_{n,t,i} - \theta_{n',t,i}) - P_{n,n',t,i}^L + (1 - u_{n,n',t,i}) \cdot N \geq 0$$
$$n, n' \in Line_l, \ \forall l \quad (6.12)$$

$$B_{n,n'} \cdot (\theta_{n,t,i} - \theta_{n',t,i}) - P_{n,n',t,i}^L - (1 - u_{n,n',t,i}) \cdot N \leq 0$$
$$n, n' \in Line_l, \ \forall l \quad (6.13)$$

$$P_{n,n'}^{L,\min} \cdot u_{n,n',t,i} \leq P_{n,n',t,i}^L \leq P_{n,n'}^{L,\max} \cdot u_{n,n',t,i}$$
$$n, n' \in Line_l, \ \forall l \quad (6.14)$$

where (6.12) and (6.13) represent the physical relations between voltage angles and power flows through transmission lines. With a sufficiently large N, (6.12) and (6.13) are redundant when lines are outages. Equation (6.14) shows the limits of transmission lines. The models of power flows through transformers are similar to (6.12)–(6.14).

5. Upper and lower limits of the outputs of generators. During the implementation of generation redispatch in state $S_{i,t}$ at time t, the upper and lower limits of generators should be satisfied by

$$P_j^{\min} \cdot o_{j,t,i} \leq P_{j,t,i}^{G} \leq P_j^{\max} \cdot o_{j,t,i}, \ \forall j \tag{6.15}$$

6. Load limits. When performing generation redispatch in state $S_{i,t}$ at time t, load shedding might be conducted to ensure power balance when considering the ramping rates of the generators, power flows through the lines, and so on. When conducting load shedding, the following constraints should be involved as follows:

$$0 \leq \Delta L_{n,t,i} \leq L_{n,t,i}, \ \forall n \tag{6.16}$$

where $L_{n,t,i}$ is the forecasted load of the bus n at t. Its value can be predicted based on existing load forecasting methods.

7. Voltage limits. The following constraint regarding voltage in state $S_{i,t}$ at time t should be satisfied:

$$\theta_n^{\min} \leq \theta_{n,t,i} \leq \theta_n^{\max}, \ \forall n \tag{6.17}$$

6.3.3 Linear Scalarization of the Model

Based on the Markov decision process, it is necessary to ensure the minimum expected cost from the current decision epoch to the terminal decision epoch. For example the optimal strategy for the state S_1 in Figure 6.4 should ensure the minimum expected cost of the current state and the future possible states, i.e. the states in the largest triangle. From the perspective of the mathematical model, it is necessary to find a solution for each state to guarantee that the multiple objectives are optimal. However, if the ramping rates are considered, it is not easy to find the optimal solutions for some scenarios to ensure optimal multiple objectives simultaneously.

The problem can be considered as an optimization with multiple objectives when considering the generators' ramping rates. This chapter uses a linear scalarization method to transform the original multiobjective optimization problem into a single-objective optimization problem. The critical point for the linear scalarization method is to determine the weight of each objective. The

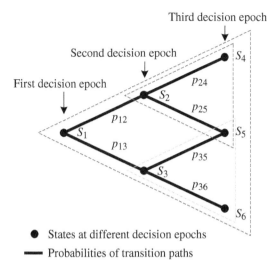

Figure 6.4 A scenario with three decision epochs.

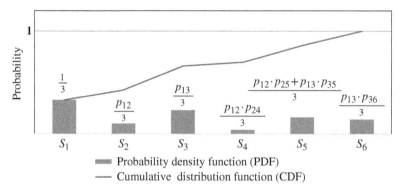

Figure 6.5 Probabilities of states.

occurrence probability of each state is considered as the weight of the objective of the corresponding state. The probability of each state can be expressed as follows:

$$P_i^{(\text{PDF})} = \frac{1}{N_T} \sum_{r \in \Omega_i^{\text{Path}}} \left(p_{r,i,t} \right) \tag{6.18}$$

Take the scenario in Figure 6.4 as an example, the probabilities of S_1, S_2, S_3, S_4, S_5, and S_6 are as shown in Figure 6.5.

With the linear scalarization method, the original problem can be expressed as a mixed integer linear programming model. The reformulated model can be expressed as follows:

$$\min \sum_{t\in\Omega_T} \sum_{i\in\Omega_{S,t}} \left\{ P_i^{(PDF)} \cdot v_t(S_{i,t}, A_{a,t}) \right\} \tag{6.19}$$

subject to (6.2)–(6.4), (6.6)–(6.13), and (6.15)–(6.18).

The CPLEX solver is used to solve the reformulated mixed integer linear programming model.

6.3.4 Case Studies

In this section, the IEEE 30-bus system and a practical power grid system are used to validate the proposed model.

6.3.4.1 IEEE 30-Bus System
Data Description Figure 6.6 shows the IEEE 30-bus system and the trajectory of the typhoon. For the sake of exposition, it is assumed that the decisions are made

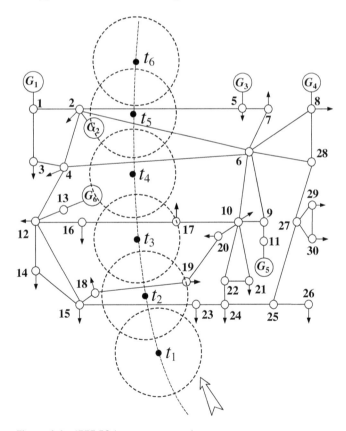

Figure 6.6 IEEE 30-bus system topology.

at discrete time t_1, t_2, t_3, t_4, and t_5. Six transmission lines, two generators, and two partial loads may be in failure on the trajectory of the typhoon. The failure probabilities of these components are listed in Table 6.1. In practice, the generic fragility curves can be used to achieve the failure probabilities. Generator parameters are listed in Table 6.2.

Mapping States to Strategies Based on the proposed model, the optimal redispatch strategy for each state is calculated in advance. The strategies at the decision epoch t_1 can be considered as preventive actions before the typhoon. During the typhoon, the system operators observe the system topology, i.e. the state, at one decision epoch, and then map the state to the optimal strategy. The same process is repeated

Table 6.1 Probabilities of component failure.

Component number	Components	Failure probabilities
C_1	Line 15-23	0.25
C_2	Line 18-19	0.22
C_3	50% load 19	0.10
C_4	Line 16-17	0.20
C_5	60% load 17	0.10
C_6	G_6	0.08
C_7	Line 4-6	0.18
C_8	G_2	0.08
C_9	Line 2-6	0.15
C_{10}	Line 2-5	0.12

Table 6.2 Parameters of generators.

Unit	Bus	P_{min}	P_{max}	Ramping rates	D_i^{ON}	D_i^{OFF}
G_1	1	30	120	120	3	3
G_2	2	35	140	120	3	3
G_3	5	10	50	72	3	3
G_4	8	5	30	60	3	3
G_5	11	10	55	72	3	3
G_6	13	15	40	60	3	3

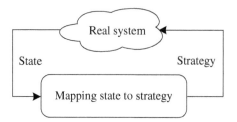

Figure 6.7 The process of mapping a state to a strategy.

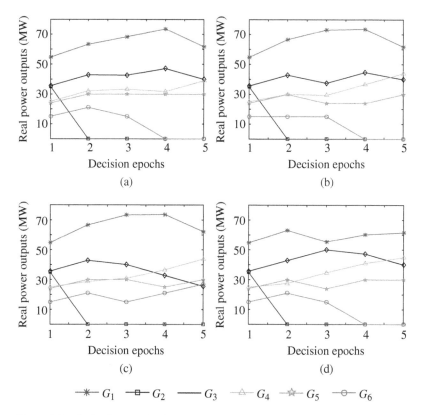

Figure 6.8 Optimal strategies for (a) scenario 1, (b) scenario 2, (c) scenario 3, and (d) scenario 4.

at the next decision epoch, as shown in Figure 6.7. Table 6.3 shows four scenarios with uncertain failure components, and Figure 6.8 shows the corresponding strategies. The optimal strategies are selected according to system states in real systems.

Table 6.3 Different failure Scenarios.

Scenario number	t_1	t_2	t_3	t_4	t_5
1	No	No	No	C_6, C_7	C_6, C_7, C_9, C_{10}
2	No	C_3	C_3, C_5	C_3, C_5, C_6, C_7	$C_3, C_5, C_6, C_7, C_9, C_{10}$
3	No	C_1	C_1, C_4	C_1, C_4, C_7	C_1, C_4, C_7, C_9
4	No	C_2, C_3	C_2, C_3, C_4, C_5	$C_2, C_3, C_4, C_5, C_6, C_7$	$C_2, C_3, C_4, C_5, C_6, C_7, C_9, C_{10}$

Because the generators G_2 and G_6 might be in failure on the typhoon's trajectory, disconnection or generation reduction are potential strategies in the process of generation dispatch. For example the generator G_6 has small outputs so that the other generators can cover its generation loss due to the typhoon, as shown in Figure 6.8a,b,d.

Potential damaged load caused by the typhoon should be also considered when making decisions. For example, the total generation at the first decision epoch is 189.2 MW. At the second decision epoch, the total generation becomes 184.45 MW if 50% load at the bus 19 is damaged due to the typhoon in Figure 6.8b, while the total generation retain 189.2 MW if the load is not damaged. Similarly, the total generation becomes 179.05 MW if 60% load at the bus 17 is damaged at the third decision epoch, while retaining 189.2 MW if not. The strategies for possible states at each decision epoch are represented in Figure 6.9. The optimal strategy for the observed state at each decision epoch can be selected based on the mapping relations calculated from the proposed model.

Effectiveness of the Proposed Method This section shows the effectiveness of the proposed method by comparing the following methods:

M1: The proposed method.
M2: Non-proactive strategies. In this case, system operators will not proactively perform generation redispatch beforehand. They only take actions after some events, i.e. line faults, to minimize LoL, in consideration of operation constraints.

The differences of LoL with M1 and M2 are presented in Figure 6.10. Most failure scenarios correspond to smaller LoL by using M1. There are a few failure scenarios with larger LoL by using M1, because M1 synthetically considers possible scenarios on the trajectory of the typhoon, and some scenarios may sacrifice LoL. From the perspective of all scenarios; however, the proposed method is more effective.

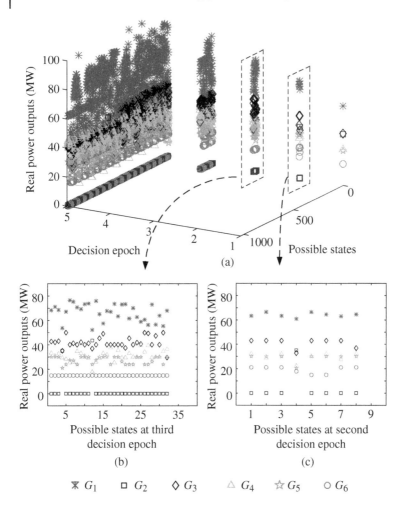

Figure 6.9 (a) Mapping states to strategies over all decision epochs. (b) Mapping states to strategies at the second decision epoch. (c) Mapping states to strategies at the third decision epoch.

Influences of Scenario Reduction The influences of state reduction are presented in this section. The results of different maximum failure component scenarios are listed in Table 6.4. It is observed that more maximum failure components have higher accuracy but result in longer CPU time.

6.3.4.2 A Practical Power Grid System

There are 116 generating units, 1117 buses, and 1771 transmission lines in the practical power grid system. The typhoon Mujigae, which struck the Guangdong

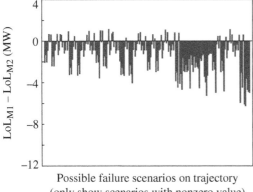

Figure 6.10 Differences of LoL with M1 and M2.

Table 6.4 Different maximum failure component scenarios.

Maximum failure components	Objective value ($)	Error (%)	Total CPU time (s)
2	322.54	47.92	29
3	476.42	23.07	72
4	569.55	8.03	185
5	603.90	2.49	542
6	616.06	0.52	1035
7	618.94	0.06	1935
8	619.27	0.0048	3021
9	619.28	0.0032	3935
10	619.30	0	4235

province of China in 2015, is used as an example. According to the trajectory, seven critical transmission lines and one critical generator undergo the typhoon sequentially. The duration between two adjacent decision epochs is about one hour. The lower and upper limits of the critical generator are 1059 and 110 MW, and the corresponding ramping rate is 100 MW/h. The failure probabilities of the seven transmission lines and the generator on the trajectory are 0.2, 0.18, 0.12, 0.1, 0.08, 0.07, and 0.1, respectively.

The benchmark strategy is based on the M2 method. The differences of LoL with M1 and M2 are presented in Figure 6.11. It is observed that the proposed

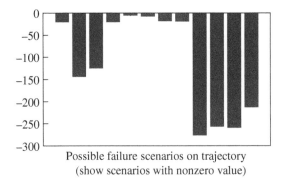

Possible failure scenarios on trajectory
(show scenarios with nonzero value)

Figure 6.11 Differences of LoL with M1 and M2.

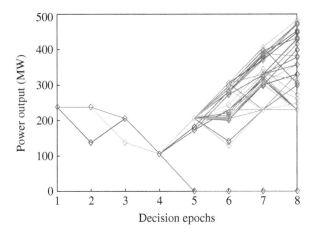

Figure 6.12 Outputs of the critical generator on the trajectory.

strategy (M1) can ensure a better performance compared with non-proactive strategies (M2).

The outputs of the critical generator on the trajectory are listed in Figure 6.12. Because the critical generator may be in failure at the fifth decision epoch, its outputs remain close to its lower limit from the first to fourth decision epochs to avoid a large LoL due to power unbalance considering operation constraints. The costs of expected LoL with different numbers of available generators are listed in Figure 6.13. Each value is the average value based on 50 simulations, which

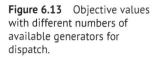

Figure 6.13 Objective values with different numbers of available generators for dispatch.

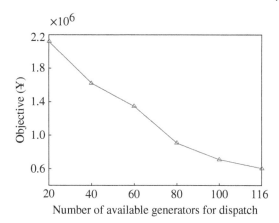

randomly select generators for dispatch. Results show that more generators for dispatch tend to be less costs of expected LoL.

6.4 Summary and Conclusions

A Markov model was proposed to construct sequentially proactive generation redispatch strategies based on system states, in consideration of operating constraints. Since trajectories and the failures of components on the trajectory due to extreme weather events are with uncertainties, a recursive value function that included the current cost and the future cost for each state was established. To address the computational difficulties caused by the constraints of ramping rates of generators, the recursive value function was transformed into a mixed integer linear programming model, with the probability of each state used as the weight of each objective. Simulations were carried out on the IEEE 30-bus system, the IEEE 118-bus system, and a simplified Guangdong power grid system to validate the proposed model. The major findings are listed as follows: (i) The Markov model can adequately describe sequential decision processes with consideration for the uncertainties during an event unfolding. (ii) The proposed strategies tend to redispatch the power flows of lines on the trajectory of the extreme weather event to reduce potential power loss. (iii) The proposed model provides insight into proactive operation strategies under extreme weather events.

Nomenclature

Indices and Sets

i, i', i''	index of Markov states
t, t'	index of time periods
a, a'	index of actions
l	index of lines
k	index of electrical devices
j	index of generators
n, n'	index of nodes
r	index of paths between states
$\Omega_{C,t}$	set of components that might be in failure at t
$\Omega_{S,t}$	set of Markov states at t
Ω_A	set of actions
Ω_T	set of time periods
Ω_N	set of nodes
Ω_G	set of generators
$\Omega_{i,t+1}^S$	set of states at $t+1$ following the state $S_{i,t}$ at t
$\Omega_{i,t+}^S$	set of states
Ω_n^G	set of generators connected with node n
Ω_n^D	set of loads connected with node n
Ω_n^N	set of nodes connected with node n
Ω_i^{Path}	set of paths from initial state to state $S_{i,t}$

Notation for Failure Rate

λ_t^w	failure rate due to typhoon, windstorms, etc.
w	a given wind speed
w_t	wind speed at t
α	a given parameter
λ	failure rate under normal weather conditions
λ_t^I	failure rate due to ice storm
$M_{t,L}$	total load in kN on line
$\overline{M}_L$	maximum load in kN on line
m_{Li}	ice load on lines
m_{Lw}	wind load on lines
$f_L(\cdot)$	joint probability density function of ice load and wind load on lines

Notation for Optimization Model

$S_{i,t}$	Markov state at t
$S_{i',t+1}$	Markov state at $t+1$
$S_{i'',t'}$	Markov state at t'
$s_{k,t}$	status of component k at t and $t+1$
$\lambda_{k,t+1}$	failure rate of component k at $t+1$
$R_t(\cdot)$	immediate cost at t
$\Pr(\cdot)$	transition probability
$\gamma_{L,t}$	penalty due to loss of load
$v_t(\cdot)$	value function at t
$v_{t+1}(\cdot)$	value function at $t+1$
$v_t^*(\cdot)$	optimal value function
$L_{n,t,i}$	load demand of node n in state $S_{i,t}$ at t
$P_j^{\min}$	lower generation of generator j
$P_j^{\max}$	upper generation of generator j
R_j^{UP}	ramp-up rate limit of generator j
R_j^{DN}	ramp-down rate limit of generator j
$P_{n,n'}^{L,\max}$	max capacity of line $n - n'$
$P_{n,n'}^{L,\min}$	min capacity of line $n - n'$
$\theta_n^{\max}$	max limit of phase angle of node n
$\theta_n^{\min}$	min limit of phase angle of node n
$A_{a,t}$	action a at t
$B_{n,n'}$	electrical susceptance of line $n - n'$
D_j^{ON}	min on time of generator j
D_j^{OFF}	min off time of generator j
ΔT	duration of each period
$P_{j,t,i}^G$	generation of generator j in state $S_{i,t}$
$P_{j,t+1,i'}^G$	generation of generator j in state $S_{i',t+1}$
$P_{n,n',t,i}^L$	power from node n to n' in state $S_{i,t}$
$\Delta L_{n,t,i}$	load shedding of node n in state $S_{i,t}$
$\theta_{n,t,i}$	phase angle of node n in state $S_{i,t}$
$o_{j,t,i}$	binary variable to indicate status of generator j in state $S_{i,t}$
$o_{j,t+1,i'}$	binary variable to indicate status of generator j in state $S_{i',t+1}$
$o_{j,t',i''}$	binary variable to indicate status of generator j in state $S_{i'',t'}$

$u_{n,n',t,i}$ binary outage indicator to indicate status of line $n - n'$ in state $S_{i'',t'}$

N_T number of time periods

N a large number

$P_i^{(PDF)}$ probability of state $S_{i,t}$

$p_{r,i,t}$ probability from initial state to state $S_{i,t}$ via path r

References

1 H. Liu, R. A. Davidson, D. V. Rosowsky, and J. R. Stedinger, "Negative binomial regression of electric power outages in hurricanes," *Journal of Infrastructure Systems*, vol. 11, no. 4, pp. 258–267, 2005.

2 S. D. Guikema, R. Nateghi, S. M. Quiring, A. Staid, A. C. Reilly, and M. Gao, "Predicting hurricane power outages to support storm response planning," *IEEE Access*, vol. 2, pp. 1364–1373, 2014.

3 R. Nateghi, S. D. Guikema, and S. M. Quiring, "Forecasting hurricane-induced power outage durations," *Natural Hazards*, vol. 74, p. 1795–1811, 2014.

4 A. Arab, A. Khodaei, S. K. Khator, K. Ding, V. A. Emesih, and Z. Han, "Stochastic pre-hurricane restoration planning for electric power systems infrastructure," *IEEE Transactions on Smart Grid*, vol. 6, no. 2, pp. 1046–1054, 2015.

5 W. Sun, C. Liu, and S. Liu, "Black start capability assessment in power system restoration," in *2011 IEEE Power and Energy Society General Meeting*, pp. 1–7, 2011.

6 F. Qiu, J. Wang, C. Chen, and J. Tong, "Optimal black start resource allocation," *IEEE Transactions on Power Systems*, vol. 31, no. 3, pp. 2493–2494, 2016.

7 C. Wang, Y. Hou, Z. Qin, C. Peng, and H. Zhou, "Dynamic coordinated condition-based maintenance for multiple components with external conditions," *IEEE Transactions on Power Delivery*, vol. 30, no. 5, pp. 2362–2370, 2015.

8 S. Zhang and V. Vittal, "Wide-area control resiliency using redundant communication paths," *IEEE Transactions on Power Systems*, vol. 29, no. 5, pp. 2189–2199, 2014.

9 S. Zhang and V. Vittal, "Design of wide-area power system damping controllers resilient to communication failures," *IEEE Transactions on Power Systems*, vol. 28, no. 4, pp. 4292–4300, 2013.

10 M. Panteli, D. N. Trakas, P. Mancarella, and N. D. Hatziargyriou, "Boosting the power grid resilience to extreme weather events using defensive islanding," *IEEE Transactions on Smart Grid*, vol. 7, no. 6, pp. 2913–2922, 2016.

11 Z. Wang and J. Wang, "Self-healing resilient distribution systems based on sectionalization into microgrids," *IEEE Transactions on Power Systems*, vol. 30, no. 6, pp. 3139–3149, 2015.

12 M. M. Adibi and L. H. Fink, "Overcoming restoration challenges associated with major power system disturbances - restoration from cascading failures," *IEEE Power and Energy Magazine*, vol. 4, no. 5, pp. 68–77, 2006.

13 Y. Hou, C. Liu, K. Sun, P. Zhang, S. Liu, and D. Mizumura, "Computation of milestones for decision support during system restoration," *IEEE Transactions on Power Systems*, vol. 26, no. 3, pp. 1399–1409, 2011.

14 C. Chen, J. Wang, F. Qiu, and D. Zhao, "Resilient distribution system by microgrids formation after natural disasters," *IEEE Transactions on Smart Grid*, vol. 7, no. 2, pp. 958–966, 2016.

15 L. H. Fink, K.-L. Liou, and C.-C. Liu, "From generic restoration actions to specific restoration strategies," *IEEE Transactions on Power Systems*, vol. 10, no. 2, pp. 745–752, 1995.

16 W. Sun, C. Liu, and L. Zhang, "Optimal generator start-up strategy for bulk power system restoration," *IEEE Transactions on Power Systems*, vol. 26, no. 3, pp. 1357–1366, 2011.

17 C. Liu, S. J. Lee, and S. S. Venkata, "An expert system operational aid for restoration and loss reduction of distribution systems," *IEEE Transactions on Power Systems*, vol. 3, no. 2, pp. 619–626, 1988.

18 S. Toune, H. Fudo, T. Genji, Y. Fukuyama, and Y. Nakanishi, "Comparative study of modern heuristic algorithms to service restoration in distribution systems," *IEEE Transactions on Power Delivery*, vol. 17, no. 1, pp. 173–181, 2002.

19 Y. Wang, C. Chen, J. Wang, and R. Baldick, "Research on resilience of power systems under natural disasters–a review," *IEEE Transactions on Power Systems*, vol. 31, no. 2, pp. 1604–1613, 2016.

7

Markov Decision Process-Based Resilience Enhancement for Distribution Systems

Because consumers' outages in distribution systems can be directly caused by failures due to extreme weather events, a state-based, decision-making model is proposed in this chapter to enhance the distribution system resilience throughout the unfolding events. The varying system topologies are modeled as Markov states, and the probabilities between different Markov states depend on the component failure caused by the unfolding events. A Markov decision process (MDP)-based model is proposed to make state-based actions, i.e. system reconfiguration, at each decision time. To overcome the curse of dimensionality, an approximate dynamic programming (ADP) approach with post-decision states is employed to optimize the proposed model, and the proposed model is validated by two test systems.

7.1 Real-Time Response Against Unfolding Disasters

Commercial and residential customers are connected to distribution systems with radial topologies, and any failures in distribution systems will result in power outages. Severe weather is one of major causes of distribution system failures. For example, weather events caused roughly 679 power outages, each of which affected at least 50 000 customers, between 2003 and 2012. Although transmission system outages did occur, a major portion of outages occurred along distribution systems [1]. The severe consequences have required distribution systems to have resilience against these extreme weather events, and this has been identified by the US Electric Power Research Institute (EPRI) [2] and the North American Electric Reliability Corporation (NERC) [3].

A power system may reside in different stages when it is imposed to natural disasters or extreme weather events. It is necessary to define these stages to enable

Power Grid Resilience against Natural Disasters: Preparedness, Response, and Recovery, First Edition.
Shunbo Lei, Chong Wang, and Yunhe Hou.
© 2023 John Wiley & Sons Ltd. Published 2023 by John Wiley & Sons Ltd.

systematic enhancement power systems against to these events. Gholami et al. [4] analyzes the notion of resilience in power systems from a fundamental viewpoint and thoroughly examines its practical implications. Usually, three critical stages, i.e. prior to events, during events, and after events, need to be included [5, 6]. Different stages have different requirements.

Prior to weather events, system hardening makes physical infrastructural changes to systems so that they are less susceptible to extreme events. For example, a coordinated hardening and distributed generator (DG) allocation strategy has been developed in [7]. An analytical method is proposed in [8] to offer a quick way for getting knowledge about adverse impacts of an approaching windstorm and taking preventive measures accordingly. Amirioun et al. [9] and [10] propose proactive scheduling for resilience enhancement of microgrids (MGs) [11–14] ahead of extreme windstorms and floods, respectively. Even though many preventive actions are performed prior to events, it is not possible to avoid outages completely. When outages occur after events, it is necessary to recover outages as quickly as possible to improve the system resilience. A conventional power system restoration includes three stages, i.e. preparation, system restoration, and load restoration [15–17]. Some algorithms such as expert systems [18] and heuristic approaches [19] are proposed to accelerate load recovery. However, there are unique characteristics associated with outages caused by weather-related events, leading to different restoration strategies such as MG-based restoration strategies [20] and decentralized restoration schemes [21].

The above studies mainly focus on the strategies prior to events and after events. There are also some studies on operational strategies during events. A concept of operational resilience is proposed in [22], but the detailed models are not established. In [23], generator re-dispatch, topology switching and load shedding are considered as emergency responses, which are modeled as a two-stage robust mixed-integer optimization model. In consideration of high controllability of MGs, Gholami et al. [24] presents a two-stage stochastic programming approach to obtain the optimal scheduling of a resilient MG. With the development of integrated natural gas and power grids, Yan et al. [25] establishes a two-stage robust model to accommodate random outages caused by natural disasters in both natural gas and power systems. A stochastic programming approach for increasing resilience of a distribution system exposed to wildfires is investigated in [26]. The uncertainties associated with wind speeds and wind directions that affect the progression of the wildfire are represented by simulated scenarios. The commonly used scenario-based stochastic programming [24, 26] and the robust stochastic programming [25] are not suitable for mapping sequentially real-time varying states to optimal strategies. To address this difficulty, MDPs can be employed to make state-based decisions on a stochastic environment caused by weather events. Some applications of MDP in power systems have been investigated [27, 28]. For the resilience enhancement, Wang et al. [29] proposes sequentially proactive MDP-based strategies to improve the transmission system resilience and a linear scalarization method based on the state tree to solve the proposed model. However, distribution systems and transmission systems differ in topologies and allowable actions, and the developed model and the solution in [29] cannot be applied to distribution systems directly. To fill the gap, it is necessary to develop state-based decision-making models for distribution systems considering their own characteristics.

7.2 Disasters' Influences on Distribution Systems

The concept of MDP is first presented in this section, and then Markov states on the trajectories of extreme events and transition probabilities between different Markov states are presented.

7.2.1 Markov States on Disasters' Trajectories

MDP provides a mathematical framework for modeling real-time, state-based decision-making in situations where there are sequential uncertainties. The processes of MDP are illustrated in Figure 7.1. At t_1, the state S_1 is observed by the decision-maker, and the action A_1 is implemented, resulting in a reward R_1. With the uncertainties caused by environments, the state S_1 reaches the state S_2 at t_2, and the decision process is repeated. During this process, the decision-maker expects the optimal action for each state considering the current cost and the future cost. The goal is to minimize the expected loss of load by means of system

Figure 7.1 Markov decision processes.

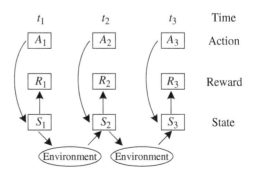

reconfiguration based on the observed states, considering the operational costs and the uncertainties due to extreme weather events. The Markov state is the system topology. The environment denotes the uncertain impacts of extreme weather events on power systems, and the uncertain impacts are represented by transition probabilities.

Usually, the sequential trajectory of a weather-related event has a great impact on a distribution system, and in consequence, the components in the different locations of the system may be in failure in different time periods. This could result in sequential changes of line states. The system topologies are modeled as Markov states. Define $\mathcal{H}_t$ as the set of disconnected lines due to system configuration but not in failure caused by the unfolding event at t. Define F_t as the set of all possible failure components due to the unfolding event at t, and $\tilde{F}_t$ as the set of the actual failure components at t, and we have $\tilde{F}_t \subseteq F_t$. Take the scenario in Figure 7.2 as an example: $F_{t_2} = \{b_{1-2}, b_{1-4}\}$ and $\tilde{F}_{t_2} = \{b_{1-4}\}$. The failure scenarios of the lines b_{1-2} and b_{1-4} are uncertain before the time period t_2, and the actual state can only be observed at t_2 and the actual failure on b_{1-4} occurs. The set of the disconnected lines is used to describe the different Markov states.

$$S_{i,t} = \mathcal{H}_t \bigcup_{\tau=1}^{t} \left(\tilde{F}_\tau - \tilde{R}_\tau \right) \tag{7.1}$$

where (7.1) shows that the Markov state $S_{i,t}$ at t. $\tilde{R}_\tau$ is the set of components repaired at time τ.

Power outages will occur with the action of proactive load shedding, so proactive load shedding will not be implemented during system reconfiguration. It means that the allowable actions include system reconfiguration. For the load level, it is based on the predicted load curve, which is assumed to be known in this chapter.

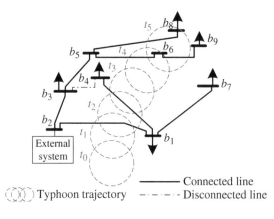

Figure 7.2 An example of a distribution system under an unfolding event.

7.2.2 Transition Probability Between Markov States

The current Markov state at t could reach any Markov state at $t+1$ on the trajectory of the event, and the probability from one Markov state to another Markov state is defined as transition probabilities. The transition probabilities depend on component failure rates caused by weather-related events.

$$\Pr(S_{j,t+1}|S_{i,t}, a_t, \xi_t) = \prod_{c \in F_{t+1}} \Pr(s_{c,t+1}|s_{c,t}, a_t, \xi_t) \tag{7.2}$$

where $\Pr(S_{j,t+1}|S_{i,t}, a_t, \xi_t)$ means the probability from the state $S_{i,t}$ to the state $S_{j,t+1}$ under the action a_t with the uncertainty ξ_t, and $\Pr(s_{c,t+1}|s_{c,t}, a_t, \xi_t)$ represents the probability from the cth component's on-off state $s_{c,t}$ to the on-off state $s_{c,t+1}$ under the action a_t with the uncertainty ξ_t.

7.3 Markov Decision Processes-Based Optimization Model

A recursive model, mapping Markov state to its optimal strategy, with the operational constraints is established in this section.

7.3.1 Markov Decision Processes-based Recursive Model

As different operational costs will be caused by different states, not only current states but also future states impacted by uncertainty should be considered. For example, because the line b_{1-2} will be impacted by the typhoon at t_1 in Figure 7.2, the line b_{1-2} might be in failure at t_1. If the line b_{1-2} is disconnected and the line b_{3-4} is connected before the line b_{1-2} is impacted by the typhoon, power outages could be avoided. For the above case, only one time period ahead and only one line at t_1 are considered. If the sequential time periods and numerous components impacted in each time period are considered, the scenarios become complicated, and a model for making decisions to ensure the minimum operational cost should be developed. A recursive model with the current cost and the expected future cost is established.

$$v_t(S_{i,t}) = \min_{a_t \in A} \left(C_t(S_{i,t}, a_t) + \sum_{S_{j,t+1} \in S} \Pr(S_{j,t+1}|S_{i,t}, a_t, \xi_t) \cdot v_{t+1}(S_{j,t+1}) \right) \tag{7.3}$$

where $v_t(S_{i,t})$ is the value function of the state $S_{i,t}$ at t. The second term on the right side of (7.3) shows the expected future cost. $C_t(S_{i,t}, a_t)$ is the current cost caused by the action a_t under the state $S_{i,t}$ at t, and this cost in this study is defined as the sum of the cost of loss of load and the operational cost of controllable lines. It is expressed as follows:

$$C_t(S_{i,t}, a_t) = \sum_{k \in B} \left(\eta_t \cdot \Delta L^p_{b,i,t} \cdot \Delta T \right) + \sum_{l \in \mathcal{L}^d_{i,t}} (\beta_{l,i,t} \cdot C_l) \tag{7.4}$$

where the first term on the right side of (7.4) denotes the cost of loss of load, and the second term is the operational cost of controllable lines.

7.3.2 Operational Constraints

The operational constraints, e.g. power balance, power flow, radial topologies, and line capacity, need to be satisfied when dispatching the distribution systems:

7.3.2.1 Radiality Constraint

The spanning tree constraint is employed to guarantee the network radiality of the distribution system [30].

$$o_{kk',i,t} + o_{k'k,i,t} = \beta_{l,i,t}, \quad l \in \mathcal{L}_{i,t}, \ t \in \mathcal{T} \tag{7.5a}$$

$$\sum_{k' \in \mathcal{N}_{k,i,t}} o_{kk',i,t} = 1, \quad k \in \overline{B}_{i,t}, \ t \in \mathcal{T} \tag{7.5b}$$

$$o_{kk',i,t} = 0, \quad k \in \tilde{B}, \ k' \in \mathcal{N}_{k,i,t}, \ t \in \mathcal{T} \tag{7.5c}$$

where (7.5a) and (7.5b) constrain that the two terminals of a connected line only have one parent bus. In practice, islanded buses, to which power cannot be supplied by the grid, maybe exist due to component failures caused by extreme events, and these islanded buses are not included in the spanning tree constraint. Equation (7.5c) indicates that the substation bus (i.e. the bus connected to the external system) has no parent buses.

In practice, it is possible that only some lines can be remotely dispatched, and a constraint with regard to non-dispatched lines can express this.

$$\beta_{l,i,t} = 1, \quad l \in \mathcal{L}^{nd}_{i,t}, \ t \in \mathcal{T} \tag{7.6}$$

7.3.2.2 Repair Constraint

Repair activities need to consider equipment/crew constraints [31, 32]. Because this chapter mainly focuses on dispatch strategies during the extreme weather event and repair scheduling belongs to a task after the extreme weather event, the repair is simply constrained by the repair duration as follows:

$$\beta_{c,i,t} = 0, \quad c \in C^f_{i,t}, \ T^f_c \le t \le T^f_c + \Delta T^r_c \tag{7.7}$$

where ΔT^r_c represents the repair duration for the component c. Equation (7.7) means that the state of the failure component c is set to 0 during the repair periods, and it is a generic constraint.

7.3.2.3 Power Flow Constraint
Power flow constraints should be satisfied during the system reconfiguration.

$$U_{k,i,t} - U_{k',i,t} \leq (1 - \beta_{l,i,t}) \cdot M + 2(r_{kk'} \cdot F^p_{kk',i,t} + x_{kk'} \cdot F^q_{kk',i,t}), \quad l \in \mathcal{L}_{i,t}, \ t \in \mathcal{T}$$
(7.8a)

$$U_{k,i,t} - U_{k',i,t} \geq (\beta_{l,i,t} - 1) \cdot M + 2(r_{kk'} \cdot F^p_{kk',i,t} + x_{kk'} \cdot F^q_{kk',i,t}), \quad l \in \mathcal{L}_{i,t}, \ t \in \mathcal{T}$$
(7.8b)

where the DistFlow model (7.8a) and (7.8b) is used to express power flow in the distribution systems. The quadratic terms in the accurate power flow model are ignored [33]. The big M is a disjunctive parameter. By means of a sufficiently large M, (7.8a) and (7.8b) are redundant when distribution lines are disconnected or outages occur. Non-islanded buses are included in these constraints.

7.3.2.4 Power Balance Constraint
When reaching the state $S_{i,t}$ at t, power balance at each bus should be satisfied. The constraint can be expressed as follows:

$$L^p_{k,t} + \sum_{k' \in \mathcal{N}_{k,i,t}} F^p_{kk',i,t} = 0, \quad k \in \overline{B}_{i,t}, \ t \in \mathcal{T}$$
(7.9a)

$$L^q_{k,t} + \sum_{k' \in \mathcal{N}_{k,i,t}} F^q_{kk',i,t} = 0, \quad k \in \overline{B}_{i,t}, \ t \in \mathcal{T}$$
(7.9b)

where (7.9a) and (7.9b) are real power balance and reactive power balance, respectively. Only non-islanded buses are included in the constraint. The load connected to the islanded buses in the system is directly considered as loss of load in (7.4).

7.3.2.5 Line Capacity Constraint
The power through each line should be within the limit for the state $S_{i,t}$ with the action a_t. The constraint can be expressed as follows:

$$(F^p_{kk',i,t})^2 + (F^q_{kk',i,t})^2 \leq \beta_{l,i,t} \cdot (F^s_{kk'})^2, \quad l \in \mathcal{L}_{i,t}, \ t \in \mathcal{T}$$
(7.10)

where (7.10) is a non-linear constraint, resulting in computational intractability. To facilitate the model solution, the constraint (7.10) is relaxed to a group of linear constraints [34], and is re-written as follows:

$$-\beta_{l,i,t} \cdot F^s_{kk'} \leq F^p_{kk',i,t} \leq \beta_{l,i,t} \cdot F^s_{kk'}, \quad l \in \mathcal{L}_{i,t}, \ t \in \mathcal{T}$$
(7.11a)

$$-\beta_{l,i,t} \cdot F^s_{kk'} \leq F^q_{kk',i,t} \leq \beta_{l,i,t} \cdot F^s_{kk'}, \quad l \in \mathcal{L}_{i,t}, \ t \in \mathcal{T}$$
(7.11b)

$$-\sqrt{2}\beta_{l,i,t} \cdot F^s_{kk'} \leq F^p_{kk',i,t} + F^q_{kk',i,t} \leq \sqrt{2}\beta_{l,i,t} \cdot F^s_{kk'}, \quad l \in \mathcal{L}_{i,t}, \ t \in \mathcal{T}$$
(7.11c)

$$-\sqrt{2}\beta_{l,i,t} \cdot F^s_{kk'} \leq F^p_{kk',i,t} + F^q_{kk',i,t} \leq \sqrt{2}\beta_{l,i,t} \cdot F^s_{kk'}, \quad l \in \mathcal{L}_{i,t}, \ t \in \mathcal{T}$$
(7.11d)

7.3.2.6 Voltage Constraint

The voltage limits under the state $S_{i,t}$ with the action a_t should be satisfied.

$$\underline{V}_k^2 \leq U_{k,i,t} \leq \overline{V}_k^2, \quad k \in \overline{B}_{i,t}, \ t \in \mathcal{T} \tag{7.12}$$

7.4 Solution Algorithms – Approximate Dynamic Programming

The challenge of solving the proposed model is first introduced, and the ADP to solve the proposed model is then presented.

7.4.1 Solution Challenges

One common approach used to address the conventional stochastic programming is to produce some discrete scenarios, representing the uncertainties, and then a model with an expected objective based on these generated scenarios is established. Figure 7.3a–c illustrates the decision-making processes at t_1, t_2, and t_3, respectively. At t_1, the state transition tree under the impacts of actions is constructed, as shown in Figure 7.3a, and the optimal action can be obtained by optimizing the recursive model (7.3). After performing the optimal action, the state reaches a new state at t_2 during uncertainty, as shown in Figure 7.3b. The state transition tree for this new state will be updated accordingly, and the recursive model (7.3) is used again to obtain the optimal action. Based on the decision processes, establishing the state transition tree with uncertainty is one critical point to optimize the MDP-based model. However, it is not easy to

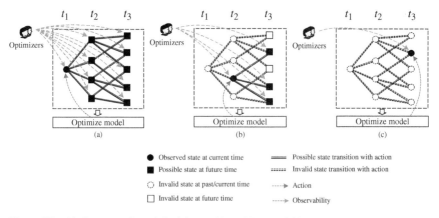

Figure 7.3 Markov state-based decision at (a) t_1, (b) t_2, and (c) t_3.

construct the state transition tree in consideration of state transitions and various actions, which could result in "curse of dimensionality." There are three curses of dimensionality: (i) the decision space $\mathcal{A}$ is too large to obtain the optimal action for each state; (ii) the state space S may be too huge to calculate the value function $v_t(S_{i,t})$ for each state within acceptable time; (iii) the outcome space may be too large to calculate the expectation of future cost.

ADP is a modeling framework that can deal with the curses of dimensionality in large, stochastic, and multiperiod models. Two important techniques in ADP, the post-decision states and the forward dynamic algorithm, are used to address the curses of dimensionality.

- Post-decision states. A post-decision state is defined as a state immediately after the action but before the arrival of a new state at the next decision time due to uncertainties. The information embedded in the post-decision states can be used to estimate the downstream future cost. For example, S_1', S_2', and S_3' in Figure 7.4 are the postdecision states with regard to S_1, S_2, and S_3 after the actions A_1, A_2, and A_3, respectively. If one estimated future cost can be assigned to the post-decision states, the stochastic and multi-period MDP-based model becomes a one-period deterministic model for each state in each decision period.
- Forward dynamic algorithm. In each decision period, the estimated value of each post-decision state needs to be known to solve the one-period deterministic model for each state. A sample path-based forward algorithm is employed to solve the recursive MDP-based model with initial values, and iteratively repeat this procedure to achieve the estimated value of each post-decision state.

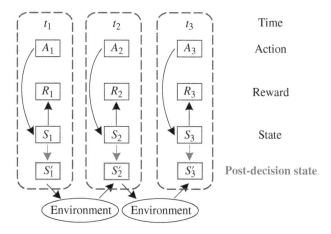

Figure 7.4 Decision processes with post-decision states.

7.4.2 Post-decision States

The post-decision state, defined as $S_{i,t}^{a_t}$, is a state immediately after the action a_t but before the arrival of a new state in consideration of uncertainties. A generic form of the proposed model in (7.3) is an expectation form listed as follows:

$$v_t(S_{i,t}) = \min_{a_t \in \mathcal{A}} \left(C_t(S_{i,t}, a_t) + \mathbb{E}\{v_{t+1}(S_{j,t+1}|S_{i,t}, a_t, \xi_t)\} \right) \tag{7.13}$$

where (7.13) can be rewritten as (7.14) with the post-decision state $S_{i,t}^{a_t}$.

$$v_t(S_{i,t}) = \min_{a_t \in \mathcal{A}} \left(C_t(S_{i,t}, a_t) + \mathbb{E}\{v_{t+1}(S_{j,t+1}|S_{i,t}^{a_t}, \xi_t)\} \right) \tag{7.14}$$

Define $\mathbb{E}\{v_{t+1}(S_{j,t+1}|S_{i,t}^{a_t}, \xi_t)\}$ by $v_t^{a_t}(S_{i,t}^{a_t})$, we have the following optimality equations.

$$v_t(S_{i,t}) = \min_{a_t \in \mathcal{A}} \left(C_t(S_{i,t}, a_t) + v_t^{a_t}(S_{i,t}^{a_t}) \right) \tag{7.15a}$$

$$v_{t-1}^{a_{t-1}}(S_{j',t-1}^{a_{t-1}}) = \mathbb{E}\left\{ v_t \left(S_{i,t}|S_{j',t-1}^{a_{t-1}}, \xi_{t-1} \right) \right\} \tag{7.15b}$$

The optimality equations of the post-decision states can be obtained by substituting (7.15a) into (7.15b).

$$v_{t-1}^{a_{t-1}} \left(S_{j',t-1}^{a_{t-1}} \right) = \mathbb{E}\left\{ \min_{a_t \in \mathcal{A}} \left(C_t(S_{i,t}, a_t) + v_t^{a} \left(S_{i,t}^{a}|S_{j',t-1}^{a_{t-1}}, \xi_{t-1} \right) \right) \right\} \tag{7.16}$$

where (7.16) can be rewritten as the form at t as follows:

$$v_t^{a_t}(S_{i,t}^{a_t}) = \mathbb{E}\left\{ \min_{a_{t+1} \in \mathcal{A}} \left(C_{t+1}(S_{j,t+1}, a_{t+1}) + v_{t+1}^{a_{t+1}} \left(S_{j,t+1}^{a_{t+1}}|S_{i,t}^{a_t}, \xi_t \right) \right) \right\} \tag{7.17}$$

7.4.3 Forward Dynamic Algorithm

If we know the value of $v_t^{a_t}(S_{i,t}^{a_t})$ in (7.15a), the optimization model (7.15a) could be solved easily. So the forward dynamic algorithm uses a deterministic optimization model (7.15a) with an initial estimation of $\tilde{v}_t^{a_t}(S_{i,t}^{a_t})$ of $v_t^{a_t}(S_{i,t}^{a_t})$ to optimize the model, and then employs the resulting observations to update an estimation $\tilde{v}_t^{a_t}(S_{i,t}^{a_t})$ thereby approximating the expected value in (7.17). To represent iterations, the superscripts n and $n-1$ are added to the model, and (7.15a) can be expressed as follows:

$$v_t^n(S_{i,t}) = \min_{a_t \in \mathcal{A}} \left(C_t(S_{i,t}, a_t) + \tilde{v}_t^{a_t,n-1}(S_{i,t}^{a_t}) \right) \tag{7.18}$$

where the decision that minimizes (7.18) at nth iteration is shown as follows:

$$a_t = \arg\min_{a_t \in \mathcal{A}} \left(C_t(S_{i,t}, a_t) + \tilde{v}_t^{a_t,n-1}(S_{i,t}^{a_t}) \right) \tag{7.19}$$

The estimated values of the post-decision states $\tilde{v}_t^{a_t,n}(S_{i,t}^{a_t})$ in the nth iteration are updated by

$$\tilde{v}_t^{a_t,n}(S_{i,t}^{a_t}) = (1 - \epsilon) \cdot \tilde{v}_t^{a_t,n-1}(S_{i,t}^{a_t}) + \epsilon \cdot v_{t+1}^n(S_{j,t+1}) \tag{7.20}$$

where the first term on the right side of (7.20) represents the estimate of the post-decision state $S_{i,t}^{a_t}$ at the $(n-1)$th iteration, and the second term represents the value of the resulting observations from the post-decision state $S_{i,t}^{a_t}$ at the nth iteration.

Based on the estimated values of the post-decision states at the $(n-1)$th iteration, the estimated value of the states $S_{i,t}$ can be optimized by means of (7.18) with the operational constraints. This is a one-period deterministic optimization model. The estimated values of the post-decision states $S_{i,t}^{a_t}$ at the nth iteration can be achieved based on (7.20). When having enough iterations, the converged estimated values of the post-decision states can be obtained. This process is offline. With the converged estimated values of the post-decision states, the online one-period deterministic optimization can be implemented to obtain the optimal strategies.

7.4.4 Proposed Model Reformulation

In the model, the term $C_t(S_{i,t}, a_t)$ is an explicit objective (7.4) with regard to the constraints (7.5)–(7.12); however, the term $\tilde{v}_t^{a_t,n}(S_{i,t}^{a_t})$ is just a value with regard to the post-decision state $S_{i,t}^{a_t}$ but has no relations to the strategies and variables. In this case, it is not easy to solve the optimization model (7.18). Therefore, it is necessary to relate $\tilde{v}_t^{a_t,n}(S_{i,t}^{a_t})$ to the strategies and variables.

System reconfiguration is the cause of changing the current state $S_{i,t}$ to $S_{i,t}^{a_t}$. For example, we have two reconfigurable lines, and the corresponding binaries are x_1 and x_2. There will be four post-decision states listed in Table 7.1. In this case, the second term $\tilde{v}_t^{a_t,n}(S_{i,t}^{a_t})$ in (7.18) can be expressed as $(1 - x_1)(1 - x_2)V_1 + (1 - x_1)x_2V_2 + x_1(1 - x_2)V_3 + x_1x_2V_4$ which relates the values of post-decision states to decision variables.

Therefore, a generic form can be expressed to rewrite the second term $\tilde{v}_t^{a_t,n}(S_{i,t}^{a_t})$ in (7.18). For the state $S_{i,t}$, there are m reconfigurable lines, resulting in 2^m post-decision states associated with the corresponding estimates (represented as $V_1, V_2, \ldots, V_{2^m}$) at the nth iteration. The 2^m post-decision states are binary-coded with a $2^m \times m$ matrix $\mathbf{b}_i$, in which the (i', l)th entry denotes the on-off state of the line l under the i'th post-decision states. The generic form of the term $\tilde{v}_t^{a_t,n}(S_{i,t}^{a_t})$ at the nth iteration is listed as follows:

$$\sum_{i' \in S_i^{\text{post}}} \left\{ \prod_{l \in \mathcal{L}_{i,t}^d} (1 - \beta_{l,i',t} - \mathbf{b}_i(i', l))(1 - 2\mathbf{b}_i(i', l))V_{i'} \right\} \tag{7.21}$$

Table 7.1 Four Post-decision states.

Post-decision states	x_1	x_2	Estimated values
$S_{1,t}^{a_t}$	0	0	$\tilde{v}_t^{a_t,n}(S_{1,t}^{a_t}) \triangleq V_1$
$S_{2,t}^{a_t}$	0	1	$\tilde{v}_t^{a_t,n}(S_{2,t}^{a_t}) \triangleq V_2$
$S_{3,t}^{a_t}$	1	0	$\tilde{v}_t^{a_t,n}(S_{3,t}^{a_t}) \triangleq V_3$
$S_{4,t}^{a_t}$	1	1	$\tilde{v}_t^{a_t,n}(S_{4,t}^{a_t}) \triangleq V_4$

where $\beta_{l,i',t}$ is a binary representing on-off states of the line l under post-decision states. $\mathbf{b}_i(i',l)$ is a known value with 0 or 1, making (7.21) a sum of multi-linear functions. The optimization model (7.18) in forward dynamic algorithm can be re-written as follows:

$$\text{min } (7.4) + (7.21)$$
$$\text{s.t. } (7.5), (7.6), (7.7), (7.8), (7.9), (7.10), (7.11), (7.12) \tag{7.22}$$

McCormick relaxation is used to connect additional variables and constraints and transform the multilinear functions in (7.21) into four binary equations. In (7.21), the multilinear function with most variables is $\beta_{1,i',t}\beta_{2,i',t} \cdots \beta_{m,i',t}$, which can be represented by additional variables

$$y_{2,i',t} = \beta_{1,i',t}\beta_{2,i',t}$$
$$y_{3,i',t} = y_{2,i',t}\beta_{3,i',t}$$
$$\cdots \tag{7.23}$$
$$y_{m,i',t} = y_{m-1,i',t}\beta_{m,i',t}$$

and additional constraints

$$y_{2,i',t} \geq \beta_{2,i',t} + \beta_{1,i',t} - 1$$
$$y_{2,i',t} \leq \beta_{1,i',t}$$
$$y_{l,i',t} \geq 0 \quad (l = 2, \ldots, m)$$
$$y_{l,i',t} \leq \beta_{l,i',t} \quad (l = 2, \ldots, m) \tag{7.24}$$
$$y_{l,i',t} \geq \beta_{l,i',t} + y_{l-1,i',t} - 1 \quad (l = 3, \ldots, m)$$
$$y_{l,i',t} \leq y_{l-1,i',t} \quad (l = 3, \ldots, m)$$

With the additional constraints and variables, the optimization problem can be established as a mixed integer linear programming model, which can be solved by many solvers such as CPLEX and GUROBI.

7.4.5 Iteration Process

With the estimation of the post-decision state, the recursive MDP-based model is solved only for one state in each time period. For each iteration, the optimization model is a mixed integer linear programming model. The detailed procedures are listed in Algorithm 7.1.

Algorithm 7.1 ADP algorithm

1: Step 1. Set the iteration counter $n = 1$ and the maximum number of iterations N, respectively.

2: Step 2. Set the initial approximation $\bar{v}_t^{a_t,n}(S_{i,t}^{a_t})$ for each state.

3: Step 3. Do for $t = 1, \ldots, T$

4:      Step 3.1. Solve (19) and (20) to get $v_t^n(S_{i,t})$ and a_t at the iteration n.

5:      Step 3.2. Update the approximation $\bar{v}_t^{a_t,n}(S_{i,t}^{a_t})$ for the post-decision $S_{i,t}^{a_t}$ with (21).

6:      Step 3.3. Obtain the post-decision $S_{i,t}^{a_t}$ from the state $S_{i,t}$ under a_t.

7:      Step 3.4. According to uncertainties of the extreme event, generate a new state $S_{j,t+1}$ at $t+1$ from the post-decision state $S_{i,t}^{a_t}$ at t.

8: Step 4. Set $n = n + 1$. If $n \leq N$ go to Step 3.

9: Step 5. If $n = N$, return $\bar{v}_t^{a_t,N}(S_{i,t}^{a_t})$ and the corresponding action a_t for the state $S_{i,t}$.

7.5 Case Studies

In this section, the proposed model and the algorithm are validated by two test systems.

7.5.1 IEEE 33-Bus System

7.5.1.1 Data Description

The topology of the IEEE 33-bus system and the typhoon trajectory are shown in Figure 7.5. It is assumed that the trajectory and the duration of the typhoon are known. The duration between two decisions in the case study is assumed to be 15 minutes, and six decisions need to be made on the typhoon trajectory. In practice, the duration of a typhoon depends on its real development. For the original topology, the lines 25–29, 1–18, 9–15, 12–22, and 8–21 are offline to ensure the radial topology. The lines 1–18, 14–15, 12–22, 8–21, 9–15, 10–11, 12–13, and 25–29 are

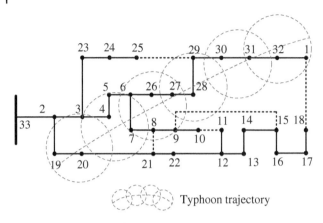

Figure 7.5 Topology of IEEE 33-bus system.

assumed to be dispatchable, and the other lines are nondispatchable. The penalty cost for loss of load is 35 000 $/kW in each period. The operational costs for the lines 10–11, 12–13, 25–29, 1–18, 14–15, 12–22, 8–21, and 9–15 are 1000, 1200, 1300, 1300, 1400, 1400, 1400, and 1400 $/period, respectively.

7.5.1.2 Estimated Values of Post-Decision States

The estimations of post-decision states are important to optimize the problem. In this case, we implemented 1500 iterations to achieve the estimated values of post-decision states. We only show the estimated values of some post-decision states for the sake of exposition in consideration of a large number of post-decision states. Figure 7.6 shows the estimated values of four post-decision states $S_{1,2}$, $S_{2,2}$, $S_{3,2}$, and $S_{4,2}$, shown in Table 7.2, in the second decision period. The estimated values of the post-decision states $S_{1,2}$, $S_{2,2}$, $S_{3,2}$, and $S_{4,2}$ converge to $\$1.29 \times 10^6$, $\$1.37 \times 10^6$, $\$1.45 \times 10^6$, and $\$1.24 \times 10^6$, respectively. The estimated value of each post-decision state can be interpreted as the expected cost for the post-decision state.

When having the estimated values of the post-decision states, the term $v_t^{a_t}(S_{i,t}^{a_t})$ is known when solving the optimization model (7.15a). In this case, the problem is transformed into an online one-period deterministic optimization problem.

Different failure rates of components can be caused by different intensity of severe weather. From the perspective of the mathematical model, different failure rates cause different estimated values of each post-decision state. The estimated values of several post-decision states with different failure rates are listed in Figure 7.7. For example, the estimated values of the post-decision state $S_{1,1}$ in the first period are $\$1.01 \times 10^6$, $\$1.58 \times 10^6$, and $\$1.81 \times 10^6$ when the failure rates are 0.02, 0.05, and 0.08, respectively. It is observed that a higher failure rate causes a larger estimated value of a post-decision state. When updating the estimated

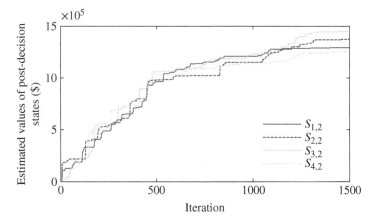

Figure 7.6 Iterations for estimated values of post-decision states.

Table 7.2 Post-decision states $S_{1,2}$, $S_{2,2}$, $S_{3,2}$, and $S_{4,2}$.

Time period	Post-decision states	Disconnected lines
2	$S_{1,2}$	10–11, 12–13, 25–29, 1–18, 8–21
2	$S_{2,2}$	10–11, 25–29, 1–18, 14–15, 8–21
2	$S_{3,2}$	10–11, 25–29, 1–18, 12–22, 8-21
2	$S_{4,2}$	25–29, 1–18, 14–15, 12–22, 8–21
1	$S_{1,1}$	10–11, 12–13, 25–29, 1–18, 8–21

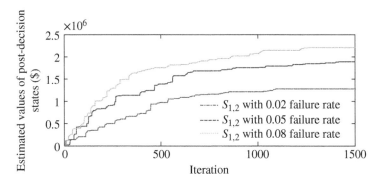

Figure 7.7 Iterations for estimated values of post-decision states with different failure rates.

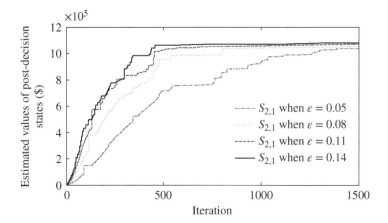

Figure 7.8 Iterations for estimated values of post-decision states with different values of ϵ.

values of post-decision states by using (7.20), ϵ is artificially set. Figure 7.8 shows the impacts of different values of ϵ on the estimated values of the postdecision state $S_{2,1}$. It is observed that the estimated values are close even when ϵ has different values.

7.5.1.3 Dispatch Strategies with Estimated Values of Post-Decision States

The strategy corresponding to one observed real-time state can be obtained by a one-period deterministic optimization problem based on the estimated values of each post-decision state. Tables 7.3 and 7.4 show the state-based strategies, and the original topology has the disconnected lines 10–11, 8–21, 9–15, 1–18, and 25–29. It is observed that the strategies make that the feeders impacted by the typhoon

Table 7.3 The first case of state-based strategy.

Time period	Observed state (component failure)	Strategy	
		Open lines	**Close lines**
1	—	19–20	10–11
2	6–7	10–11, 6–26	19–20, 8–21, 1–18
3	—	—	—
4	—	—	—
5	—	—	—
6	—	—	—

Table 7.4 The second case of state-based strategy.

Time period	Observed state (component failure)	Strategy	
		Open lines	**Close lines**
1	—	19–20	10–11
2	—	6–26	19–20, 1–18
3	—	—	—
4	—	—	—
5	—	—	—
6	—	—	—

are downstream. This is reasonable because downstream feeders cause smaller outages even they are in failure due to the typhoon. When the system operators do nothing during the unfolding event, the values of post-decision states $S_{1,2}$, $S_{2,2}$, $S_{3,2}$, and $S_{4,2}$ are $\$2.13 \times 10^6$, $\$2.56 \times 10^6$, $\$2.71 \times 10^6$, and $\$2.01 \times 10^6$, respectively. They are larger than the values $\$1.27 \times 10^6$, $\$1.89 \times 10^6$, and $\$2.19 \times 10^6$ by using the proposed method, respectively. Because the values of post-decision states are expected costs in the subsequent periods, smaller values indicate better strategies.

7.5.2 IEEE 123-Bus System

7.5.2.1 Data Description
The topology of the IEEE 123-bus system and the typhoon trajectory are shown in Figure 7.9. The lines 101–119, 67–117, 56–76, 39–57, 42–120, 52–53, 46–65, 60–57, 60–117, 63–64, 16–96, 92–120, 51–108, 71–85, and 38–43 are dispatchable. The original topology has the disconnected lines 42–120, 38–43, 39–57, 92–120, 115–116, 46–65, 51–108, 71–85, 16–96, and 56–76. The operational costs for the dispatchable lines are 1200 \$/period. The penalty cost for loss of load is 35 000 \$/kW in each period.

7.5.2.2 Simulated Results
The estimated values of post-decision states can be obtained, and then the state-based strategies can be optimized. A total of 1500 iterations were implemented to obtain the estimated values of the post-decision states. Table 7.5 shows the state-based strategies on the trajectory of the typhoon, and the states on the trajectory are assumed to be generated stochastically based on failure rates caused by the typhoon. Figure 7.10a,b shows the topologies after implementing

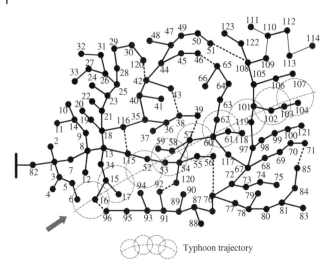

Figure 7.9 Topology of IEEE 123-bus system.

Table 7.5 State-based strategy for IEEE 123-bus system.

Time period	Observed state (component failure)	Strategy	
		Open lines	**Close lines**
1	—	—	—
2	15–17	—	—
3	—	52–53	46–65
4	—	—	—
5	58–57	—	—
6	—	57–60, 60–117,	52–53, 56–76
6	—	101–119	51–108
7	—	—	—
8	102–103	—	—
9	—	—	—

the state-based strategies in the third and sixth periods, respectively. In the third period, the line 52–53 is disconnected and the line 46–65 is connected to avoid blackout of downstream feeders if the typhoon fails the line 52–53. In the sixth period, three lines (57–60, 60–117, 101–119) are disconnected and three lines (52–53, 56–76, 51–108) are connected to reduce possible blackout areas.

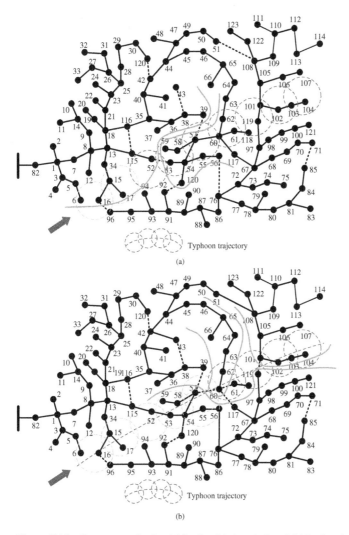

Figure 7.10 System topologies (a) in the third period and (b) in the sixth period.

7.6 Summary and Conclusions

A Markov state-based decision-making strategy, implementing system topology reconfiguration, is proposed to improve the distribution system resilience throughout the unfolding events. Markov states represent the sequential states of system topologies that may be changed by the unfolding events. Transition probabilities represent the uncertainties between different Markov states, and they depend on the component failure rates caused by the unfolding events. An MDP-based optimization model, including the current cost and the expected cost in the future, is proposed to make state-based actions at each decision time. An ADP approach with the post-decision state is employed to address the "curse of dimensionality" caused by uncertainties. With the estimated values of post-decision states, the stochastic problem with sequential multi-period stochastic optimization problem is transformed into a one-period deterministic problem. Case studies demonstrate that the state-based strategies try to make the feeders on the trajectory locate the terminal of the whole network to reduce potential loss of load, and in consequence to improve system resilience.

Nomenclature

Indices and Sets

c	index of components
l	index of lines
k, k'	index of terminal buses of line l
t, τ	index of time periods
i, i', j, j'	index of states
$\mathcal{A}$	set of actions
$\mathcal{B}$	set of buses
$\overline{\mathcal{B}}_{i,t}$	set of nonislanded buses under the state $S_{i,t}$
$\tilde{\mathcal{B}}$	set of substation nodes
$\mathcal{C}^f_{i,t}$	set of repaired components under state $S_{i,t}$
$\mathcal{F}_t$	set of all possible failure components at t
$\tilde{\mathcal{F}}_t$	set of actual failure components at t
$\mathcal{L}_{i,t}$	set of nonislanded lines under state $S_{i,t}$
$\mathcal{L}_{i',t}$	set of dispatchable lines under postdecision state $S^{a_i}_{i',t}$
$\mathcal{L}^{nd}_{i,t}$	set of nondispatchable lines under state $S_{i,t}$
$\mathcal{L}^d_{i,t}$	set of dispatchable lines under state $S_{i,t}$
$\tilde{\mathcal{R}}_\tau$	set of repaired components at τ
$\mathcal{S}$	set of Markov states
S^{post}_i	set of postdecision states of state $S_{i,t}$
$\mathcal{T}$	set of time periods

Notation for Markov Decision Process-Based Model

a_t	action at t
C_t	immediate cost at t
C_l	operational cost of line l at t
$F^p_{kk',i,t}$	active power flow on line $k - k'$ under $S_{i,t}$ at t

$F^q_{kk',i,t}$	reactive power flow on line $k - k'$ under $S_{i,t}$ at t
$F^s_{kk'}$	apparent power capacity of line $k - k'$
$\Delta L^q_{k,i,t}$	loss of reactive load of bus k under state $S_{i,t}$
$\Delta L^p_{k,i,t}$	loss of active load of bus k under state $S_{i,t}$
$L^p_{k,t}$	active load of bus k at t
$L^q_{k,t}$	reactive load of bus k at t
M	large positive number
$o_{kk',i,t}$	binary variable, the value is 1 if bus k' is the parent bus for bus k under state $S_{i,t}$, otherwise 0
$r_{kk'}/x_{kk'}$	resistance/reactance of line $k - k'$
$s_{c,t}$	on-off state of component c at t
$S_{i,t}, S_{j,t+1}$	Markov state i and j at t and $t + 1$, respectively
ΔT	duration of each time period
T^f_c	time period from normal state to failure state of component c
ΔT^{τ}_c	repair duration for component c
$U_{k,i,t}$	squared voltage magnitude of bus k
$\underline{V}_k, \overline{V}_k$	low/upper limits of voltage value of bus k
v_t, v_{t+1}	values functions at t and $t + 1$
Pr	transition probability
$\beta_{l,i,t}, \beta_{c,i,t}$	binary variables representing on-off states of line l and c, respectively. 1 denotes on state, and 0 denotes off state
ξ_t	uncertainty of extreme event at t
η_t, η_{t-1}	penalty due to loss of load at t and $t - 1$ (\$/MWh)

Notation for Solution Method

$\mathbf{b}_i$	binary-coded matrix for postdecision states
$\mathbb{E}$	expected value
m	number of dispatchable lines
n	number of iterations
N	maximum number of iterations
$S^{a_t}_{i,t}$	post-decision after $S_{i,t}$ with action a_t
$S^{a_{t-1}}_{j',t-1}$	post-decision after $S_{j',t-1}$ with action a_{t-1}
T	number of decision periods
$v^{a_t}_t$	value function of post-decision state
v^n_t	value function of state at nth iteration at t
$\bar{v}^{a_t,n}_t$	approximated value function of postdecision state at nth iteration
$V_{i'}$	known values of post-decision states
x_1, x_2	binary variables
$y_{l,i',t}$	binary variable
ϵ	a coefficient

References

1 Executive Office of the President, USA, "Economic benefits of increasing electric grid resilience to weather outages," Aug. 2013.

2 Electric Power Research Institute, "Enhancing distribution resiliency: Opportunities for applying innovative technologies." 2013.

3 North American Electric Reliability Corporation, "Severe impact resilience: Considerations and recommendations." 2012.

4 A. Gholami, T. Shekari, M. H. Amirioun, F. Aminifar, M. H. Amini, and A. Sargolzaei, "Toward a consensus on the definition and taxonomy of power system resilience," *IEEE Access*, vol. 6, pp. 32035–32053, 2018.

5 M. Panteli and P. Mancarella, "The grid: Stronger, bigger, smarter?: Presenting a conceptual framework of power system resilience," *IEEE Power and Energy Magazine*, vol. 13, pp. 58–66, May 2015.

6 National Academy of Sciences USA, *National Research Council: Disaster Resilience: A National Imperative*. Washington, DC, USA: National Academies Press, 2012.

7 W. Yuan, J. Wang, F. Qiu, C. Chen, C. Kang, and B. Zeng, "Robust optimization-based resilient distribution network planning against natural disasters," *IEEE Transactions on Power Systems*, vol. 7, pp. 2817–2826, 2016.

8 M. Amirioun, F. Aminifar, H. Lesani, and M. Shahidehpour, "Metrics and quantitative framework for assessing microgrid resilience against windstorms," *International Journal of Electrical Power & Energy Systems*, vol. 104, pp. 716–723, 2019.

9 M. H. Amirioun, F. Aminifar, and H. Lesani, "Resilience-oriented proactive management of microgrids against windstorms," *IEEE Transactions on Power Systems*, vol. 33, pp. 4275–4284, 2018.

10 M. H. Amirioun, F. Aminifar, and H. Lesani, "Towards proactive scheduling of microgrids against extreme floods," *IEEE Transactions on Smart Grid*, vol. 9, pp. 3900–3902, 2018.

11 L. Zhang, K. Sun, Y. W. Li, X. Lu, and J. Zhao, "A distributed power control of series-connected module-integrated inverters for PV grid-tied applications," *IEEE Transactions on Power Electronics*, vol. 33, pp. 7698–7707, 2018.

12 F. Iqbal and A. S. Siddiqui, "Optimal configuration analysis for a campus microgrid–a case study," *Protection and Control of Modern Power Systems*, vol. 2, pp. 1–12, 2017.

13 W. Guo and L. Mu, "Control principles of micro-source inverters used in microgrid," *Protection and Control of Modern Power Systems*, vol. 1, pp. 1–7, 2016.

14 C. Wang, B. Cui, and Z. Wang, "Analysis of solvability boundary for droop-controlled microgrids," *IEEE Transactions on Power Systems*, vol. 33, pp. 5799–5802, 2018.

15 M. M. Adibi and L. H. Fink, "Overcoming restoration challenges associated with major power system disturbances - restoration from cascading failures," *IEEE Power and Energy Magazine*, vol. 4, pp. 68–77, 2006.

16 Y. Hou, C. C. Liu, K. Sun, P. Zhang, S. Liu, and D. Mizumura, "Computation of milestones for decision support during system restoration," *IEEE Transactions on Power Systems*, vol. 26, pp. 1399–1409, 2011.

17 W. Sun, C. C. Liu, and L. Zhang, "Optimal generator start-up strategy for bulk power system restoration," *IEEE Transactions on Power Systems*, vol. 26, pp. 1357–1366, 2011.

18 C. C. Liu, S. J. Lee, and S. S. Venkata, "An expert system operational aid for restoration and loss reduction of distribution systems," *IEEE Transactions on Power Systems*, vol. 3, pp. 619–626, 1988.

19 S. Toune, H. Fudo, T. Genji, Y. Fukuyama, and Y. Nakanishi, "Comparative study of modern heuristic algorithms to service restoration in distribution systems," *IEEE Transactions on Power Delivery*, vol. 17, pp. 173–181, 2002.

20 A. Castillo, "Microgrid provision of blackstart in disaster recovery for power system restoration," in *IEEE International Conference on Smart Grid Communications*, pp. 534–539, Oct. 2013.

21 F. Ren, M. Zhang, D. Soetanto, and X. Su, "Conceptual design of a multi-agent system for interconnected power systems restoration," *IEEE Transactions on Power Systems*, vol. 27, pp. 732–740, 2012.

22 M. Panteli, P. Mancarella, D. N. Trakas, E. Kyriakides, and N. D. Hatziargyriou, "Metrics and quantification of operational and infrastructure resilience in power systems," *IEEE Transactions on Power Systems*, vol. 32, pp. 4732–4742, 2017.

23 G. Huang, J. Wang, C. Chen, J. Qi, and C. Guo, "Integration of preventive and emergency responses for power grid resilience enhancement," *IEEE Transactions on Power Systems*, vol. 32, pp. 4451–4463, 2017.

24 A. Gholami, T. Shekari, F. Aminifar, and M. Shahidehpour, "Microgrid scheduling with uncertainty: The quest for resilience," *IEEE Transactions on Smart Grid*, vol. 7, pp. 2849–2858, 2016.

25 M. Yan, Y. He, M. Shahidehpour, X. Ai, Z. Li, and J. Wen, "Coordinated regional-district operation of integrated energy systems for resilience enhancement in natural disasters," *IEEE Transactions on Smart Grid*, vol. 10, pp. 4881–4892, 2019.

26 D. N. Trakas and N. D. Hatziargyriou, "Optimal distribution system operation for enhancing resilience against wildfires," *IEEE Transactions on Power Systems*, vol. 33, pp. 2260–2271, 2018.

27 E. Byon and Y. Ding, "Season-dependent condition-based maintenance for a wind turbine using a partially observed Markov decision process," *IEEE Transactions on Power Systems*, vol. 25, pp. 1823–1834, 2010.

28 S. K. Abeygunawardane, P. Jirutitijaroen, and H. Xu, "Adaptive maintenance policies for aging devices using a Markov decision process," *IEEE Transactions on Power Systems*, vol. 28, pp. 3194–3203, 2013.

29 C. Wang, Y. Hou, F. Qiu, S. Lei, and K. Liu, "Resilience enhancement with sequentially proactive operation strategies," *IEEE Transactions on Power Systems*, vol. 32, pp. 2847–2857, 2017.

30 S. Lei, C. Chen, Y. Song, and Y. Hou, "Radiality constraints for resilient reconfiguration of distribution systems: Formulation and application to microgrid formation," *arXiv preprint: 1907.04951*, p. 1, 2019.

31 A. Arif, Z. Wang, J. Wang, and C. Chen, "Power distribution system outage management with co-optimization of repairs, reconfiguration, and DG dispatch," *IEEE Transactions on Smart Grid*, vol. 9, pp. 4109–4118, 2018.

32 S. Lei, C. Chen, Y. Li, and Y. Hou, "Resilient disaster recovery logistics of distribution systems: Co-optimize service restoration with repair crew and mobile power source dispatch," *IEEE Transactions on Smart Grid*, vol. 10, pp. 6187–6202, 2019.

33 J. A. Taylor and F. S. Hover, "Convex models of distribution system reconfiguration," *IEEE Transactions on Power Systems*, vol. 27, pp. 1407–1413, 2012.

34 X. Chen, W. Wu, and B. Zhang, "Robust restoration method for active distribution networks," *IEEE Transactions on Power Systems*, vol. 31, pp. 4005–4015, 2016.

Part IV

Recovery After a Natural Disaster

8

Microgrids with Flexible Boundaries for Service Restoration

Network reconfiguration is an efficient strategy of microgrids (MGs) and distribution systems (DSs) for different purposes including resilience improvement. Specifically, MG technology development, distributed generation (DG) integration, and DS automation, etc., are enabling much more flexible operation and reconfiguration of power networks, e.g. MGs and DSs with flexible boundaries. Nonetheless, the formulation of optimization problems considering MG and DS reconfiguration and including such new flexibilities is not trivial, especially for the topology issue. That is, the topology needs to be radial. In fact, formulating radiality constraints with existing methods can lead to underutilization of MG and DS flexibilities. This chapter presents a new method for formulating radiality constraints to fully include topological and related flexibilities in reconfiguration-related MG and/or DS optimization problems, whose feasibility and optimality thus are extended and enhanced, respectively. The method's theoretical validity is certified by graph theoretic analyses. As it involves integer variables, this chapter further analyzes the tightness and compactness issues. The new radiality constraints are particularly used for post-disaster MG formation that is involved in many resilience-enhancing service restoration and infrastructure recovery problems of MGs and DSs. The new MG formation model allows much more flexible separation and/or merge of sub-grids with flexible boundaries. The model also establishes superiority over the models in the literature. Two test systems are used to conduct demonstrative case studies.

8.1 Using Microgrids in Service Restoration

Network reconfiguration is a multifunction and effective strategy of DSs [1–3]. For example, it can serve as an electric service restoration strategy used after natural disasters. Its optimization, therefore, has been studied extensively. Note that to decrease short-circuit currents and facilitate involved protection and

Power Grid Resilience against Natural Disasters: Preparedness, Response, and Recovery, First Edition.
Shunbo Lei, Chong Wang, and Yunhe Hou.
© 2023 John Wiley & Sons Ltd. Published 2023 by John Wiley & Sons Ltd.

coordination [4], most DSs need to maintain a radial topology. Thus, works such as [4–8] specifically study the mathematical formulation of radiality constraints. Using their approaches, the radiality requirement is included in most optimization problems of DSs [4], including post-disaster recovery and normal-state operation [1–3, 7, 9–14]. Generally, the formulation of radiality constraints in conventional DS reconfiguration optimization problems has been resolved. However, with added flexibilities of MG components, DGs, and automation equipment, etc., DSs now can have much more adaptive operation and reconfiguration. Specifically, with respect to the topology issue, now MGs and DSs can have flexible boundaries [15–17]. For instance, in [17], the DS is separated into a to-be-optimized number of MGs. Actually, more resilient DS and MG reconfiguration is empowered by such added flexibilities.

However, using existing approaches of formulating radiality constraints, the resulting DS and MG reconfiguration optimization models do not fully include the new flexibilities [18]. That is, these formulations give feasible regions that are only the actual feasible region's subsets. For example, existing formulations do not allow the optimization algorithm to select decisions taking advantages of DS and MG flexible boundaries. Flexibilities of DSs and MGs, therefore, will be less coordinated and underutilized, imposing especially adverse impacts on resilience-oriented reconfiguration [19]. For instance, as numerous faults are caused by natural disasters, post-disaster reconfiguration can only use quite limited flexibilities [20]. In such cases, the failures also make decision-making more complex and even require different restoration efforts to be co-optimized [11]. In fact, co-optimizing reconfiguration with the repairing of damaged grid components involves reconfiguring MG and DS networks with physical structures depending on repairing variables. Current approaches of formulating radiality constraints are not capable of properly handling such situations, including the operation mode with flexible boundaries.

As mentioned before, *resilience* can be defined as the ability to withstand and reduce the magnitude and/or duration of disruptive events, which includes the capability to anticipate, absorb, adapt to, and/or rapidly recover from such an event caused by a natural disaster, etc. In this regard, to improve MG and DS resilience, an enhanced utilization of topological flexibilities (such as flexible boundaries of MGs) is essential [21]. For instance, the use of DGs to assist power supply recovery or continuity in MGs and DSs greatly relies on the topological flexibilities [19].

A new formulation of radiality constraints is proposed in this chapter for fully enabling topological and related flexibilities. The new formulation can be included in related optimization models to get post-disaster MG and DS reconfiguration strategies that are more resilient. The new approach is superior to other attempts with the same or similar goals in the literature, e.g. in [12, 18] (see later parts of this chapter for comparisons). Its analytical validity is affirmed by graph-theoretic justifications. Since it involves binary variables, this chapter also analyzes its compactness and tightness, which are critical issues in formulating grid optimization models including integer variables [22–24]. In general, using the new radiality constraints in MG and DS reconfiguration-related optimization models can achieve extended feasibility and hence enhanced optimality, partly owing to the flexible boundaries of DSs and MGs.

For demonstration, the presented radiality constraints are used to build a new model for optimizing post-disaster MG formation that reconfigures a DS to form MGs powered by DGs and/or other energy resources. It is an essential restoration strategy in most resilience-oriented recovery problems of MGs and DSs [11, 12, 18–20]. In [18, 19, 25–27], this strategy, which also has to meet the radiality requirement, is proved efficient and effective for post-disaster recovery of critical loads in DSs and MGs. The presented MG formation optimization model shows superiority over the literature's two types of MG formation formulations built in [25] and [26], respectively, as the model here better utilizes the flexible boundary operation approach. As shown by the comparison in Figure 8.1, while their formulations exclude some flexibilities of DSs and MGs, the model here enables more adaptive separation/merge of sub-grids, and more flexible pick-up of loads, etc. (See later parts of this chapter for more details.) As a result, the formulation here attains MG formation strategies that have better recovery performance, implying that a higher level of resilience is achieved.

Note that though most MGs and DSs need to operate radially [4, 8, 28], some MGs and DSs (especially in urban areas) may operate in a meshed or ring manner to maintain higher levels of reliability. With increasingly more MGs and DGs, which can lead to bidirectional power flow, in the future, there will be more meshed or ring DSs supported by equipment including soft open points and bidirectional protective relays [29]. Those are beyond the scope of this chapter, which focuses on the most common cases requiring radial DS and MG topologies. Generally, the presented approach for formulating radiality constraints can be

I: Search region of the model in [25]
II: Search region of the model in [26]
III: Search region of the model here

Figure 8.1 The search region (i.e. included flexibilities) of the MG formation model in this chapter and the models in [25, 26]. (Note that in this chapter all comparisons assume the same model other than the topology formulation. For instance, the model in [26] considers dispatchable loads. They are altered to be non-dispatchable as in [25] and here.)

used in different DS and MG reconfiguration problems that have the radiality requirement. And based on the new radiality formulation, the new post-disaster MG formation model can produce recovery strategies utilizing flexible boundaries of MGs and DSs.

The main features of approaches presented in this chapter are as follows: (i) Topological and related flexibilities of MGs and DSs in reconfiguration-related optimization problems are fully enabled by the new formulation of radiality constraints. (ii) The proposed formulation's analytical validity is certified by in-depth graph-theoretic justifications and analyses. (iii) The compactness and tightness of the new radiality constraints, which influence related optimization problems' computational efficiency, are discussed in depth. (iv) Based on the presented radiality constraints, the new post-disaster MG formation model attains a higher resilience level, compared to existing models it has extended feasibility and enhanced optimality, partly owing to the better consideration of flexible boundary-based operation of DSs and MGs.

In this chapter, Section 8.2 presents the new approach of formulation of radiality constraints. Note that, in Section 8.2, the new formulation of radiality constraints is conceptually presented with discussions on the flexible boundary issue. Explicit formulation of the new radiality constraints is introduced in Section 8.3. Section 8.4 uses the new radiality constraints to construct a new post-disaster MG formation optimization model. Section 8.4 also emphasizes on the radiality constraint formulation's compactness and tightness issues. MG formation case studies are then discussed in Section 8.5. Section 8.6 briefly concludes this chapter.

8.2 Dynamically Formed Microgrids

8.2.1 Flexible Boundaries in Microgrid Formation Optimization

As mentioned earlier, now DSs and MGs can have much more adaptive operation and reconfiguration with flexibilities added by MG components, DGs, and automation equipment, etc. In particular, with respect to the topology issue, those technologies and flexibilities enable DSs and MGs to operate with flexible boundaries [15–17]. Such operation strategies are especially effective in enhancing resilience of DSs and MGs after natural disasters, e.g. by splitting a DS into a flexible number of MGs [17]. Therefore, more resilient reconfiguration of MGs and DSs is empowered by those added flexibilities.

In this regard, MG formation has been recently explored in [25, 26], etc. To recover critical loads after a natural disaster, a DS is resiliently reconfigured into multiple DG-energizing MGs. Chen et al. [25] first propose this strategy. Ding et al. [26] present a different model to reformulate the MG formation model introduced in [25]. Those two optimization models in [25] and [26] for this problem are adopted in most of the relevant studies, e.g. [19, 20, 27, 30–32]. However, their models do not fully include the flexibilities related to flexible boundaries of DSs and MGs.

Specifically, to optimize post-disaster MG formation, the formulation in [25] is based on clustering nodes. It assigns every node to at most one DG and enforces branch-node constraints, connectivity constraints, and other conditions involved in node clustering. The formulation in [26] uses single-commodity flow-based radiality constraints. It specifies that the number of closed lines equals the number of all nodes minus a fixed number of subgraphs and requires each subgraph to be connected. For their detailed modeling, refer to [25] and [26]. In general, the topological and related flexibilities are not fully utilized, as their radiality constraint formulations implicitly impose some unnecessary restrictions. Those restrictions disallow some post-disaster reconfiguration decisions involving flexible boundaries of DSs and MGs.

8.2.2 Radiality Constraints and Topological Flexibility

Generally, incorporating radiality constraints in MG and DS reconfiguration-related optimization models is to ensure that the system has a radial topology and each sub-grid has at most one substation node. Nevertheless, some other restrictions on MG and DS topologies are also implicitly imposed by existing radiality formulations in the literature. For example, forming de-energized load islands is not allowed, and the allocation of power resources into sub-grids and the number of sub-grids are partly or completely fixed. Such implicit limitations restrict topological and relevant flexibilities to be fully used, especially for cases with increasingly more MG and DG technologies, which enable more adaptive and flexible operation and reconfiguration of MGs and DSs. In contrast, the radiality model presented in this chapter enables more flexible DS and MG reconfiguration after disastrous events (e.g. more flexible boundaries), because it does not enforce such unneeded restrictions.

This chapter proposes to model radiality constraints using two simple concepts in graph theory and their relationships.

First, the definition of *spanning tree*, an extensively used concept, is provided here so that this chapter is self-contained:

Definition 8.1 A spanning tree is a graph that connects all the nodes and contains no loops [33].

Second, *spanning forest*, the other concept which is less common, is defined in the following definition.

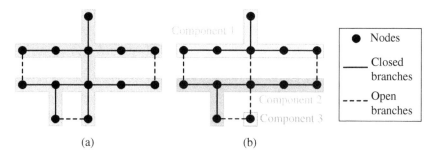

Figure 8.2 (a) A spanning tree. (b) A spanning forest. (Demonstrated on the modified IEEE 13-node test feeder [34].)

Definition 8.2 A spanning forest is a graph having no cycles [33].

Prior to further discussions, here we explain two other concepts in graph theory. First, a *subgraph* of a graph is composed of a subset of the graph's nodes and links. Second, a *component* or a *connected component* in a graph, is a subgraph where any two nodes are directly or indirectly connected to each other and that is linked to no additional nodes in the graph.

As Figure 8.2 indicates, a spanning forest is a graph that has spanning trees as its components. In fact, a spanning forest that has κ connected components is called a *κ-tree*. For instance, a 3-tree is shown in Figure 8.2b. Note that a spanning tree, e.g. Figure 8.2a, is a 1-tree. Apparently, spanning forests and spanning trees are both radial.

In a DS, let κ_1 be the number of substation nodes. Considering Definitions 8.1 and 8.2, we can observe that common DS reconfiguration for supply capacity improvement [2] and loss reduction [1] is essentially to form a κ_1-tree. Thus, if $\kappa_1 = 1$, a spanning tree is constructed, while if $\kappa_1 > 1$, a spanning forest is constructed. It also requires that every connected component includes a substation node.

In terms of resilient reconfiguration of DSs (e.g. post-disaster MG formation with flexible boundaries for service recovery), with $\kappa_2 \geq \kappa_1$, it forms a κ_2-tree. It also requires that every connected component includes at most one substation. In fact, resilient reconfiguration should also optimize the value of κ_2. If it is prefixed, the actual optimal solution may not be attained. For relevant cases, mostly we have $\kappa_2 > 1$, indicating that a spanning forest is formed. Thus, in resilient reconfiguration of MGs and DSs, the topology issues can be handled by enforcing that the network forms a spanning forest. Therefore, this chapter proposes to formulate radiality constraints as Eqs. (8.1) and (8.2) that are inspired by the following remark.

Remark 8.1 For a spanning tree, an arbitrary subgraph is a spanning forest.

For instance, Figure 8.2b shows a subgraph of the graph in Figure 8.2a. Based on Definitions 8.1 and 8.2, Remark 8.1 is certainly true. Therefore, the proposed formulation of radiality constraints is as follows:

$$\beta \in \Omega \tag{8.1}$$

$$\alpha_{ij} \leq \beta_{ij}, \ \forall (i,j) \in \mathcal{L} \tag{8.2}$$

Specifically, $\mathcal{L}$ is the set of DS lines; $\alpha \triangleq \{\alpha_{ij}, \ \forall (i,j) \in \mathcal{L}\}$, with α_{ij} being the connection status of line (i,j) (0 if open, and 1 if closed); $\beta \triangleq \{\beta_{ij}, \ \forall (i,j) \in \mathcal{L}\}$ with β_{ij} being the *fictitious* connection status of line (i,j) (0 if open, and 1 if closed). By "fictitious," it means that β are only auxiliary variables that do not actually determine the system's network topology. It is still variables α determining the topology. The set Ω includes all incidence vectors of spanning tree topologies that can be formed by network reconfiguration.

Therefore, constraint (8.1) requires that β forms a fictitious spanning tree. Then constraint (8.2) enforces that the DS closes a subset of the spanning tree's (determined by β) closed lines. In other words, α constructs a subgraph of the fictitious spanning tree. Remark 8.1 implies that the resulting system topology specified by α is a spanning forest. Note that here constraint (8.1) is only conceptually expressed to enforce that β is a characteristic vector of a spanning tree. Constraint (8.1) is not to be formulated by listing all spanning trees in Ω. (In fact, it is not NP-hard to find all spanning trees of a graph [35]. Still, generally getting all spanning trees of a DS is only used in DS reconfiguration with heuristic or meta-heuristic algorithms.) The explicit formulation of constraint (8.1) is elaborated on in Section 8.3, where constraints (8.3)–(8.9) are used to formulate constraint (8.1). Particularly, any β belonging to Ω satisfies constraints (8.3)–(8.9), and any β that satisfies constraints (8.3)–(8.9) belongs to the set Ω. Note that constraints (8.1) and (8.2) assume implicitly that the system includes only one substation node. Section 8.4 will shortly explain the reason and present a simple technique enabling the proposed radiality formulation to fully deal with a DS including more than one substation nodes. For uncontrollable lines with faulted switches or without switches, constraints may be included to fix their connection status (see Section 8.4).

Remark 8.1 just indicates that any α meeting constraints (8.1) and (8.2) is topologically feasible. The theorem in the following specifies that any topologically feasible α meets constraints (8.1) and (8.2). Since it is less obvious, we provide its proof. Theorem 8.1 and Remark 8.1 together support the proposed radiality model's validity.

Theorem 8.1 A spanning forest subgraph of a connected graph is also the subgraph of a spanning tree subgraph of the connected graph. (See the proof in Appendix 8.A.)

The proposed radiality constraints (8.1) and (8.2) can be seen as a two-step approach to constrain DS topology in post-disaster reconfiguration with flexible boundaries of DSs and MGs. In the first step, constraint (8.1) assures topology radiality by fixing a spanning tree to be the system topology's super-graph that is decided by the lines' fictitious connection status. (If G_A is a subgraph of G_B, G_B is a *super-graph* of G_A.) In the second step, constraint (8.2) allows more flexible reconfiguration by letting the DS choose a subgraph of the fictitious spanning tree to be the system's actual topology.

In contrary, other approaches of radiality formulation in the literature (e.g. [4, 8]) use a one-step model directly enforcing the distribution topology to be a spanning forest or a spanning tree. Some of their formulations (e.g. the single-commodity flow model [4]) may be adopted here to formulate constraint (8.1). Still, Section 8.3 will present a new formulation. Above all, many more DS reconfiguration flexibilities (e.g. more flexible assignment of power resources among sub-grids, and more adaptive separation or merge of sub-grids) are enabled by the presented two-step technique for formulating radiality constraints. In Sections 8.4 and 8.5, using the new radiality constraints for resilient post-disaster MG formation with flexible boundaries, these benefits will be elaborated on.

8.3 Mathematical Formulation of Radiality Constraints

Constraint (8.1) is just conceptually expressed in Section 8.2. As mentioned, normal models or approaches of formulating radiality constraints in MG and DS reconfiguration-related studies may be adopted here for its explicit formulation. This chapter revisits these models and approaches. A new model is introduced, too. The compactness and tightness issues are briefly discussed in this section and explored in more detail in Section 8.4.2.

DS and MG reconfiguration essentially is a mixed-integer programming (MIP) problem commonly solved by branch-and-cut (B&C) algorithms, which have to solve *linear programming (LP) relaxations* (i.e. relaxing binary/integer constraints) of the MIP problem. An MIP problem's computational complexity depends on its *compactness* and *tightness*, etc. An MIP model is *tight* if its feasible region is similar to its LP relaxation, resulting in a smaller gap between the optimal solutions of the MIP problem and its LP relaxation, and less-explored nodes in the B&C search tree (i.e. less iterations and quicker convergence). A *compact* MIP model has a small number of constraints and variables, resulting in shorter computation time of each iteration (associated with each explored node) in the B&C search tree. (*Compactness* just implies the number of constraints and variables in this chapter, while it may have other meanings in other areas,

Figure 8.3 Illustrating the *tightness* concept.

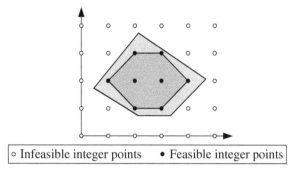

| ○ Infeasible integer points | • Feasible integer points |

e.g. in real mathematical analysis.) In formulating a MIP model, compactness and tightness are usually conflicting objectives [22].

Figure 8.3 gives an instance to illustrate the concept of *tightness*. There are infinite formulations to define a set including feasible integer points and excluding infeasible integer points. Different formulations are associated with different sets with different tightness levels. The smaller shadowed set is defined by the *tightest* formulation that defines the convex hull of feasible integer points, and the larger shadowed set is defined by a less tight formulation. A tight model is desired when formulating optimization problems involving integer variables.

Several useful concepts are introduced in the following. Here, *incidence/characteristic vectors of spanning trees* are values of β specifying fictitious spanning tree topologies of the DS, and *spanning tree polytope* is the convex hull of those incidence vectors. For instance, Figure 8.4a shows a system having $\beta = [\beta_{12}, \beta_{13}, \beta_{23}]$. Its characteristic vectors of spanning trees include $[0, 1, 1]$, $[1, 0, 1]$, and $[1, 1, 0]$, and the shaded region in Figure 8.4b with such vectors being the vertices is its spanning tree polytope. *Incidence/characteristic vectors of spanning forests* and the *spanning forest polytope* can be defined similarly.

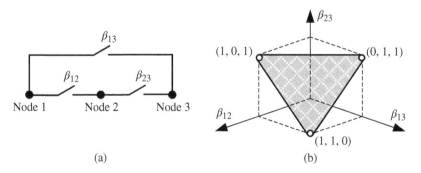

Figure 8.4 (a) An example system with three nodes. (b) The *set of spanning tree characteristic vectors* (i.e. Ω) and the *spanning tree polytope* (i.e. convex hull of Ω) demonstrated on the example system.

That is, constraint (8.1) essentially enforces that β is a characteristic vector of a spanning tree. A compact and tight formulation should be used to explicitly model this condition. LP relaxations of the tightest models actually define the spanning tree polytope.

Some models or methods of expressing radiality constraints in DS reconfiguration-related studies may be adopted for formulating the spanning tree constraint (8.1). In the following, those models and methods for explicitly formulating constraints (8.1) are revisited and compared.

8.3.1 Loop-Eliminating Model

The loop-eliminating model lists all cycles in the DS and requires that each is open [5]. Still, it is NP-hard to find all cycles in a graph. This model is actually equivalent to the spanning tree constraint's *subtour elimination formulation* in graph theory. It is one of the tightest formulations. Moreover, its LP relaxation characterizes the spanning tree polytope that has integer extreme points. It has limited applications, as it has an exponential number of constraints though it has only $|\mathcal{L}|$ variables.

8.3.2 Path-Based Model

The path-based model lists all routes between the substation and each node, and activates just one path for each node [6]. However, it is NP-hard to find all routes between two nodes in a graph. This model again has a graph-theoretic counterpart, i.e. the spanning tree constraint's *directed cutset formulation* with an exponential number of constraints and $3 \cdot |\mathcal{L}|$ variables. It is again one of the tightest formulations, and its LP relaxation has integer vertices.

8.3.3 Single-Commodity Flow-Based Model

The single-commodity flow-based model requires that $|\mathcal{N}| - 1$ lines are closed ($\mathcal{N}$ being the set of nodes) and assures connectivity by enforcing that unit fictitious commodity flows to each node from the substation [4]. It is one of the most compact models, as it has only a linear number of constraints and $3 \cdot |\mathcal{L}|$ (or decreased to $2 \cdot |\mathcal{L}|$) variables. It is also the most commonly adopted radiality formulation, owing to its simplicity. However, it is not the tightest. Its LP relaxation's projection onto the β-space is a set larger than the spanning tree polytope, and thus has fractional extreme points. Note that this formulation is actually the same as the virtual power flow-based radiality model, which is widely adopted in [18, 30], etc.

8.3.4 Parent–Child Node Relation-Based Model

The parent–child node relation-based model requires each node except the sub-station node to have one parent node [7]. Nevertheless, this formulation may lead to a disconnected topology with cycles [8]. In many DS reconfiguration works (e.g. [28]), it still works because it implicitly requires the number of closed lines to be $|\mathcal{N}| - 1$, and the connectivity requirement (i.e. requiring the system to be a connected graph) is also implicitly imposed by power flow constraints. As mentioned above, these two requirements are sufficient and necessary conditions of ensuring radiality of a DS having one substation node.

8.3.5 Primal and Dual Graph-Based Model

The primal and dual graph-based formulation enforces every primal (respectively, dual) vertex to be linked to another primal (respectively, dual) vertex, and prohibits the primal and dual spanning trees from intersecting each other [8]. It is one of the most compact models. Specifically, it has a linear number of constraints and $4 \cdot |\mathcal{L}|$ variables. It is originally presented in [36], and in fact, it has a totally unimodular constraint matrix defining a polyhedron that has integer vertices. Therefore, in the β-space its LP relaxation actually defines the spanning tree polytope. That is, it is one of the tightest models. However, it needs to model dual graphs, and it is only useful for planar graphs. As mentioned in [37], it also has a flaw to be dealt with by proper choice of primal and dual root nodes. In general, its utilization is far less convenient, and it has limited applicability.

8.3.6 Spanning Forest-Based Model

Using notations listed at the end of this chapter, a different flow-based model for the spanning tree constraint (8.1) is further introduced as follows:

$$\sum_{(j,i_r)\in\mathcal{L}} f_{ji_r}^k - \sum_{(i_r,j)\in\mathcal{L}} f_{i_rj}^k = -1, \ \forall k \in \mathcal{N}\backslash i_r \tag{8.3}$$

$$\sum_{(j,k)\in\mathcal{L}} f_{jk}^k - \sum_{(k,j)\in\mathcal{L}} f_{kj}^k = 1, \ \forall k \in \mathcal{N}\backslash i_r \tag{8.4}$$

$$\sum_{(j,i)\in\mathcal{L}} f_{ji}^k - \sum_{(i,j)\in\mathcal{L}} f_{ij}^k = 0, \ \forall k \in \mathcal{N}\backslash i_r, \forall i \in \mathcal{N}\backslash\{i_r,k\} \tag{8.5}$$

$$0 \leq f_{ij}^k \leq \lambda_{ij}, 0 \leq f_{ji}^k \leq \lambda_{ji}, \ \forall k \in \mathcal{N}\backslash i_r, \forall(i,j) \in \mathcal{L} \tag{8.6}$$

$$\sum_{(i,j)\in\mathcal{L}} (\lambda_{ij} + \lambda_{ji}) = |\mathcal{N}| - 1 \tag{8.7}$$

$$\lambda_{ij} + \lambda_{ji} = \beta_{ij}, \forall(i,j) \in \mathcal{L} \tag{8.8}$$

$$\lambda_{ij}, \lambda_{ji} \in \{0,1\}, \forall(i,j) \in \mathcal{L} \tag{8.9}$$

The model above is called the *directed multi-commodity flow-based formulation* of the spanning tree constraint. The model defines a *fictitious* commodity for every node $k \neq i_r$, and forces that unit commodity k flows to node k from the substation node i_r. Particularly, for every node $k \neq i_r$, constraint (8.4) requires that unit commodity k flows to node k, and constraint (8.3) states that unit commodity k flows out from i_r, i.e. the substation node. Constraint (8.5) prohibits each commodity k from flowing to nodes except node k. Constraint (8.6) states that every commodity may flow on a link only if the directed spanning tree (determined by variables λ_{ij}) includes the link. Equation (8.7) fixes the number of links in the spanning tree. Constraint (8.8) attains the fictitious connection status β_{ij} of lines based on the inclusion of links in the spanning tree. Constraint (8.9) enforces that some variables are binary. Though β_{ij} and λ_{ij} are both binary variables, it is not needed to have binary constraints for variables β_{ij}, because constraint (8.8) already enforces β_{ij} to be binary.

Because post-disaster network reconfiguration with flexible boundaries is critically important in enhancing DS and MG resilience, many works also introduce new approaches of assuring radiality to enable undiminished adaptivities and flexibilities of MGs and DSs in optimization. Particularly, studies [12] and [18] have proposed two applicable approaches. Here, they are briefly compared with the proposed method of formulating radiality constraints, which are conceptually expressed by constraints (8.1) and (8.2). In general, Remark 8.1 and Theorem 8.1 here theoretically certify the validity of constraints (8.1) and (8.2). Studies [12] and [18] do not provide analytical proofs for their models. Particularly, the approach in [12] is partially built on the spanning tree constraints presented in [7], which actually do not guarantee radiality as explained in [8]. It may still be valid; however, a proof is critically needed. More comparisons are provided in later parts of this chapter.

8.4 Adaptive Microgrid Formation for Service Restoration

8.4.1 Formulation and Validity

The proposed radiality model (8.1) and (8.2) can be used in different optimization problems involving MGs and/or DSs with flexible boundaries [15–17]. In this chapter, to verify its advantages, it is applied to regulate the topology of DSs in resilient post-disaster MG formation.

Post-disaster MG formation has been studied extensively in [25, 26], etc. To attain a resilient DS by quickly recovering critical loads after a natural disaster, it reconfigures a DS into multiple DG-energizing MGs. As mentioned before,

Chen et al. [25] first propose this strategy. Ding et al. [26] then present a different model for reformulating the MG formation model presented in [25]. These two formulations in [25] and [26] for this problem are adopted in most of the relevant studies, e.g. [19, 20, 27, 30–32]. Based on the new radiality model in this chapter, a new resilient MG formation optimization model for better utilization of flexible boundaries of DSs and MGs is constructed as follows:

$$\max \sum_{i \in \mathcal{N}} \delta_i \cdot \omega_i \cdot p_i^c \tag{8.10}$$

s.t. (8.1) and (8.2)

$$p_i^g - \delta_i \cdot p_i^c + \sum_{(j,i) \in \mathcal{L}} P_{ji} - \sum_{(i,j) \in \mathcal{L}} P_{ij} = 0, \ \forall i \in \mathcal{N} \tag{8.11}$$

$$q_i^g - \delta_i \cdot q_i^c + \sum_{(j,i) \in \mathcal{L}} Q_{ji} - \sum_{(i,j) \in \mathcal{L}} Q_{ij} = 0, \ \forall i \in \mathcal{N} \tag{8.12}$$

$$p_i^g = q_i^g = 0, \ \forall i \in \mathcal{N} \backslash \{\mathcal{N}_r, \mathcal{N}_g\} \tag{8.13}$$

$$0 \le p_i^g \le \overline{p}_i^g, 0 \le q_i^g \le \overline{q}_i^g, \ \forall i \in \{\mathcal{N}_r, \mathcal{N}_g\} \tag{8.14}$$

$$v_i - v_j \ge 2 \cdot (P_{ij} \cdot r_{ij} + Q_{ij} \cdot x_{ij}) + (\alpha_{ij} - 1) \cdot M, \ \forall (i,j) \in \mathcal{L} \tag{8.15}$$

$$v_i - v_j \le 2 \cdot (P_{ij} \cdot r_{ij} + Q_{ij} \cdot x_{ij}) + (1 - \alpha_{ij}) \cdot M, \ \forall (i,j) \in \mathcal{L} \tag{8.16}$$

$$\underline{v}_i \le v_i \le \overline{v}_i, \ \forall i \in \mathcal{N} \tag{8.17}$$

$$P_{ij}^2 + Q_{ij}^2 \le \alpha_{ij} \cdot \overline{S}_{ij}^2, \ \forall (i,j) \in \mathcal{L} \tag{8.18}$$

$$\alpha_{ij} = 0, \forall (i,j) \in \mathcal{L}_o; \ \alpha_{ij} = 1, \ \forall (i,j) \in \mathcal{L}_c \tag{8.19}$$

$$\delta_i = 0, \forall i \in \mathcal{N}_o; \ \delta_i \ge \varepsilon_i, \ \forall i \in \mathcal{N}_c \tag{8.20}$$

$$\varepsilon_i = 1, \ \forall i \in \{\mathcal{N}_r, \mathcal{N}_g\} \tag{8.21}$$

$$\left(\sum_{(i,j) \in \mathcal{L}} \varepsilon_j \cdot \alpha_{ij} + \sum_{(j,i) \in \mathcal{L}} \varepsilon_j \cdot \alpha_{ji} \right) / \vartheta_i \le \varepsilon_i$$
$$\le \sum_{(i,j) \in \mathcal{L}} \varepsilon_j \cdot \alpha_{ij} + \sum_{(j,i) \in \mathcal{L}} \varepsilon_j \cdot \alpha_{ji}, \ \forall i \in \mathcal{N} \backslash \{\mathcal{N}_r, \mathcal{N}_g\} \tag{8.22}$$

Notations are listed in the Nomenclature section at the end of this chapter.

The objective function (8.10) is to maximize the sum of recovered loads weighted by their priorities [19, 31]. Other objectives, e.g. to minimize the cost or number of switch actions, may be included, too [7, 12]. Here, the post-disaster MG formation model follows many related works, especially Refs. [25] and [26] that are the major benchmarks, to only maximize the restored loads. Constraints (8.1) and (8.2) are included to assure radiality and allow topological flexibilities and more flexible boundaries of MGs, etc. Constraint (8.1) can be modeled using the approaches discussed in Section 8.3. Constraints (8.11) and (8.12) state the active and reactive power balance conditions, respectively. Constraint (8.13) enforces zero power output of nodes that have no power sources. Constraint (8.14) expresses active and reactive power limits of DGs and substations. Normally DGs may absorb or inject reactive power [9], which helps avoid overvoltage and balance the MG/DS [28]. However, the power factor of some DGs can have a restricted range and only injecting reactive power is allowed. Following some benchmarks such as [18] and [25], the model here considers the DGs' positive injection of reactive power. Constraint (8.14) can be easily adjusted to consider DGs' absorption of reactive power. Constraints (8.15) and (8.16), which are relaxed for open branches, express the power flow equations for closed lines based on the LinDistFlow model. Baran and Wu [38] first proposed the DistFlow model. By ignoring quadratic line loss terms that are much smaller, it is simplified in [1] to obtain the LinDistFlow model that is used extensively in the literature (e.g. [25] and [26]) and proved to have acceptable accuracy [39]. To adopt power flow models that are more accurate, semidefinite programming (SDP) or second-order cone programming (SOCP) convexification/relaxation approaches in [9] and [40] may be used. Constraint (8.17) states the voltage magnitude's range limits. Constraint (8.18) is nonlinear but convex. It may be linearized using the method in [3], etc. It is to limit each line's apparent power by its capacity. Constraint (8.19) fixes the connection status of faulted closed and faulted open lines. Load shedding is considered, and a load is either completely shed or completely picked up [25]. For picking up a load, the associated node has to be energized and the load switch has to be closed. For shedding a load, the associated node needs to be de-energized or the load switch needs to be open. Constraint (8.20) forces picking up loads with faulted closed switches if their associated nodes are energized and forbids picking up loads with faulted open switches. The faulted closed or faulted open load and branch switches are due to the disastrous event [25]. Constraint (8.21) states that DG and substation nodes are energized by default. By examining whether a node is linked with an energized node, constraint (8.22) derives the node's energization status. Using the McCormick envelopes [41], the nonconvex and nonlinear terms, i.e. $\varepsilon_j \cdot \alpha_{ji}$ and $\varepsilon_j \cdot \alpha_{ij}$, can be linearized equivalently.

Note that the optimization model also considers regions with surviving main grid power from substations. Full restoration of those regions using power from substations may be prohibited by operational constraints after a natural disaster, and forming MGs with flexible boundaries may help attain better recovery [20]. For simplicity in explanations, in this chapter a sub-grid supplied by a substation is counted as an MG, too.

8.4.2 Tightness and Compactness

In this section, the compactness and tightness of the spanning tree constraints (8.3)–(8.9) and spanning forest constraints (8.1) and (8.2), which greatly impact the tightness and compactness of the new post-disaster MG formation model, are discussed.

By a polynomial number of constraints and variables, the LP relaxation of the spanning tree constraints (8.3)–(8.9) defines the spanning tree polytope (i.e. conv(Ω)) in the β-space. Table 8.1 compares the number of constraints and variables in different types of spanning tree constraints, including the formulation introduced in this chapter. Note that the comparison in Table 8.1 is with respect to different models or approaches of explicitly formulating the spanning tree constraint (8.1). As indicated in Table 8.1, compared to the directed cutset formulation and the subtour elimination formulation, both the tightest but with an exponential number of constraints, our presented formulation only involves a polynomial number of constraints and is also the tightest. Compared with the single-commodity flow-based model, our presented model is tighter. Compared to the primal and dual graph-based model, which only applies to planar graphs, our presented model applies to both nonplanar and planar graphs. In short, among the formulations that are applicable for both nonplanar and planar graphs and are also the tightest, our presented one is the most compact. In fact, the formulations compared in Table 8.1 cover a wide spectrum of compactness and tightness

Table 8.1 Comparing different types of spanning tree constraints.

Type of formulation	Number of binary variables	Number of continuous variables	Number of constraints	Spanning tree polytope or not
Subtour elimination	$\|\mathcal{L}\|$	0	$\mathcal{O}(2^{\|\mathcal{N}\|})$	Yes
Directed cutset	$2 \cdot \|\mathcal{L}\|$	$\|\mathcal{L}\|$	$\mathcal{O}(2^{\|\mathcal{N}\|-1})$	Yes (extended formulation)
Single-commodity flow	$\|\mathcal{L}\|$	$2 \cdot \|\mathcal{L}\|$	$\|\mathcal{N}\| + 2 \cdot \|\mathcal{L}\|$	No
Primal and dual graph-based	$4 \cdot \|\mathcal{L}\|$	0	$2 \cdot \|\mathcal{L}\|$	Yes (extended formulation)
Directed multi-commodity flow	$2 \cdot \|\mathcal{L}\|$	$2 \cdot \|\mathcal{N}\| \cdot \|\mathcal{L}\| - \|\mathcal{L}\|$	$\|\mathcal{N}\|^2 + 2 \cdot \|\mathcal{N}\| \cdot \|\mathcal{L}\| - \|\mathcal{N}\| - \|\mathcal{L}\| + 1$	Yes (extended formulation)

levels in formulating the spanning tree constraints. Practitioners and researchers can select a proper formulation based on their needs or preferences. Note that Table 8.1 just compares different spanning tree constraints for formulating constraint (8.1). Constraint (8.2) includes $|\mathcal{L}|$ more constraints and $|\mathcal{L}|$ more binary variables (i.e. $\alpha \triangleq \{\alpha_{ij}, \forall (i,j) \in \mathcal{L}\}$) in the proposed radiality model.

Next, the compactness and tightness of constraints (8.1) and (8.2), rather than solely constraint (8.1), are further discussed using the following proposition.

Proposition 8.1 *If the LP relaxation of constraint (8.1)'s explicit formulation defines the spanning tree polytope in the β-space, the proposed radiality constraints (8.1) and (8.2)'s LP relaxation defines the spanning forest polytope in the α-space. (See the proof in Appendix 8.A.)*

Proposition 8.1 indicates that the compactness and tightness characteristics of constraints (8.1) and (8.2) actually follow the explicit formulation of constraint (8.1). Therefore, with different methods or models to formulate constraint (8.1), the proposed radiality constraints (8.1) and (8.2) also cover a wide spectrum of compactness and tightness levels of formulating the spanning forest polytope. An appropriate formulation may be chosen based on one's needs, etc.

Particularly, we assume that the DS has just one substation node, so as to exploit the formulations' properties mentioned earlier. If a DS has two or substation nodes, one may just merge them into one node when formulating constraints (8.1) and (8.2), but still treat them as individual nodes when formulating operational constraints of the DS. Figure 8.5 provides an instance. Figure 8.5a shows the original DS with two substations in nodes 1 and 2. Figure 8.5b shows the transformed DS attained by merging the two substation nodes into one node. Figure 8.5c then shows a radial topology attained by using the presented radiality constraints on the transformed DS. Particularly, it is associated with the original system's topology shown in Figure 8.5d. (Operational constraints of the DS, e.g. power flow constraints and power balance conditions, are still based on Figure 8.5a,d, i.e. separately dealing with each substation node.) The above approach is correct due

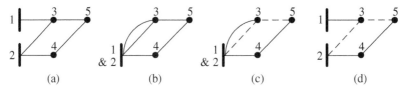

(a)   (b)   (c)   (d)

Figure 8.5 (a) The *original* example system. (b) The *transformed* system attained by *merging* substation nodes 1 and 2. (c) A radial topology of the *transformed* system obtained by enforcing the proposed radiality constraints. (d) The *original* system's topology associated with the *transformed* system's topology in (c).

to two equivalences as follows: (i) Having no circles in the transformed network is equivalent to having neither loops nor paths among substation nodes in the original network. (ii) Any topology in the original network is associated with a topology in the transformed network, and vice versa.

With constraint (8.1) appropriately formulated, the proposed formulation attains the tightest model of the system's spanning forest polytope, i.e. the convex hull of characteristic vectors of possible spanning forest topologies. The approaches in [12] and [18] result in relatively less tight models.

Constraint (8.1) may be formulated in more compact but less tight manners, too. In this regard, the resulting topology constraints (8.1) and (8.2) generally are still tighter than or as tight as those in [12] and [18], and more compact than (i.e. having fewer constraints and variables) those in [12] and [18].

8.4.3 Applicability and Application

The application of the proposed radiality model can be very straightforward, i.e. simply including constraint (8.2) to the single-commodity flow-based spanning tree constraints that are commonly used. The approaches in [18] and [12], however, need to model a virtual DC optimal power flow subproblem and its Karush–Kuhn–Tucker conditions, and introduce virtual branches and a virtual source node, respectively.

The presented radiality constraints (8.1) and (8.2) can be used in optimization problems of DSs considering reconfiguration, e.g. post-disaster MG formation. Topological and related flexibilities, therefore, are completely enabled in optimization to attain a better utilization of flexible boundaries of MGs, etc. In this regard, the approaches in [12] and [18] have some limits. For instance, the radiality constraints presented in [18] do not consider the possible separation of sub-grids in optimization, though allowing the merge of sub-grids. In fact, the approaches in [12] and [18] are not particularly constructed for post-disaster MG formation. They are more useful in co-optimizing damage repairs and DS service recovery, etc.

As aforementioned, there are two other formulations for post-disaster MG formation optimization [25, 26]. The formulation in [25] is based on the idea of clustering nodes. It designates every node to one of the DGs and imposes branch-node and connectivity constraints, etc. The formulation in [26] uses the single-commodity flow-based spanning tree constraints. It enforces every subgraph to be connected and the number of closed lines to be the number of nodes minus a given number of subgraphs.

The number of constraints and variables in different MG formation models are compared in Table 8.2. The formulation in [26] has the smallest number of constraints and variables. Compared to the formulation in [25], the presented

Table 8.2 Number of constraints and variables in different MG formation models

	Model in [25]	Model in [26]	Proposed model
Number of binary variables	$2 \cdot |\mathcal{N}| \cdot (|\mathcal{N}_r| + |\mathcal{N}_g|) + |\mathcal{N}| + |\mathcal{L}|$	$|\mathcal{N}| + |\mathcal{L}|$	$2 \cdot |\mathcal{N}| + 3 \cdot |\mathcal{L}|$
Number of continuous variables	$4 \cdot |\mathcal{N}| \cdot (|\mathcal{N}_r| + |\mathcal{N}_g|)$	$|\mathcal{N}| + 3 \cdot (|\mathcal{N}_r| + |\mathcal{N}_g|) + 3 \cdot |\mathcal{L}|$	$2 \cdot |\mathcal{N}| \cdot |\mathcal{L}| + |\mathcal{N}| + 2 \cdot (|\mathcal{N}_r| + |\mathcal{N}_g|) + |\mathcal{L}|$
Number of constraints	$(11 \cdot |\mathcal{N}| - 2 \cdot |\mathcal{N}_r| - 2 \cdot |\mathcal{N}_g| + 1) \cdot (|\mathcal{N}_r| + |\mathcal{N}_g|) + 2 \cdot |\mathcal{N}| + |\mathcal{N}_o| + |\mathcal{N}_c| + |\mathcal{L}| + |\mathcal{L}_o| + |\mathcal{L}_c|$	$3 \cdot |\mathcal{N}| + 3 \cdot (|\mathcal{N}_r| + |\mathcal{N}_g|) + |\mathcal{N}_o| + |\mathcal{N}_c| + 5 \cdot |\mathcal{L}| + |\mathcal{L}_o| + |\mathcal{L}_c| + 1$	$|\mathcal{N}|^2 + 2 \cdot |\mathcal{N}| \cdot |\mathcal{L}| + 3 \cdot |\mathcal{N}| + 2 \cdot (|\mathcal{N}_r| + |\mathcal{N}_g|) + |\mathcal{N}_o| + |\mathcal{N}_c| + 3 \cdot |\mathcal{L}| + |\mathcal{L}_o| + |\mathcal{L}_c| + 1$

Table 8.3 Comparing different MG formation models

	Model in [25]	Model in [26]	Proposed model
Applicable systems	Radial	Radial or meshed	Radial or meshed
Radiality constraints	—	Single-commodity flow-based	Proposed radiality constraints
Allocation of DGs into MGs	One DG in each MG	One DG in each MG	Flexible
Number of MGs	Fixed	Fixed	Flexible
De-energized islands	Not allowed	Not allowed	Allowed
Loads with faulted closed switches	Forced to pick up	Forced to pick up	Flexible

formulation has less binary variables but more continuous variables and more constraints. However, as case studies in Section 8.5 will show, the presented formulation has the shortest computation time on average.

Specifically, the largest difference between the presented radiality formulation and other existing formulations in the literature is that, the presented model completely enables topological flexibilities of DSs and MGs, while other models do not have this capacity. By including the presented radiality formulation in the new post-disaster MG formation optimization model, some other related flexibilities (e.g. intentional formation of de-energized islands, flexible pick-up of loads with faulted closed switches, flexible allocation of DGs into MGs, and therefore flexible boundaries of MGs) are further enabled via the complete utilization of topological flexibilities.

Table 8.3 further compares the proposed model with the two existing post-disaster MG formation models in the literature. The summaries in the table are self-explanatory. Brief explanations are given as follows:

1. The model in [25] is designed for radial systems. It is extended in [31] to deal with meshed systems. Still, the extended model does not eliminate loops. Thus, it may be used for the bulk power system, but does not apply to meshed DSs, which have to open some branches to be operated in radial topologies. Both model in [26] and the model proposed in this chapter can handle meshed DSs, as radiality constraints are included to avoid loops when reconfiguring the network. Single-commodity flow-based radiality constraints are used in [26], while the model here adopts the proposed radiality constraints to fully enable topological flexibilities and flexible boundaries of MGs, etc.

2. Both post-disaster MG formation models in [25] and [26] allocate one DG to each MG. Therefore, the number of MGs is actually fixed and equal to the number of DGs and/or other power sources. As for the proposed model here, the number of DGs designated to different MGs, and the resulting number of MGs, are flexible. Thus, it can also be used for dynamic MG formation with dynamically flexible boundaries of MGs, which involves adaptive merge and/or separation of MGs when damaged parts of the DS are sequentially repaired. Such flexibilities introduce many benefits. For example, larger MGs can be formed to better match DGs with different-sized loads, so as to enhance capacity utilization rates of DGs and the restoration of critical loads.

3. Both models in [25] and [26] energize all nodes included in $\mathcal{N}$. Thus, their models cannot consider the nodes in load islands without power sources and isolated by faulted open branches. Discrete/non-dispatchable loads at nodes $\mathcal{N}_c$ are also forced to be picked up in their models. The model here can consider such load islands and can be easily extended to optimize the allocation of mobile power sources in such islands and the merge among islands after faulted open branches is repaired. The model here can also intentionally form de-energized islands in optimization, enabling more flexible pick-up of the loads at nodes $\mathcal{N}_c$. This flexibility can be critical, as some DSs may have a large portion of loads not equipped with switches, causing $\mathcal{N}_c$ to be a large set. In [19], with additional binary variables, the model in [26] is modified to permit more flexible allocation of DGs into MGs. However, it does not fully enable topological and some related flexibilities. For example, it still requires energizing all nodes. In general, the model proposed in this chapter enables more flexibilities, including the flexible boundaries of MGs.

Note that in some cases, the post-disaster MG formation models in [25] and [26] and the model here can be infeasible, i.e. having no feasible solutions. Still, *model* infeasibility does not indicate *system* infeasibility. Actually, the system is always feasible, since it can just turn off all generations and shed all loads in the most extreme case. However, the models do not include all flexibilities of the system. Figure 8.6 gives an example to illustrate the model infeasibility issue. In Figure 8.6, the damaged branch $(1, 2)$ disconnects nodes 2, 3, and 4 from the substation node 1. In this example, the models in [25] and [26] require nodes 2, 3, and 4 to be connected as a single MG. Node 3 thus is enforced to be energized, and its load is then enforced to be picked up as its load switch is faulted closed. This leads to infeasibility, as $\overline{p}_2^g < p_3^c$. In contrast, the proposed model is feasible in this case, as the model can choose to open branch $(2, 3)$ to shed the load at node 3. That is, the proposed model has more flexibilities in determining the boundary of the MG to be formed. Still, if branch $(2, 3)$ is faulted closed, the proposed model will also be infeasible. There can be other infeasible scenarios.

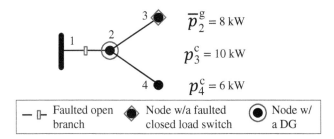

$$\overline{p}_2^g = 8 \text{ kW}$$

$$p_3^c = 10 \text{ kW}$$

$$p_4^c = 6 \text{ kW}$$

— ⊐- Faulted open branch	◆ Node w/a faulted closed load switch	⊙ Node w/ a DG

Figure 8.6 A small example to illustrate *infeasibility* of the MG formation models ($p_2^c = 0$).

The proposed post-disaster MG formation model and those in [25, 26] can consider microturbine-based DGs and renewable DGs, etc. For microturbine-based DGs, a reliable fuel supply instructed by the US Federal Emergency Management Agency [42] and other authorities can be achieved via gas networks, or by equipping DGs with fuel storage tanks [42, 43], preallocating fuel [13], and dispatching fuel trucks [44]. For renewable DGs, the adopted model can be extended to handle their intermittency and uncertainty. Specifically, the advantages of the proposed radiality model and MG formation model do not depend on the types of DGs.

In general, as indicated in Table 8.3, the proposed post-disaster MG formation model is more adaptive and allows more flexibilities including the flexible boundaries of MGs. Specifically, with the proposed radiality constraints fully enabling topological and some related flexibilities, the proposed MG formation model achieves higher effectiveness than other models.

8.5 Case Studies

In this section, the proposed post-disaster MG formation model is demonstrated on two test systems. A computer with an Intel i5-4278U processor and 8GB of memory is used. Involved MIP problems are solved by Gurobi 7.5.2 with the default settings.

8.5.1 Illustration on a Small Test System

In this and the next test systems, microturbine-based DGs are considered [45]. They can also be truck-mounted emergency generators (i.e. transportable DGs) pre-positioned in DSs [20]. Following [26, 46], etc., it is assumed that the continuous operating time of DGs is sufficiently long here. To use renewable DGs in post-disaster MG formation, robust or stochastic optimization methods may be necessary to handle their intermittency and uncertainty. It is left for future investigation.

Figure 8.7 depicts an illustrative case based on a scenario with 27 faults in total. The model in [25] does not apply to meshed DSs. Thus, as indicated in Figure 8.7a, it does not consider the normally open branches. It forms six MGs energized by the six DGs, respectively. As the model in [26] can handle meshed DSs, it gets better results via further reconfiguration involving those normally open lines. Figure 8.7b shows that it restores more loads by forming six larger MGs, which are essentially a 6-tree. By contrast, as shown in Figure 8.7c, the post-disaster MG formation model proposed in this chapter forms a 7-tree, i.e. 4 MGs and three load islands. That is, the proposed model better utilizes the flexible boundaries in forming MGs. Specifically, MG 4 and MG 6 in Figure 8.7b are merged into a single MG in Figure 8.7c, so that the loads at nodes 31 and 33 can also be restored. MG 2 in Figure 8.7b is separated into an MG and a load island in Figure 8.7c, so that node 23 is not energized and its load is not forced to be picked up. The models in [25] and [26] and the proposed model have DG capacity utilization rates of 55.6%, 67.4%, and 95.4%, respectively. They restore 1500, 1820, and 2575 kW loads, respectively. The proposed model achieves a more coordinated matching among the different-sized DGs and loads, and thus attains better service restoration. Generally, as the proposed radiality constraints can fully enable topological and some other related flexibilities of the DS, the proposed MG formation model has extended feasibility and enhanced optimality.

To establish a superiority of the proposed radiality constraints and post-disaster MG formation model, 10 000 cases based on randomly generated scenarios of DS faults are run. Table 8.4 shows that the proposed model has far less infeasible cases. In many scenarios, it finds a feasible operating point of the DS, while the models in [25] and [26] return infeasibility. Such results verify that the proposed MG formation model has extended feasibility as the proposed radiality constraints fully enable topological and some related flexibilities, including the flexible boundaries of MGs. Table 8.4 also reports the average computation time of the models in [25] and [26] and the model here. As indicated in the comparison, although the proposed model has the largest search space, its average computation time is the shortest. Thus, its enlarged feasible set may possess more computationally tractable characteristics. Nevertheless, the average computation time of all three models are close to each other in this case.

As in Table 8.5, the proposed post-disaster MG formation model has the highest average/median/minimum restored loads and the smallest standard deviation. The same maximum value of restored loads of all three models corresponds to the case(s) with all loads restored. On average, the proposed model restores 15.0% and 8.4% more loads than the models in [25] and [26], respectively.

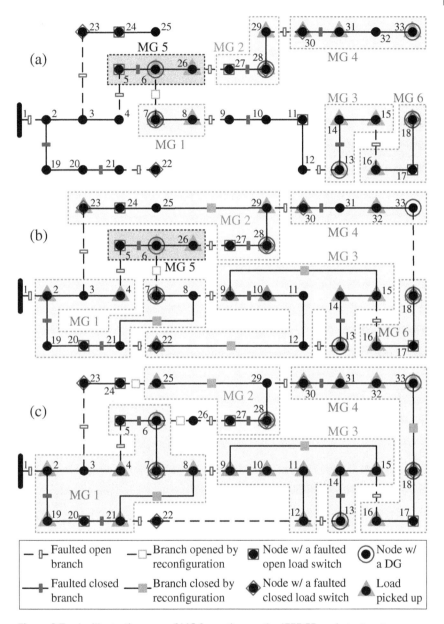

Figure 8.7 An illustrative case of MG formation on the IEEE 33-node test system: (a) Using the model in [25]. (b) Using the model in [26]. (c) Using the proposed model.

Table 8.4 Computation time and the number of infeasible cases (IEEE 33-node test system).

	Model in [25]	Model in [26]	Proposed model
Average computation time (s)	0.56	0.75	0.49
Infeasible cases	212/10 000	272/10 000	2/10 000

Table 8.5 Summary statistics for restored loads (IEEE 33-node test system).

	Restored loads (kW)					Box plots
	Average	Standard	Maximum	Median	Minimum	Box plots
1: Model in [25]	2686	531	3715	2750	1175	
2: Model in [26]	2849	495	3715	2915	1380	
3: Proposed model	3088	393	3715	3165	1425	

Actually, the proposed model performs better or equally well in all 10,000 cases. The histograms in Figure 8.8a depict the outperformance. Such results validate the enhanced optimality of the proposed model. Here, one major reason for its superiority is the more coordinated matching among DGs and loads, taking advantage of more flexible boundaries of MGs. Table 8.6 indicates that the average DG capacity utilization rate of the proposed model is much higher than those of the models in [25] and [26]. The smaller minimum and larger standard deviation of the proposed model are due to the cases with the substation contributing much power injection for service restoration.

The single-commodity flow-based model is used to formulate constraint (8.1) in all previous cases. Here, for comparison, the directed multicommodity flow-based model is used instead. Consequently, the proposed radiality constraints and post-disaster MG formation model become tighter, though less compact. The revised model is run on the same 10,000 scenarios. As indicated in Table 8.7, N_m is much smaller than N_s, both on average and in 95.8% of the cases. The histograms in Figure 8.8b also detail the extra explored nodes in the B&C search tree of the less tight model. Although T_s is slightly shorter than T_m on average, T_m is much shorter than T_s in 10.9% cases. Specifically, in those cases, T_s and T_m are 1.92 and 0.69 seconds on average, respectively; N_s and N_m are 815 and 186 on average, respectively. In general, the computation time of the revised tighter model is more consistent. It also reduces the computation time

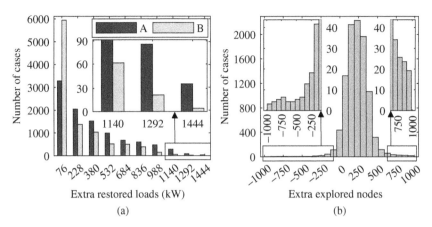

Figure 8.8 (a) Histograms of the extra restored loads of the proposed MG formation model, compared with the model in [25] (A), and compared with the model in [26] (B). (b) Histograms of the extra explored nodes in the B&C search tree of the proposed MG formation model using the single-commodity flow-based method to formulate constraint (8.1), compared with that using the directed multicommodity flow-based method (IEEE 33-node test system).

Table 8.6 Summary statistics for DG capacity utilization rate (IEEE 33-node test system).

	DG capacity utilization rate (%)					Box plots	
	Average	Standard	Maximum	Median	Minimum	Box plots	
1: Model in [25]	64.4	9.8	88.4	65.1	29.7		
2: Model in [26]	60.7	15.1	94.8	61.0	15.3		
3: Proposed model	74.1	15.7	100	74.4	6.6		

in many cases that require the less tight model to explore much more nodes in the B&C search tree. That is, the proposed radiality constraints can satisfy the need for tighter formulations of DS and MG reconfiguration-related optimization problems.

8.5.2 Results on a Large Test System

Figure 8.9 provides an illustrative case based on a 60-fault scenario using the proposed post-disaster MG formation model. It forms seven MGs powered by eight DGs, and a sub-grid powered by the substation. The DG capacity utilization

Table 8.7 Computation time and the number of explored nodes in the B&C search tree (IEEE 33-node test system).

		Number of cases with		
Average T_s	Average T_m	$T_s < T_m$	$T_s = T_m$	$T_s > T_m$
0.49 s	0.56 s	8911/10 000	0/10 000	1089/10 000
		Number of cases with		
Average N_s	Average N_m	$N_s < N_m$	$N_s = N_m$	$N_s > N_m$
407	143	402/10 000	21/10 000	9577/10 000

T_s/T_m (resp. N_s/N_m): Computation time (resp. the number of ex-plored nodes in the B&C search tree) of the proposed MG formation model using the single-commodity flow-based method/directed multi-commodity flow-based method to formulate constraint (8.1).

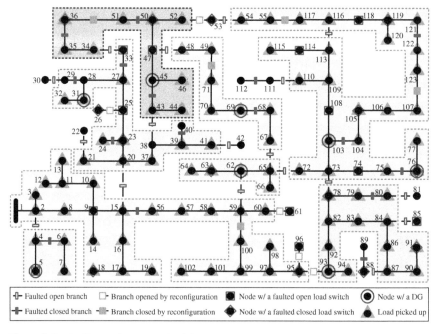

Figure 8.9 An illustrative case of MG formation on the IEEE 123-node test system (using the proposed model).

Table 8.8 Computation time and the number of infeasible cases (IEEE 123-node test system).

	Model in [25]	Model in [26]	Proposed model
Average computation time (s)	13.41	4.36	1.79
Infeasible cases	109/1 0000	196/10 000	2/10 000

Table 8.9 Summary statistics for restored loads (IEEE 123-node test system).

	Restored loads (kW)					Box plots
	Average	Standard	Maximum	Median	Minimum	Boxplots
1: Model in [25]	2917	482	4150	2915	1555	
2: Model in [26]	3008	524	4350	3010	1265	
3: Proposed model	3189	494	4450	3195	1665	

rate is 75.0%, and 3730 kW loads are restored. Again, 10,000 cases based on random scenarios of DS faults are run. Tables 8.8–8.11 and Figure 8.10 show results similar to those of the previous system. That is, the superiority of the proposed radiality constraints and MG formation model is also established on this larger system. For example, the proposed model matches DGs and loads in a more coordinated manner, thus achieving better service restoration. For space limit, this section does not go into further details. In short, the proposed MG formation model has extended feasibility and enhanced optimality, as the proposed radiality constraints fully enable topological and some related flexibilities of DSs, including more flexible boundaries in forming MGs. Besides, as shown in Table 8.8, the proposed model again has a shorter average computation time than the models in [25] and [26] for this larger system.

Table 8.10 Summary statistics for DG capacity utilization rate (IEEE 123-node test system)

	DG capacity utilization rate (%)					Box plots
	Average	Standard	Maximum	Median	Minimum	Box plots
1: Model in [25]	63.0	9.1	90.5	63.1	30.8	
2: Model in [26]	63.3	10.6	97.4	63.1	30.1	
3: Proposed model	70.9	12.4	100	70.7	24.9	

Table 8.11 Computation time and the number of explored nodes in the B&C search tree (IEEE 123-node test system).

		Number of cases with		
Average T_s	Average T_m	$T_s < T_m$	$T_s = T_m$	$T_s > T_m$
1.79 s	1.54 s	7348/10 000	0/10 000	2652/10 000

		Number of cases with		
Average N_s	Average N_m	$N_s < N_m$	$N_s = N_m$	$N_s > N_m$
496	71	108/10 000	134/10 000	9758/10 000

T_s/T_m (resp. N_s/N_m): Computation time (resp. the number of ex-plored nodes in the B&C search tree) of the proposed MG formation model using the single-commodity flow-based method/directed multi-commodity flow-based method to formulate constraint (8.1).

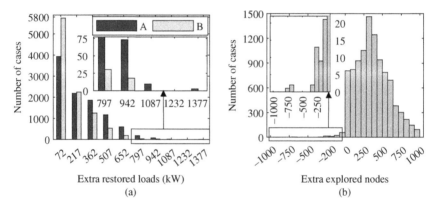

Figure 8.10 (a) Histograms of the extra restored loads of the proposed MG formation model, compared with the model in [25] (A), and compared with the model in [26] (B). (b) Histograms of the extra explored nodes in the B&C search tree of the proposed MG formation model using the single-commodity flow-based method to formulate constraint (8.1), compared with that using the directed multicommodity flow-based method. (IEEE 123-node test system).

It is worth mentioning that, as shown in Table 8.11, T_m is averagely shorter than T_s for this larger system. Specifically, in the cases with $T_m < T_s$, T_m and T_s are 1.26 and 5.59 seconds, on average, respectively; N_m and N_s are 89 and 937, on average, respectively. That is, the computation time of the tighter version of the proposed post-disaster MG formation model is not only shorter on average but also more consistent. It greatly reduces the computation time especially for cases requiring the less tight model to explore much more B&C search tree nodes.

Thus, the tighter version of the proposed radiality constraints can allow both more flexible DS operation and more efficient computation for reconfiguration-related optimization problems, e.g. MG formation with flexible boundaries here.

8.5.3 LinDistFlow Model Accuracy

Using the LinDistFlow model instead of the DistFlow model improves the computational efficiency. As the neglected higher-order power loss terms are typically much smaller than power flows, this simplification generally only introduces a small error. Still, this section validates its effectiveness using a method similar to that in [47]. Specifically, after obtaining a solution by the post-disaster MG formation model with LinDistFlow equations, the LinDistFlow equations are replaced with DistFlow equations, and the values of network topology and load pick-up decisions (i.e. α_{ij} and δ_i) are fixed according to the obtained solution. The resulting new model is called the *validation model*, which is solved using IPOPT [48] to obtain an operating point (i.e. v_i, p_i^g, q_i^g, P_{ij} and Q_{ij}). If the validation model is feasible (resp. infeasible), it means that in this case, using the LinDistFlow model attains a feasible (resp. infeasible) restoration plan.

Among all the cases (i.e. three different MG formation models tested on two systems with 10 000 scenarios of each system), only in two cases of the 33-node system and one case of the 123-node system, the proposed MG formation model with LinDistFlow equations obtains a solution that results in infeasibility of the validation model. All these three cases have an 100% DG capacity utilization rate. That is, no margins of DG capacities are available to compensate for the loss. To resolve this issue, a constraint can be added to reserve a small portion of DG capacities for loss compensation. A more fundamental solution is to develop algorithms for solving the MG formation model with nonconvex DistFlow equations. SOCP and SDP relaxation/convexification methods are potentially useful here. It can be investigated in the future. In general, the validation studies in this section affirm that using the LinDistFlow model in this chapter is effective.

8.6 Summary and Conclusions

This chapter proposes a new method for formulating radiality constraints that fully enables topological and some other related flexibilities in DS and MG reconfiguration optimization problems. It is specifically applied to resilient post-disaster formation of MGs with flexible boundaries to attain extended feasibility and enhanced optimality. As verified in case studies, compared to existing MG formation models, the proposed model based on the proposed radiality constraints achieves a higher resilience enhancement via more coordinated utilization of DS and MG flexibilities and reduces the computational complexity. Exploring the effects of increased topological flexibilities on other DS and MG performance metrics can be an interesting future research direction. Cases with unbalanced DSs are also worth of future investigations. In such cases, SOCP or SDP relaxation/convexification methods can be used [49]. It is also worth studying the application of the proposed radiality model and MG formation model to multiperiod reconfiguration of large-scale DSs, which might require developing system simplification methods or parallel computing algorithms, etc. Besides, how other widely used radiality models (e.g. the spanning tree constraints in [7]) can be modified to fully enable topological flexibilities is worth investigating, too.

Appendix 8.A

8.A.1 Proof of Theorem 8.1

Proof: Assume a κ-tree denoted as $\mathcal{G}_\kappa$ to be a subgraph of a graph $\mathcal{G}$. As $\mathcal{G}$ is connected, there exists at least one edge in $\mathcal{G}$ that can link the κth component of $\mathcal{G}_\kappa$ to another component of $\mathcal{G}_\kappa$. Adding this edge to $\mathcal{G}_\kappa$, which does not create cycles, $\mathcal{G}_\kappa$ becomes a $(\kappa - 1)$-tree denoted as $\mathcal{G}_{\kappa-1}$. Repeating this process, $\mathcal{G}_\kappa$ ends up to be a 1-tree denoted as $\mathcal{G}_1$. Obviously, the spanning forest $\mathcal{G}_\kappa$ is a subgraph of the spanning tree $\mathcal{G}_1$, which is a subgraph of $\mathcal{G}$. This completes the proof. □

8.A.2 Proof of Proposition 8.1

Proof: Assume that the LP relaxation of constraints (8.1) and (8.2) has a fractional vertex (α^*, β^*) in the (α, β)-space. Further assume that β^* has fractional entries. Thus, β^* can be represented by an integer vertex β^m and another point β^n of the spanning tree polytope: $\beta^* = \xi \cdot \beta^m + (1 - \xi) \cdot \beta^n$ with $0 < \xi < 1$. For (i,j) with $\beta^m_{ij} = 0$ or $\beta^m_{ij} = 1$, we have $\beta^*_{ij} = (1 - \xi) \cdot \beta^n_{ij}$ and $\beta^*_{ij} = \xi + (1 - \xi) \cdot \beta^n_{ij}$, respectively. As $0 \le \beta^n_{ij} \le 1$ and $0 \le \alpha^*_{ij} \le \beta^*_{ij} \le 1$, we have $0 \le \frac{\alpha^*_{ij}}{1-\xi} \le \beta^n_{ij}$ for (i,j) with $\beta^m_{ij} = 0$, and have $0 \le \frac{\alpha^*_{ij} - \xi \cdot \chi_{ij}}{1-\xi} \le \beta^n_{ij}$ with $\chi_{ij} \in \left[\frac{\alpha^*_{ij} - \beta^*_{ij} + \xi}{\xi}, \frac{\alpha^*_{ij}}{\xi} \right] \cap [0, 1]$ for (i,j) with $\beta^m_{ij} = 1$. Then, we can construct α^m and α^n to have $\alpha^m_{ij} = 0$ and $\alpha^n_{ij} = \frac{\alpha^*_{ij}}{1-\xi}$ for (i,j) with $\beta^m_{ij} = 0$, and have $\alpha^m_{ij} = \chi_{ij}$ and $\alpha^n_{ij} = \frac{\alpha^*_{ij} - \xi \cdot \chi_{ij}}{1-\xi}$ for (i,j) with $\beta^m_{ij} = 1$, so that $(\alpha^*, \beta^*) = \xi \cdot (\alpha^m, \beta^m) + (1 - \xi) \cdot (\alpha^n, \beta^n)$. This contradicts the assumption of (α^*, β^*) being a vertex. The case with β^* being a vertex of conv(Ω) can be analyzed similarly. Thus, vertices of the projection of constraints (8.1) and (8.2)'s LP relaxation onto the α-space are 0–1 incidence vectors of spanning forests of the DS. □

Nomenclature of Spanning Tree Constraints

Indices and Sets

i, j	indices of DS nodes
i_r	index of the substation node
k	index of fictitious commodities
$\mathcal{N}$	set of all DS nodes
$\mathcal{L}$	set of all DS branches

Variables

f_{ij}^k	flow of commodity k from node i to node j
λ_{ij}	binary, 1 if arc (i, j) is included in the directed spanning tree, 0 otherwise
β_{ij}	binary, 1 if the fictitious connection status of branch (i, j) is closed, 0 if open

Nomenclature of MG Formation Model

Sets

$\mathcal{N}_r, \mathcal{N}_g$	set of substation nodes and DG nodes, respectively
$\mathcal{N}_o, \mathcal{N}_c$	set of nodes with faulted open load switches or faulted closed load switches, respectively
$\mathcal{L}_o, \mathcal{L}_c$	set of faulted open branches and faulted closed branches, respectively

Parameters

p_i^c, q_i^c	real and reactive power demand of the load at node i
$\overline{p}_i^g, \overline{q}_i^g$	real and reactive power capacity of the power source at node i
$\overline{v}_i, \underline{v}_i$	maximum and minimum squared voltage magnitude of node i
r_{ij}, x_{ij}	resistance and reactance of branch (i, j)
$\overline{S}_{ij}$	apparent power capacity of branch (i, j)
ω_i	priority weight of the load at node i
ϑ_i	number of branches starting or ending with node i
M	a large enough positive number

Variables

δ_i	binary, 1 if the load at node i is picked up, 0 otherwise
ε_i	binary, 1 if node i is energized, 0 otherwise
α_{ij}	binary, 1 is branch (i, j) is closed, 0 if open

p_i^g, q_i^g real and reactive power output of the power source at node i
v_i squared voltage magnitude of node i
P_{ij}, Q_{ij} real and reactive power flow on branch (i,j)

References

1 M. E. Baran and F. F. Wu, "Network reconfiguration in distribution systems for loss reduction and load balancing," *IEEE Transactions on Power Delivery*, vol. 4, no. 2, pp. 1401–1407, 1989.

2 K. Chen, W. Wu, B. Zhang, S. Djokic, and G. P. Harrison, "A method to evaluate total supply capability of distribution systems considering network reconfiguration and daily load curves," *IEEE Transactions on Power Systems*, vol. 31, no. 3, pp. 2096–2104, 2016.

3 X. Chen, W. Wu, and B. Zhang, "Robust restoration method for active distribution networks," *IEEE Transactions on Power Systems*, vol. 31, no. 5, pp. 4005–4015, 2016.

4 M. Lavorato, J. F. Franco, M. J. Rider, and R. Romero, "Imposing radiality constraints in distribution system optimization problems," *IEEE Transactions on Power Systems*, vol. 27, no. 1, pp. 172–180, 2012.

5 J.-Y. Fan, L. Zhang, and J. D. McDonald, "Distribution network reconfiguration: Single loop optimization," *IEEE Transactions on Power Systems*, vol. 11, no. 3, pp. 1643–1647, 1996.

6 E. R. Ramos, A. G. Expósito, J. R. Santos, and F. L. Iborra, "Path-based distribution network modeling: Application to reconfiguration for loss reduction," *IEEE Transactions on Power Systems*, vol. 20, no. 2, pp. 556–564, 2005.

7 R. A. Jabr, R. Singh, and B. C. Pal, "Minimum loss network reconfiguration using mixed-integer convex programming," *IEEE Transactions on Power Systems*, vol. 27, no. 2, pp. 1106–1115, 2012.

8 H. Ahmadi and J. R. Martí, "Mathematical representation of radiality constraint in distribution system reconfiguration problem," *International Journal of Electrical Power & Energy Systems*, vol. 64, pp. 293–299, 2015.

9 M. Farivar, C. R. Clarke, S. H. Low, and K. M. Chandy, "Inverter VAR control for distribution systems with renewables," in *2011 IEEE International Conference on Smart Grid Communications (SmartGridComm)*, pp. 457–462, Brussels, Belgium, Oct. 2011.

10 S. Lei, J. Wang, and Y. Hou, "Remote-controlled switch allocation enabling prompt restoration of distribution systems," *IEEE Transactions on Power Systems*, vol. 33, no. 3, pp. 3129–3142, 2018.

11 A. Arif, Z. Wang, J. Wang, and C. Chen, "Power distribution system outage management with co-optimization of repairs, reconfiguration, and DG dispatch," *IEEE Transactions on Smart Grid*, vol. 9, no. 5, pp. 4109–4118, 2018.

12 A. Arif, S. Ma, and Z. Wang, "Dynamic reconfiguration and fault isolation for a self-healing distribution system," in *IEEE/PES Transmission and Distribution Conference & Exposition*, pp. 1–5, Denver, CO, USA, Apr. 2018.

13 H. Gao, Y. Chen, S. Mei, S. Huang, and Y. Xu, "Resilience-oriented pre-hurricane resource allocation in distribution systems considering electric buses," *Proceedings of the IEEE*, vol. 105, pp. 1214–1233, 2017.

14 S. Lei, Y. Hou, F. Qiu, and J. Yan, "Identification of critical switches for integrating renewable distributed generation by dynamic network reconfiguration," *IEEE Transactions on Sustainable Energy*, vol. 9, no. 1, pp. 420–432, 2018.

15 S. A. Arefifar, Y. A.-R. I. Mohamed, and T. El-Fouly, "Optimized multiple microgrid-based clustering of active distribution systems considering communication and control requirements," *IEEE Transactions on Industrial Electronics*, vol. 62, no. 2, pp. 711–723, 2015.

16 Y. Kim, J. Wang, and X. Lu, "A framework for load service restoration using dynamic change in boundaries of advanced MGs with synchronous-machine DGs," *IEEE Transactions on Smart Grid*, vol. 9, no. 4, pp. 3676–3690, 2018.

17 S. D. Manshadi and M. E. Khodayar, "Expansion of autonomous microgrids in active distribution networks," *IEEE Transactions on Smart Grid*, vol. 9, no. 3, pp. 1878–1888, 2018.

18 S. Ma, S. Li, Z. Wang, and F. Qiu, "Resilience-oriented design of distribution systems," *IEEE Transactions on Power Systems*, vol. 34, no. 4, pp. 2880–2891, 2019.

19 Z. Bie, Y. Lin, G. Li, and F. Li, "Battling the extreme: A study on the power system resilience," *Proceedings of the IEEE*, vol. 105, no. 7, pp. 1253–1266, 2017.

20 S. Lei, J. Wang, C. Chen, and Y. Hou, "Mobile emergency generator pre-positioning and real-time allocation for resilient response to natural disasters," *IEEE Transactions on Smart Grid*, vol. 9, no. 3, pp. 2030–2041, 2018.

21 Y. Wang, C. Chen, J. Wang, and R. Baldick, "Research on resilience of power systems under natural disasters–A review," *IEEE Transactions on Power Systems*, vol. 31, no. 2, pp. 1604–1613, 2016.

22 G. Morales-España, J. M. Latorre, and A. Ramos, "Tight and compact MILP formulation of start-up and shut-down ramping in unit commitment," *IEEE Transactions on Power Systems*, vol. 28, no. 2, pp. 1288–1296, 2013.

23 B. Yan, P. B. Luh, T. Zheng, D. A. Schiro, M. A. Bragin, F. Zhao, J. Zhao, and I. Lelic, "A systematic formulation tightening approach for unit commitment problems," *IEEE Transactions on Power Systems*, vol. 35, no. 1, pp. 782–794, 2020.

24 A. J. Miller and L. A. Wolsey, "Tight formulations for some simple mixed integer programs and convex objective integer programs," *Mathematical Programming: Series A and B*, vol. 98, no. 1, pp. 73–88, 2003.

25 C. Chen, J. Wang, F. Qiu, and D. Zhao, "Resilient distribution system by microgrids formation after natural disasters," *IEEE Transactions on Smart Grid*, vol. 7, no. 2, pp. 958–966, 2016.

26 T. Ding, Y. Lin, G. Li, and Z. Bie, "A new model for resilient distribution systems by microgrids formation," *IEEE Transactions on Power Systems*, vol. 32, pp. 4145–4147, 2017.

27 T. Ding, Y. Lin, Z. Bie, and C. Chen, "A resilient microgrid formation strategy for load restoration considering master-slave distributed generators and topology reconfiguration," *Applied Energy*, vol. 199, pp. 205–216, 2017.

28 L. Bai, J. Wang, C. Wang, C. Chen, and F. Li, "Distribution locational marginal pricing for congestion management and voltage support," *IEEE Transactions on Power Systems*, vol. 33, no. 4, pp. 4061–4073, 2018.

29 W. Cao, J. Wu, N. Jenkins, C. Wang, and T. Green, "Operating principle of soft open points for electrical distribution network operation," *Applied Energy*, vol. 164, pp. 245–257, 2016.

30 S. Lei, C. Chen, H. Zhou, and Y. Hou, "Routing and scheduling of mobile power sources for distribution system resilience enhancement," *IEEE Transactions on Smart Grid*, vol. 10, no. 5, pp. 5650–5662, 2019.

31 K. S. A. Sedzro, A. J. Lamadrid, and L. F. Zuluaga, "Allocation of resources using a microgrid formation approach for resilient electric grids," *IEEE Transactions on Power Systems*, vol. 33, no. 3, pp. 2633–2643, 2018.

32 K. S. A. Sedzro, X. Shi, A. J. Lamadrid, and L. F. Zuluaga, "A heuristic approach to the post-disturbance and stochastic pre-disturbance microgrid formation problem," *IEEE Transactions on Smart Grid*, vol. 10, no. 5, pp. 5574–5586, 2019.

33 J. A. Bondy and U. Murthy, *Graph Theory with Applications*. New York, NY USA: Elsevier, 1976.

34 IEEE PES Power System Analysis, Computing and Economics Committee, *IEEE 13 Node Test Feeder*. Sep. 2010.

35 H. N. Gabow and E. W. Myers, "Finding all spanning trees of directed and undirected graphs," *SIAM Journal on Computing*, vol. 7, no. 3, pp. 280–287, 1978.

36 J. C. Williams, "A linear-size zero-one programming model for the minimum spanning tree problem in planar graphs," *Networks*, vol. 39, no. 1, pp. 53–60, 2002.

37 H. Validi and A. Buchanan, "A note on 'a linear-size zero-one programming model for the minimum spanning tree problem in planar graphs'," *Networks*, vol. 73, no. 1, pp. 135–142, 2019.

38 M. E. Baran and F. F. Wu, "Optimal capacitor placement on radial distribution systems," *IEEE Transactions on Power Delivery*, vol. 4, no. 1, pp. 725–734, 1989.

39 K. Turitsyn, P. Sulc, S. Backhaus, and M. Chertkov, "Distributed control of reactive power flow in a radial distribution circuit with high photovoltaic penetration," in *IEEE Power and Energy Society General Meeting*, pp. 1–6, Minneapolis, MN USA, Jul. 2010.

40 S. H. Low, "Convex relaxation of optimal power flow–Part I: Formulations and equivalence," *IEEE Transactions on Control of Network Systems*, vol. 1, no. 1, pp. 15–27, 2014.

41 G. P. McCormick, "Computability of global solutions to factorable nonconvex programs: Part I–Convex underestimating problems," *Mathematical Programming*, vol. 10, no. 1, pp. 147–175, 1976.

42 Federal Emergency Management Agency, Mitigation Assessment Team Report: Hurricane Sandy in New Jersey and New York. 2013.

43 Global Power Supply, LLC. (USA), Mobile Diesel Generators for Rent or Lease. 2018.

44 G. G. Brown, C. J. Ellis, G. W. Graves, and D. Ronen, "Real-time wide area dispatching of mobil tank trucks," *Interfaces*, vol. 17, no. 1, pp. 107–120, 1987.

45 W. Huang, J. Zhang, Z. Wu, and M. Niu, "Dynamic modelling and simulation of a micro-turbine generation system in the microgrid," in *IEEE International Conference on Sustainable Energy Technologies*, pp. 345–350, Singapore, Nov. 2008.

46 S. Iwai, T. Kono, M. Hashiwaki, and Y. Kawagoe, "Use of mobile engine generators as source of back-up power," in *IEEE 31st International Telecommunications Energy Conference*, pp. 1–6, Incheon, South Korea, Oct. 2009.

47 S. Lei, C. Chen, Y. Li, and Y. Hou, "Resilient disaster recovery logistics of distribution systems: Co-optimize service restoration with repair crew and mobile power source dispatch," *IEEE Transactions on Smart Grid*, vol. 10, no. 6, pp. 6187–6202, 2019.

48 A. Wächter and L. T. Biegler, "On the implementation of an interior-point filter line-search algorithm for large-scale nonlinear programming," *Mathematical Programming*, vol. 106, no. 1, pp. 25–57, 2006.

49 R. Zafar, J. Ravishankar, J. E. Fletcher, and H. R. Pota, "Optimal dispatch of battery energy storage system using convex relaxations in unbalanced distribution grids," *IEEE Transactions on Industrial Informatics*, vol. 16, no. 1, pp. 97–108, 2020.

9

Microgrids with Mobile Power Sources for Service Restoration

Mobile power sources (MPSs), including electric vehicle (EV) fleets, truck-mounted mobile emergency generators (MEGs) and mobile energy storage systems (MESSs) have great potential to improve distribution grid resilience against natural disasters. Nevertheless, the dispatch of MPSs is not well studied. This chapter presents a two-stage framework for resilient routing and scheduling of MPSs. In the first stage, i.e. prior to the natural disaster, MPSs are pre-positioned in the distribution grid to enable prompt pre-restoration of electric service, so as to improve the survivability of power supply to critical loads. Distribution network is also proactively reconfigured into a less stressed or impacted state. The first-stage decisions are obtained by a two-stage robust optimization model solved by the column-and-constraint generation (C&CG) algorithm. In the second stage, i.e. after the disaster, MPSs are dynamically dispatched in the distribution grid to cooperate with conventional service recovery efforts in order to improve system recovery. A novel mixed-integer programming model, which resolves the dynamic formation of microgrids (MGs), the coupling of road and power networks, and different timescales of distribution grid operation and MPS dispatch, etc., is formulated for optimizing MPSs' dynamic dispatch. Numerical results based on the IEEE 33-node and 123-node test feeders demonstrate the proposed approach's effectiveness in routing and scheduling MPSs for enhancing distribution grid resilience.

9.1 Grid Survivability and Recovery with Mobile Power Sources

As mentioned earlier, in recent years, more frequent natural disasters, e.g. hurricanes, have led to severe power outages that are long-duration and large-scale [1–3]. The associated significant life risk and tremendous economic loss have highlighted the urgency and importance of improving power grid

Power Grid Resilience against Natural Disasters: Preparedness, Response, and Recovery, First Edition.
Shunbo Lei, Chong Wang, and Yunhe Hou.

resilience, which generally refers to the ability to withstand and rapidly recover from disruptive events [1–9]. However, conventional outage management is implemented in a passive manner. Proactive measures are not sufficient. Service recovery efforts are commonly initiated after the occurrence of a failure causing electric service interruptions to some customers. In general, those schemes are sufficiently efficient for expected high-frequency, low-impact outages (e.g. single-fault outages), but only have limited effect against natural disasters that are low-probability, but high-impact [10]. Much more proactive preparedness is needed. In fact, natural disasters including extreme weather events are the main causes of electric outages in the United States and around the world [1]. Considering the climate change, it is expected that the number of outages caused by such events will rise [1]. Moreover, the electrical grids remain vulnerable to natural disasters, which in many cases cause sustained damages of multiple components of the grid, leading to isolated areas which suffer prolonged outages [11]. Solely conventional recovery efforts cannot timely restore their power supply. Particularly, around 70% electric service interruptions are caused by contingencies in distribution systems (DSs) [12]. DSs have critical needs for novel resilience-enhancing strategies (e.g. dynamic formation of MGs), which utilize available flexibilities in a more coordinated and proactive manner to attain faster service recovery.

MPSs, including EV fleets, truck-mounted MEGs [13, 14], and MESSs [15–17], can provide spatial flexibilities to improve the resilience of DSs. Particularly, spatial flexibilities refer to the characteristic of MPSs being capable of delivering power and energy via road networks. They can serve as backup power and energy resources to ensure electric service of critical loads [3–5]. In this regard, MPSs can prevent outages to some extent, even if the system loses its access to the main grid power. Furthermore, when a severe outage occurs, the transportability of MPSs provides a new dimension of spatial flexibilities which can be delivered via road networks. Specifically, MPSs can be dispatched into the DS to recover loads by forming MGs [14, 18, 19]. For example in Figure 9.1, island 2 can be supplied with the main grid power (via the substation at node 1) only after

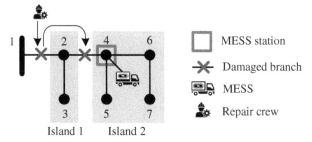

Figure 9.1 A demonstrative case illustrating resilience-enhancing effects of MPSs.

both lines $(2, 4)$ and $(1, 2)$ are repaired. For quicker service recovery, a MESS is sent and connected to node 4 for restoring critical loads in island 2, which acts as a MG. That is, while conventional restoration has to wait for the repair of damaged components and/or the recovery of the main grid, quicker restoration can be achieved by delivering MPSs to outage islands via road networks. In fact, the access to damaged components can be impeded by the natural disaster's aftermath, leading to delayed repair and prolonged electric outages [3, 4]. On the other hand, because of the finely meshed structure and redundancy, road networks are generally resilient in terms of the availability and survivability of origin-destination paths [20]. Charging stations and MESS stations normally also have more reliable access. In this regard, road networks support critical loads by providing paths for transporting spatial flexibilities to outage islands which can be restored to act as MGs.

Nevertheless, the dispatch of MPSs, which essentially is a vehicle routing and scheduling problem coupled with the service restoration problem of DSs, is insufficiently studied. For their great potential to improve reliability and resilience against natural disasters, the utilization of EVs, MEGs, and MESSs to support outage loads in restoration is reported or recommended in [5, 6, 7, 8, 21, 22], [3, 4, 23] and [15, 24], respectively. For example, in [5], emergency backup is identified as a reliability/resilience measure provided by EVs' feeding power back to the grid (e.g. via charging stations) during outages. In [8], EVs are expected to enhance electric service security by providing backup power and energy supply. To this end, using EVs, MEGs, and MESSs in restoration is investigated in [25, 26, 27, 28, 29, 30, 31], [13, 14, 32] and [17, 33], respectively. Generally, the literature concentrate on EVs' effects in improving electric service reliability against common outage events, but contain little works on the routing and scheduling of EVs for resilient response against natural disasters. Only in [30], electric buses, which can be classified as medium-duty EVs, are pre-dispatched in DSs prior to a hurricane. However, essentially it is a one-shot pre-dispatch without further routing and scheduling EVs. As for MESSs, Abdeltawab and Mohamed [16] investigate their dispatch for a variety of services in active DSs assuming the state of normal operation. In [17], the reliability improvement with MESSs is assessed. Kim and Dvorkin [33] study the installation of MESSs and considers their effects in improving DS resilience. With respect to MEGs, Refs. [13, 14] provide detailed literature reviews. In short, an efficient and integrated approach for resilient routing and scheduling of MPSs in DSs has to be designed for their better utilization. The strategy of dynamically forming MGs should also be specifically considered.

Based on the report [6] by the US Electric Power Research Institute, the resilience of DSs depends on three elements, i.e. prevention, survivability, and recovery. This chapter proposes a two-stage framework for MPSs' resilient routing and scheduling to improve both survivability and recovery of DSs against natural disasters. (Since prevention refers to avoiding the damages of components, it is out of the scope of MPS dispatch.) In the first stage, i.e. before the natural disaster, MPSs are proactively pre-positioned in the DS and co-optimized with network reconfiguration of the DS, so as to improve survivability of electricity supply of the DS. In the second stage, i.e. after the disaster strikes, MPSs are dynamically dispatched for coordinating with conventional service recovery efforts, so as to improve the effectiveness of grid recovery especially by dynamically forming MGs. Specifically, this chapter follows [30] to consider medium-duty EVs. The other two types of MPSs mentioned above, i.e. MEGs and MESSs that have different operational constraints and characteristics, are considered, too. As the damages of DSs are uncertain before the disaster, an approach based on robust optimization is developed to determine proactive actions in the first stage. Regarding the second stage, a novel mixed-integer programming model is constructed to generate decisions of MPSs' dynamic dispatch, which involves the coordination among different types of MPSs, etc. These two stages use two different models to improve two different resilience metrics. Later sections provide more details. Note that this chapter is fundamentally different from Chapter 3 regarding the investigated problem, modeling, problem formulations, and solution algorithms, etc.

The rest of this chapter is organized as follows: Section 9.2 presents the proposed two-stage framework for resilient routing and scheduling of MPSs. Discussions on some practical issues are provided in Section 9.3. Section 9.4 introduces the optimization methods for MPS pre-positioning in the first stage and MPS dynamic dispatch in the second stage. Solution methods are presented in Section 9.5. Section 9.6 provides the case studies. Section 9.7 shortly concludes this chapter.

9.2 Routing and Scheduling Mobile Power Sources in Microgrids

In many cases, natural disasters including extreme weather events cause a partial or even complete loss of power supply from the main grid to a DS. Main reasons include broken laterals or feeders of the DS, faults in the substation, and outages of the transmission system, which greatly limit conventional service recovery strategies' effects. Customers therefore experience prolonged power outages. In such cases, MPSs are effective response resources to achieve enhanced electric service survivability and power supply recovery of the grid, e.g. by forming MGs, as their utilization is much less influenced by sustained damages of DS components.

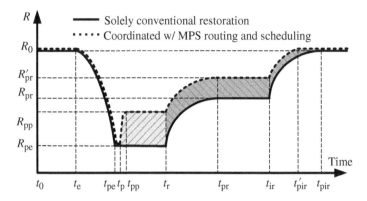

Figure 9.2 Conceptual resilience curves associated with an outage event.

Here we use the conceptual resilience curve in [10] for clearer explanations. In Figure 9.2, R is a metric of the grid's resilience level. Associated with a natural disaster, a DS experiences the states as follows: resilient state $t_0 \sim t_e$, disaster progress $t_e \sim t_{pe}$, post-disaster degraded state $t_{pe} \sim t_r$, restorative state $t_r \sim t_{pr}$, post-restoration state $t_{pr} \sim t_{ir}$, and infrastructure recovery $t_{ir} \sim t_{pir}$. Note that post-disaster restoration and recovery of power grids is a process of multiple stages. Prior to the stages of conventional restoration and recovery efforts, prompt pre-restoration can be implemented right after the disaster, e.g. to connect emergency backup generators and form MGs. With pre-restoration, the post-disaster state $t_{pe} \sim t_r$ can be separated into three stages: post-disaster state $t_{pe} \sim t_p$, pre-restorative state $t_p \sim t_{pp}$ (when conducting pre-restoration), and post-pre-restoration state $t_{pp} \sim t_r$ (the state after pre-restoration).

This chapter proposes a two-stage framework for implementing resilient MPS routing and scheduling. In the first stage, i.e. before the natural disaster $t_0 \sim t_e$, MPSs are proactively pre-positioned to selected DS nodes for enhancing survivability of the system. Specifically, the survivability indicated by the resilience level of the grid during $t_{pe} \sim t_r$ is evaluated. Fast pre-restoration of electric service is enabled by the pre-positioned MPSs. Shortly after the disaster, at t_p, the DS can be connected with MPSs to quickly improve the resilience level from R_{pe} to R_{pp}. DS network reconfiguration is also co-optimized to shift the grid into a state less stressed and impacted by the disaster [34]. With these measures, the resilience level of the grid in the post-disaster state, i.e. the survivability level, is improved from R_{pe} to R_{pp} during $t_{pp} \sim t_r$, as shown in Figure 9.2. By contrary, if MPSs are unavailable or not used, the grid's post-disaster resilience level remains on R_{pe} until t_r, as strategies including resilient MG formation are not used.

In the second stage, i.e. after the natural disaster strikes, MPSs are dynamically dispatched to cooperate with service restoration and infrastructure recovery measures, in order to improve grid recovery during $t_r \sim t_{pir}$. The recovery here is highly relevant to the Π metric proposed in [9], which assesses how fast the grid recovers. Power scheduling and the transportation of MPSs, power dispatch, and dynamic network reconfiguration of the DS, etc., are co-optimized. Therefore, the resilience level of the grid in the post-restoration state is improved from R_{pr} to R'_{pr}, and the loads in the system can be completely recovered at t'_{pir} that is earlier than t_{pir}, by forming MGs and other involved strategies.

As aforementioned, in the above process, MGs powered by MPSs may be formed. The service recovery strategy of forming MGs with EVs, MEGs, and MESSs actually has been implicitly or explicitly considered in [27, 28, 29, 30, 31, 35], [14, 32] and [33, 36], respectively. For instance, Gao et al. [30] allocate electric buses to charging stations for supplying critical loads in electrical islands after a natural disaster. Pang et al. [31] use EVs to power buildings on a feeder islanded from the main grid by a fault. Sun et al. [35] use EVs for building islanded networks, forming, and operating islands in the black start of power systems. In [28] studying the self-healing process, EVs act as backup power supplied to customers in the faulted zone. In [27, 29], during outages, EVs supply power to the owners' houses and inject power surplus back to the grid for serving other customers in the local isolated grid.

Forming MGs with EVs in service recovery is actually recommended by relevant authorities or organizations. For example, a consensus study report by the US National Academies of Sciences, Engineering and Medicine instructs the Department of Energy (DoE) to work with manufacturers to fully enable such strategies, e.g. utilizing EVs to energize isolated feeders during long-duration and large-area outages, especially after natural disasters [7]. Such strategies are also validated by many relevant industry practices. For instance, in the Fukishima disaster, 7000 Nissan Leafs were used to supply three to five days of power for households [5, 21]. In a project funded by the US DoE, the United Parcel Service (UPS) and the Oak Ridge National Laboratory worked together to develop technologies enabling EVs to energize the power grid during outages [22]. With companies such as Enel and Endesa producing operational vehicle-to-grid chargers, vehicle-to-business, vehicle-to-neighborhood, and vehicle-to-street projects have been demonstrated in many countries [37, 38].

Note that the two stages have two different objectives (i.e. improving the survivability and recovery, respectively). The first stage is not a preparedness for the second stage. The second stage is not the continuation of the first-stage task. The two shadowed areas in Figure 9.2 conceptually represent the survivability and recovery improvement of the grid, respectively. Since the damages of grid components are uncertain before the natural disaster, DS proactive reconfiguration and MPS pre-positioning decisions are generated using a two-stage robust optimization model presented in Section 9.4. Regarding the problem of coordinated and dynamic dispatch of MPSs, the issues of multiple timescales, coupling of power and road networks, and the formulation of dynamic MG formation for service restoration need to be addressed in its modeling, which is also introduced in Section 9.4.

9.3 Mobile Power Sources and Supporting Facilities

9.3.1 Availability

As mentioned above, this chapter considers MPSs owned by utilities. Particularly, some electric trucks may be unavailable to serve as MPSs, as post-disaster repairing efforts also require supports from them. In such cases, the utility may just use its duty EVs of other kinds (e.g. electric buses [30, 37]) and MPSs of other kinds (i.e. MEGs and MESSs). For instance, in [30], electric buses that have large charging/discharging power rates and battery capacities are dispatched for improving grid resilience. In fact, because of the importance of fast service recovery and the large impact of natural disasters, the utility may also get additional resources such as MPSs provided by volunteering organizations [39] and government entities [2–4, 7], or shared by other utilities under agreements of mutual assistance [1–3, 6, 7]. And considering that utilities are instructed by relevant authorities to assure sufficient resources for recovering electric service as fast as possible [1–4, 6–8], this chapter follows relevant works and presume that MPSs are available in DSs for assisting recovery, especially by forming MGs.

If MPSs of other profit-seeking individuals or agents (e.g. private-owned EVs) are to be utilized to further improve DS resilience against natural disasters, appropriate incentives have to be developed based on the behaviors and wills of the individuals and agents, and the strategies of committing them for DS service recovery have to be studied. In fact, to use private-owned EVs, a decentralized dispatch approach with privacy-protecting distributed algorithms is much more proper. Those issues are out of the scope of this chapter and can be studied in the future.

With respect to the supporting facilities, MESS stations are operated by the utility to support the use of MESSs. The connection of MEGs can also be supported. Regarding charging stations, this chapter considers utility-owned ones, which include public charging stations operated by the utility, fleet charging stations for the utility's duty EVs, and workplace charging stations for EVs of utility employees [37]. As mentioned in [37, 38], utilities are installing a rising number of charging stations in DSs. Since resilient responses against natural disasters require coordinated efforts of different organizations and infrastructures, the utility may also be temporarily authorized to use charging stations owned by others, e.g. under the requested or streamlined coordination of the government and relevant authorities including the Red Cross Organization [3, 4, 39]. Nevertheless, this chapter only considers utility-own facilities. More specifically, we follow [28, 30, 35] and consider a small number of charging stations owned and operated by the utility in the DS for electric service restoration.

9.3.2 Grid-Forming Functions

Years of industry practices have well demonstrated the grid-forming functions of MESSs, MEGs, and their supporting facilities including MESS stations [3, 4, 15, 24]. Enabled by operational vehicle-to-grid chargers, etc., the grid-forming functions of charging stations and EVs have been demonstrated in the industry, too [22, 38, 40]. For instance, vehicle-to-grid's capability of helping real-time frequency regulation is verified by PJM in [40]. In the Netherlands, companies including Nuvve and New Motion have partnered to utilize EVs for providing ancillary services [38]. In fact, frequency regulation and voltage regulation are identified as the ancillary services that charging stations and EVs can provide [5, 37]. In this regard, utilizing EVs as emergency power backup for post-disaster service restoration are practiced and studied in [5, 21, 22] and [29–31, 35], respectively, which essentially consider forming islanded MGs of different voltage levels.

To implement the proposed post-disaster service restoration strategy in practice, multiple technical issues of different system levels (or component levels) and timescales are involved. This chapter follows [27–31, 33, 35, 36] to concentrate on the dispatch of available flexibilities, which in this chapter refers to the optimal MPS routing and scheduling coordinated with the MG formation and load pick-up of the DS, etc. Relevant DS operational constraints are also handled. Some other issues, e.g. design of the physical system and control of the inverters for real-time frequency and voltage regulations, are essentially important, too. Note that charging stations and EVs are able to help achieve improved power quality via some advanced control schemes in real time. The results in this chapter act as a start for further studies on the power quality control [30]. Particularly, the loads' nominal values are used in order to save sufficient capacities of the MPSs for frequency and voltage regulations in real time [41]. In short, this chapter focuses on more relevant issues and leave those beyond-scope issues for future investigation. Interested readers can refer to [26], etc.

9.3.3 Cost-Effectiveness

Industry practices have proven that using MESSs and MEGs in service recovery is cost-effective [3, 4, 15, 23, 24]. Regarding using medium-duty EVs in MGs, note that the power interruption cost is much higher than the energy storage-levelized cost [16, 33, 36, 42]. Massive extra cost or payment is also avoided by the use of facilities owned by the utility [37, 38]. In fact, many other post-disaster recovery activities rely on the power supply, and regulation authorities have set up benefits or penalty for encouraging fast electric service recovery. In general, the proposed strategy's cost-effectiveness is supported by the high value of power supply reliability [12, 25, 29] and huge economic benefits of improved power system resilience [1, 7]. Therefore, to improve DS resilience with EVs [5, 7], this chapter follows [28–31, 35] and presumes that the resulting benefit of reduced outage duration and scale is larger than the possible cost, and generally follows those works on the modeling of cost issues in the formulations.

9.4 A Two-Stage Dispatch Framework

This section provides the problem formulations. Supported by the industry practices of utilizing EVs [5, 21, 22], MEGs [3, 4, 23], and MESSs [15, 24] in DS electric service restoration, MPSs are dispatched and connected to charging stations or MESS stations for supplying critical loads. DS reconfiguration and MG formation, etc., are also involved. In general, the considered post-disaster restoration efforts are implemented by utilities using facilities and resources owned by them [16, 24, 37].

9.4.1 Proactive Pre-Dispatch

Before the natural disaster, MPSs are pre-positioned in the DS to support prompt pre-restoration for enhancing DS survivability, which is assessed by the survived electric service to critical loads. A two-stage robust optimization model is proposed to obtain pre-positioning decisions. The objective is maximizing the worst-case weighted sum of survived loads at $t = 0$ (i.e. t_{pp} here):

$$\max_{\alpha_i^m, \lambda_{ij,0}, f_{ij,0}, g_{i,0}} \left[\min_{u_{ij}} \left(\max_{p_{i,0}, q_{i,0}, P_{i,0}, Q_{i,0}, pf_{ij,0}, qf_{ij,0}, v_{i,0}} \sum_{i \in \mathcal{N}} \omega_i \cdot p_{i,0} \right) \right] \tag{9.1}$$

The outermost level co-optimizes the DS topology (i.e. $\lambda_{ij,0}$) and the MPSs' pre-positioning nodes (i.e. α_i^m). In other words, network reconfiguration is also considered and coordinated with the pre-positioning of MPSs, so as to improve the utilization of MPSs' capacities in improving DS survivability. With MEGs fully refueled and MESSs/EVs fully charged prior to the disaster, the pre-positioned MPSs are quickly connected to the grid to provide power shortly after the disaster at t_p. In such a pre-restoration process, MGs might be formed. Involved variables in the outermost level are subject to the constraints as follows:

$$\sum_{i \in \mathcal{N}_m} \alpha_i^m = 1, \ \forall m \in \mathcal{M} \tag{9.2}$$

$$\sum_{m \in \mathcal{M}_i} \alpha_i^m \leq CAP_i, \ \forall i \in \bigcup_{m \in \mathcal{M}} \mathcal{N}_m \tag{9.3}$$

$$\sum_{(i,j) \in \mathcal{L}} \lambda_{ij,0} = |\mathcal{N}| - |\mathcal{N}^{\text{sub.}}| \tag{9.4}$$

$$\sum_{(j,i) \in \mathcal{L}} f_{ji,0} - \sum_{(i,j) \in \mathcal{L}} f_{ij,0} = l_{i,0}, \ \forall i \in \mathcal{N} \backslash \mathcal{N}^{\text{sub.}} \tag{9.5}$$

$$-\lambda_{ij,0} \cdot K_1 \leq f_{ij,0} \leq \lambda_{ij,0} \cdot K_1, \ \forall (i,j) \in \mathcal{L} \tag{9.6}$$

Constraint (9.2) forces that before the natural disaster, each MPS is pre-positioned to one of the candidate nodes for its connection. For the choice of candidate nodes, please refer to [4, 14, 23]. EV fleets can be connected to nodes with charging stations [37], while MESSs and MEGs are connected to nodes with MESS stations [16, 23, 24]. Constraint (9.3) restricts the number of MPSs connected to each station. The limit, i.e. CAP_i, may be set according to the station's capacity, or just set as 1 for all stations for assuring more dispersed pre-positioning, which normally is more resilient. Constraints (9.3)–(9.6) together assure radiality of the DS. Specifically, as mentioned in Chapter 8, the DS topology has to meet two requirements: (i) It has $|\mathcal{N}| - |\mathcal{N}^{\text{sub.}}|$ closed lines; (ii) Every load node is linked to one substation node. Equation (9.4) represents the first requirement. Let load nodes and substation nodes be the sinks and sources for a fictitious flow, respectively. Then, the second requirement is met by forcing every load node to get unit fictitious flow into it from one of the source nodes. Equation (9.5) assures that all load nodes meet the fictitious flow balance condition. Particularly, the fictitious load $l_{i,0}$, which denotes the amount of fictitious flow into a load node, is set as 1 for all nodes. Constraint (9.6) enforces the fictitious flow on open lines to be zero. Setting $K_1 = |\mathcal{N}| - |\mathcal{N}^{\text{sub.}}|$ is sufficiently large. For more details on adopting the fictitious flow model to formulate radiality constraints, please refer to Chapter 8 or [43]. To avoid islanding a single substation node, the constraints as follows are also included:

$$\sum_{(i,j) \in \mathcal{L}} f_{ij,0} - \sum_{(j,i) \in \mathcal{L}} f_{ji,0} = g_{i,0}, \ \forall i \in \mathcal{N}^{\text{sub.}} \tag{9.7}$$

$$g_{i,0} \geq 1, \ \forall i \in \mathcal{N}^{\text{sub.}} \tag{9.8}$$

For a DS that has no distributed generations or zero-load nodes, the fictitious flow needs not to be used, since the power flow constraints will naturally force the satisfaction of the second requirement mentioned above [43]. In this chapter, there are MPSs in the DS, so the fictitious flow and related concepts are introduced.

The midlevel of the two-stage robust optimization model, with the first-stage decisions of DS reconfiguration and MPS pre-positioning given by the outer-most level, finds the set of damaged lines that minimize the weighted sum of post-disaster survived loads supported by pre-restoration. Line damages are subject to the constraints as follows:

$$\sum_{(i,j)\in\mathcal{L}^{\text{vul.}}} (1 - u_{ij}) \leq \Gamma_1 \tag{9.9}$$

$$\sum_{(i,j)\in\mathcal{L}\backslash\mathcal{L}^{\text{vul.}}} (1 - u_{ij}) \leq \Gamma_2 \tag{9.10}$$

Lines are separated into two groups, i.e. the set of vulnerable lines $\mathcal{L}^{\text{vul.}}$ and its complement $\mathcal{L}\backslash\mathcal{L}^{\text{vul.}}$. Some may consider only the damages of vulnerable lines, e.g. in [34]. A commonly used approach for finding vulnerable components is to compare the failure probability of each component to a given threshold. That is, lines in $\mathcal{L}\backslash\mathcal{L}^{\text{vul.}}$ is less or far less vulnerable, but not invulnerable. In fact, esti-mations of failure probabilities can be inaccurate. Moreover, the damage of some lines in $\mathcal{L}\backslash\mathcal{L}^{\text{vul.}}$, though less or far less probable, may have more critical influ-ences. Therefore, failures of both vulnerable and less vulnerable lines are taken into account here. Particularly, constraints (9.9) and (9.10) restrict the number of damages for lines in the sets $\mathcal{L}^{\text{vul.}}$ and $\mathcal{L}\backslash\mathcal{L}^{\text{vul.}}$, respectively, by an estimated maximum number. In [44], etc., which also use robust optimization for improving resilience of power grids, all components are implicitly assumed to have an equal vulnerability level. Compared to those publications, here, with $\Gamma_1/|\mathcal{L}^{\text{vul.}}|$ larger than $\Gamma_2/(|\mathcal{L}| - |\mathcal{L}^{\text{vul.}}|)$, the vulnerability level of the lines in $\mathcal{L}^{\text{vul.}}$ is set larger than that of the lines in $\mathcal{L}\backslash\mathcal{L}^{\text{vul.}}$, which is more practical. In fact, a higher resolution of line vulnerability levels can be modeled if separating the lines into more than two groups. Generally, with those mid-level constraints, the proactive network recon-figuration in the outermost level is apt to open vulnerable lines whose failures have larger influences, and close lines that are less vulnerable or have smaller influences if damaged by the natural disaster.

The innermost level is normally called the second-stage recourse problem. With first-stage decisions from the outermost level and the grid damage situation from the middle level, the weighted sum of survived loads is maximized by the fast pre-restoration right after the natural disaster. It is subject to the constraints as follows:

$$\sum_{(j,i)\in\mathcal{L}} pf_{ji,0} - \sum_{(i,j)\in\mathcal{L}} pf_{ij,0} = P_{i,0} - P_{i,0}, \ \forall i \in \mathcal{N} \tag{9.11}$$

$$\sum_{(j,i)\in\mathcal{L}} qf_{ji,0} - \sum_{(i,j)\in\mathcal{L}} qf_{ij,0} = q_{i,0} - Q_{i,0}, \ \forall i \in \mathcal{N} \tag{9.12}$$

$$-\sum_{m\in\mathcal{M}_i\cap\{S,\mathcal{V}\}} \alpha_i^m \cdot P_m^{\max} \leq P_{i,0} \leq \sum_{m\in\mathcal{M}_i} \alpha_i^m \cdot P_m^{\max}, \ \forall i \in \bigcup_{m\in\mathcal{M}} \mathcal{N}_m \tag{9.13}$$

$$- \sum_{m \in \mathcal{M}_i} \alpha_i^m \cdot Q_m^{\max} \leq Q_{i,0} \leq \sum_{m \in \mathcal{M}_i} \alpha_i^m \cdot Q_m^{\max}, \ \forall i \in \bigcup_{m \in \mathcal{M}} \mathcal{N}_m \tag{9.14}$$

$$P_{i,0} = Q_{i,0} = 0, \ \forall i \in \mathcal{N} \backslash \bigcup_{m \in \mathcal{M}} \mathcal{N}_m \tag{9.15}$$

$$0 \leq p_{i,0} \leq \overline{p}_{i,0}, \ \forall i \in \mathcal{N} \tag{9.16}$$

$$q_{i,0} = \frac{\overline{q}_{i,0}}{\overline{p}_{i,0}} \cdot p_{i,0}, \ \forall i \in \mathcal{N} \tag{9.17}$$

$$pf_{ij,0}^2 + qf_{ij,0}^2 \leq \lambda_{ij,0} \cdot u_{ij} \cdot \left(s_{ij}^{\max}\right)^2, \ \forall (i,j) \in \mathcal{L} \tag{9.18}$$

$$v_{i,0} - v_{j,0} \leq (1 - \lambda_{ij,0} \cdot u_{ij}) \cdot K_2 + 2 \cdot (r_{ij} \cdot pf_{ij,0} + x_{ij} \cdot qf_{ij,0}), \ \forall (i,j) \in \mathcal{L} \tag{9.19}$$

$$v_{i,0} - v_{j,0} \geq (\lambda_{ij,0} \cdot u_{ij} - 1) \cdot K_2 + 2 \cdot (r_{ij} \cdot pf_{ij,0} + x_{ij} \cdot qf_{ij,0}), \ \forall (i,j) \in \mathcal{L} \tag{9.20}$$

$$\left(V_i^{\min}\right)^2 \leq v_{i,0} \leq \left(V_i^{\max}\right)^2, \ \forall i \in \mathcal{N} \tag{9.21}$$

Equations (9.11) and (9.12) force all nodes to meet the real and reactive power balance requirements, respectively. Constraints (9.13) and (9.14) limit the real and reactive power injection or extraction of candidate nodes for MPS connection, respectively. For any candidate node, its real/reactive power injection or extraction limit is the sum of real/reactive power capacities of the MPSs connected to it. With the natural disaster leading to multiple damages splitting the system into islands that potentially can operate as MGs, the MPSs commonly just inject power in power pre-restoration [30], especially for islands which lose the main grid power. Constraint (9.15) specifies zero power output of nodes which are not for connecting MPSs. In this chapter, end-use consumers are simply modeled as loads. Constraint (9.16) expresses a range for the survived load of each node affected. Equation (9.17) fixes the power demands' power factors. Constraint (9.18) limits the apparent power on each line by its capacity, and forces both real and reactive power flow on open lines to be zero. Constraints (9.19) and (9.20) represent the DistFlow model-based power flow equations [45]. The far smaller quadratic terms are neglected [45]. A sufficiently big positive number K_2 is used for relaxing these constraints for open lines. Constraints (9.18)–(9.20) include terms $\lambda_{ij,0} \cdot u_{ij}$. They need not to be linearized. Since $\lambda_{ij,0}$ and u_{ij} are outermost-level and midlevel variables of the two-stage robust optimization model, respectively, these terms actually do not bring non-linearity to the solving algorithm in Section 9.5, and constraint (9.21) represents limits on the voltage magnitude.

9.4.2 Dynamic Routing and Scheduling

After the natural disaster, outages and damages of the grid are evaluated, and then the service restoration and infrastructure recovery process begin. In this process, MPSs are dispatched dynamically to cooperate with conventional restoration efforts, so as to decrease the outage duration and scale by forming MGs, etc. The MPS dynamic dispatch objective is

$$
\max \sum_{t \in \mathcal{T}} \sum_{i \in \mathcal{N}} \omega_i \cdot p_{i,t} - \sum_{t \in \mathcal{T}} \sum_{m \in \mathcal{M}} \delta_m \cdot \gamma_t^m - \sum_{t \in \mathcal{T}} \sum_{m \in \{\mathcal{S}, \mathcal{V}\}} \left| \frac{\zeta_m}{100} \right| \cdot C_m^{\mathrm{B}} \cdot (cp_t^m + dp_t^m)
$$

$$
\tag{9.22}
$$

In (9.22), the first term represents the weighted sum of recovered loads during $t = 1 \sim |\mathcal{T}|$ (i.e. $t_r \sim t'_{\mathrm{pir}}$), the second term represents the transportation cost associated with the traveling of MPSs [30, 46], and the third term represents MESSs and EVs' battery lifecycle degradation cost [33, 47]. The second term is included for two main reasons [46]: (i) If all loads are recovered at some time $t < |\mathcal{T}|$, it is no longer needed to dispatch MPSs during time $t \sim |\mathcal{T}|$. In such cases, with the second term included, the traveling of MPSs during time $t \sim |\mathcal{T}|$ is limited. (ii) There may exist multiple strategies of MPS dispatch achieving the same optimal recovery effect. The second term is included to assure choosing the strategy attaining the optimal performance with the smallest MPS transportation cost, as redundant MPS travels lead to extra cost. Regarding the issue of battery lifecycle degradation, it is not considered in many relevant works because of the low frequency of contingency events and high value of improved grid resilience or reliability [12, 25, 29]. Particularly, while it raises concerns on battery wear and tear due to frequent charging and discharging when utilizing vehicle-to-grid for frequency regulation, etc., generally it does not result in extra or accelerated battery degradation when utilizing EVs for post-disaster service recovery, since the number of such outages is small [25, 29]. In this regard, related works including [27, 28, 30, 31] do not consider the issue of battery lifecycle degradation. Still, the above objective function includes this term to avoid unneeded charging and discharging of batteries. In general, the objective of maximizing recovered loads is dominating.

To cooperate the post-disaster MPS dynamic dispatch with DS service recovery, three interdependent sub-tasks are involved: (i) The routing of MPSs. (ii) The power scheduling of MPSs. (iii) The involved operations of the DS, including power dispatch, network reconfiguration, and MG formation. Here, they are co-optimized in the MPS dynamic dispatch problem. Note that the dispatch of repair crews can also be included in the co-optimization. It is left for detailed studies in Chapter 10, and this chapter assumes a given repair plan.

The remainder of this section introduces the constraints in the MPS dynamic dispatch problem. First, the following constraints, which are similar to constraints

(9.2) and (9.3), are imposed:

$$\sum_{i \in \mathcal{N}_m} \beta_{i,t}^m \leq 1, \ \forall m \in \mathcal{M}, \ \forall t \in \mathcal{T} \tag{9.23}$$

$$\sum_{m \in \mathcal{M}_i} \beta_{i,t}^m \leq CAP_i, \ \forall i \in \bigcup_{m \in \mathcal{M}}, \ \forall t \in \mathcal{T} \tag{9.24}$$

Constraint (9.23) states that every MPS can be connected to at most one candidate node at each time step during post-disaster recovery. Constraint (9.24) restricts the number of MPSs connected to each station at each time step. Note that traveling on the road network and being connected to the grid are mutually exclusive and collectively exhaustive states of a MPS at each time step. This feature is expressed by the equation as follows:

$$\gamma_t^m = 1 - \sum_{i \in \mathcal{N}_m} \beta_{i,t}^m, \ \forall m \in \mathcal{M}, \ \forall t \in \mathcal{T} \tag{9.25}$$

Regarding the routing of MPSs, the following constraint is enforced:

$$\beta_{i,t+\tau}^m + \beta_{j,t}^m \leq 1, \ \forall i,j \in \mathcal{N}_m, \ \forall \tau \leq tr_{ij}^m, \ \forall t + \tau \leq |\mathcal{T}| \tag{9.26}$$

Note that vehicle routing, which essentially is an NP-hard combinational optimization problem, is challenging per se. For a problem involving vehicle routing, e.g. the problem in [46], many extra integer variables are normally added to model begin-at-depot, end-at-depot, and path-flow balance constraints, etc., based on the transportation network that may be further simplified into an undirected or a directed graph including only the links and vertices of interest. In this chapter, the post-disaster routing of MPSs is formulated in a concise and novel manner without introducing new integer variables. That is, the proposed formulation just needs constraint (9.26) to assure that the traveling of MPSs between different nodes meets the needed traveling time. Other involved constraints, including the request satisfaction constraints, path-flow balance, and travel distance limits here, are implicitly reflected or satisfied owing to the interdependencies between MPS routing and other constraints and decisions. A simple instance is used to explain constraint (9.26): If it takes two time steps of MPS 1 to travel between node 1 and node 2, the following constraints are enforced:

$$\beta_{2,t+1}^1 + \beta_{1,t}^1 \leq 1, \ \forall t + 1 \leq |\mathcal{T}| \tag{9.27}$$

$$\beta_{2,t+2}^1 + \beta_{1,t}^1 \leq 1, \ \forall t + 2 \leq |\mathcal{T}| \tag{9.28}$$

Thus, if $\beta_{1,t}^1 = 1$ (i.e. MPS 1 is connected to node 1 at time t), then $\beta_{2,t+1}^1 = \beta_{2,t+2}^1 = 0$ (i.e. it cannot be connected to node 2 in the following two time periods due to the necessary travel time from node 1 to node 2) and vice versa. It is straightforward that constraint (9.26) can also be equivalently transformed into

$$\sum_{\tau=t+1}^{\min\{t+tr_{ij}^m,|\mathcal{T}|\}} \beta_{j,\tau}^m \leq (1 - \beta_{i,t}^m) \cdot \min\{tr_{ij}^m, |\mathcal{T}| - t\}, \ \forall m \in \mathcal{M}, \ \forall i,j \in \mathcal{N}_m,$$

$$\forall t + 1 \leq |\mathcal{T}| \tag{9.29}$$

The number of constraints thus is reduced.

The following constraints are on the power scheduling of MPSs:

$$SoC_t^m = SoC_{t-1}^m + (cp_t^m \cdot \eta_m^c - dp_t^m/\eta_m^d) \cdot \Delta t, \ \forall m \in S, \ \forall t \geq 1 \tag{9.30}$$

$$SoC_t^m = SoC_{t-1}^m + (cp_t^m \cdot \eta_m^c - dp_t^m/\eta_m^d - \gamma_t^m \cdot tp^m) \cdot \Delta t, \ \forall m \in \mathcal{V}, \ \forall t \geq 1 \tag{9.31}$$

$$SoC_m^{\min} \leq SoC_t^m \leq SoC_m^{\max}, \ \forall m \in \{S, \mathcal{V}\}, \ \forall t \in \mathcal{T} \tag{9.32}$$

$$0 \leq cp_t^m \leq c_t^m \cdot cp_m^{\max}, \ \forall m \in \{S, \mathcal{V}\}, \ \forall t \in \mathcal{T} \tag{9.33}$$

$$0 \leq dp_t^m \leq d_t^m \cdot dp_m^{\max}, \ \forall m \in \{S, \mathcal{V}\}, \ \forall t \in \mathcal{T} \tag{9.34}$$

$$c_t^m + d_t^m \leq \sum_{i \in \mathcal{N}_m} \beta_{i,t}^m, \ \forall m \in \{S, \mathcal{V}\}, \ \forall t \in \mathcal{T} \tag{9.35}$$

$$0 \leq gp_t^m \leq \sum_{i \in \mathcal{N}_m} \beta_{i,t}^m \cdot P_m^{\max}, \ \forall m \in \mathcal{G}, \ \forall t \in \mathcal{T} \tag{9.36}$$

$$0 \leq gq_t^m \leq \sum_{i \in \mathcal{N}_m} \beta_{i,t}^m \cdot Q_m^{\max}, \ \forall m \in \mathcal{M}, \ \forall t \in \mathcal{T} \tag{9.37}$$

Equations (9.30) and (9.31) represent the state of charge (SoC) variations over time of MESSs and EV fleets, respectively. A MESS's SoC is determined by its charging and discharging behaviors, while an EV fleet's SoC is also determined by its traveling behavior. Constraint (9.32) imposes SoC ranges for MESSs and EV fleets. Constraints (9.33) and (9.34) specify their charging power and discharging power limits, respectively. Constraint (9.35) ensures that in each time period charging and discharging are mutually exclusive states for a MESS or an EV fleet, and if it is not connected to a MG or subgrid in the DS, it can neither charge nor discharge. Constraint (9.36) restricts real power outputs of MEGs by their capacities and enforces zero power outputs for MEGs that are not connected to the DS. Constraint (9.37) restricts MPS reactive power outputs similarly. Note that the fuel for MEGs can be pre-allocated in the DS [30]. Portable or towable fuel tanks and optimally dispatched fuel trucks can refuel MEGs, too [4]. Thus, this chapter follows [23], etc., and assumes adequate fuel for MEGs in post-disaster service restoration.

The following constraints are on DS operation:

$$\lambda_{ij,t} = 0, \ \forall t \in \mathcal{T}, \ \forall (i,j) \in \mathcal{L}_t^{\text{off}} \tag{9.38}$$

$$\lambda_{ij,t} = \lambda_{ij,0}, \ \forall t \in \mathcal{T}, \ \forall (i,j) \in \mathcal{L}\backslash\{\mathcal{L}^{\text{RCS}}, \mathcal{L}_t^{\text{off}}\} \tag{9.39}$$

$$\sum_{(i,j)\in\mathcal{L}} \lambda_{ij,t} = |\mathcal{N}| - I_t, \ \forall t \in \mathcal{T} \tag{9.40}$$

$$\sum_{(j,i)\in\mathcal{L}} f_{ji,t} - \sum_{(i,j)\in\mathcal{L}} f_{ij,t} = l_{i,t}, \ \forall t \in \mathcal{T}, \ \forall i \in \mathcal{N}\backslash\mathcal{N}_t^{\text{src.}} \tag{9.41}$$

$$\sum_{(i,j)\in\mathcal{L}} f_{ij,t} - \sum_{(j,i)\in\mathcal{L}} f_{ji,t} = g_{i,t}, \ \forall t \in \mathcal{T}, \ \forall i \in \mathcal{N}_t^{\text{src.}} \tag{9.42}$$

$$-\lambda_{ij,t} \cdot K_1 \leq f_{ij,t} \leq \lambda_{ij,t} \cdot K_1, \ \forall t \in \mathcal{T}, \ \forall (i,j) \in \mathcal{L} \tag{9.43}$$

$$\sum_{(j,i)\in\mathcal{L}} pf_{ji,t} - \sum_{(i,j)\in\mathcal{L}} pf_{ij,t} = p_{i,t} - P_{i,t}, \ \forall i \in \mathcal{N}, \ \forall t \in \mathcal{T} \tag{9.44}$$

$$\sum_{(j,i)\in\mathcal{L}} qf_{ji,t} - \sum_{(i,j)\in\mathcal{L}} qf_{ij,t} = q_{i,t} - Q_{i,t}, \ \forall i \in \mathcal{N}, \ \forall t \in \mathcal{T} \tag{9.45}$$

$$P_{i,t} = \sum_{m\in\mathcal{M}_i\cap\{S,\mathcal{V}\}} \beta_{i,t}^m \cdot (dp_t^m - cp_t^m) + \sum_{m\in\mathcal{M}_i\cap\mathcal{G}} \beta_{i,t}^m \cdot gp_t^m,$$
$$\forall i \in \bigcup_{m\in\mathcal{M}} \mathcal{N}_m, \ \forall t \in \mathcal{T} \tag{9.46}$$

$$Q_{i,t} = \sum_{m\in\mathcal{M}_i} \beta_{i,t}^m \cdot gq_t^m, \ \forall i \in \bigcup_{m\in\mathcal{M}} \mathcal{N}_m, \ \forall t \in \mathcal{T} \tag{9.47}$$

$$P_{i,t} = Q_{i,t} = 0, \ \forall i \in \mathcal{N}\backslash \bigcup_{m\in\mathcal{M}} \mathcal{N}_m, \ \forall t \in \mathcal{T} \tag{9.48}$$

$$0 \leq p_{i,t} \leq \overline{p}_{i,t}, \ \forall i \in \mathcal{N}, \forall t \in \mathcal{T} \tag{9.49}$$

$$\frac{p_{i,t-1}}{\overline{p}_{i,t-1}} \leq \frac{p_{i,t}}{\overline{p}_{i,t}}, \ \forall i \in \mathcal{N}, \forall t \geq 1 \tag{9.50}$$

$$q_{i,t} = \frac{\overline{q}_{i,t}}{\overline{p}_{i,t}} \cdot p_{i,t}, \ \forall i \in \mathcal{N}, \ \forall t \in \mathcal{T} \tag{9.51}$$

$$pf_{ij,t}^2 + qf_{ij,t}^2 \leq \lambda_{ij,t} \cdot \left(s_{ij}^{\max}\right)^2, \ \forall (i,j) \in \mathcal{L}, \ \forall t \in \mathcal{T} \tag{9.52}$$

$$v_{i,t} - v_{j,t} \leq (1 - \lambda_{ij,t}) \cdot K_2 + 2 \cdot (r_{ij} \cdot pf_{ij,t} + x_{ij} \cdot qf_{ij,t}), \ \forall (i,j) \in \mathcal{L}, \ \forall t \in \mathcal{T} \tag{9.53}$$

$$v_{i,t} - v_{j,t} \geq (\lambda_{ij,t} - 1) \cdot K_2 + 2 \cdot (r_{ij} \cdot pf_{ij,t} + x_{ij} \cdot qf_{ij,t}), \ \forall (i,j) \in \mathcal{L}, \ \forall t \in \mathcal{T} \tag{9.54}$$

$$\left(V_i^{\min}\right)^2 \leq v_{i,t} \leq \left(V_i^{\max}\right)^2, \ \forall i \in \mathcal{N}, \ \forall t \in \mathcal{T} \tag{9.55}$$

The above constraints can be roughly divided into two interdependent groups. Constraints (9.38)–(9.43) are mainly on post-disaster network reconfiguration and MG formation, implemented by connection status changes of branches with switches. Constraint (9.38) restricts a branch to be open if it is damaged and still unrepaired at time t. Constraint (9.39) indicates that undamaged branches without switches remain in their initial status. Constraints (9.40)–(9.43), similar to constraints (9.4)–(9.8) in the MPS pre-positioning problem, ensure radial topology of the DS in the dynamic reconfiguration and MG formation process. Specifically, all $l_{i,t}$ are set as 1, and constraint (9.40) states that the number of closed branches at time t is less than the number of nodes minus the number of islands caused by damaged and unrepaired branches. Constraints (9.44)–(9.55), similar to constraints (9.11)–(9.21) in the problem of MPS prepositioning, are mainly on power dispatch. Specifically, Eqs. (9.46) and (9.47) calculate the real and reactive power outputs of each candidate node for MPS connection by summing the real and reactive power outputs of MPSs connected to it, respectively. Constraint (9.50) enforces that each load is non-decreasingly restored to reach 100% electric service recovery, that is, it prevents de-energizing loads that are restored.

9.5 Solution Method

9.5.1 Column-and-Constraint Generation Algorithm

The MPS pre-positioning model is a two-stage robust optimization problem. This chapter uses the C&CG algorithm to solve it [48]. For conciseness and clarity, the MPS pre-positioning model is recast into a compact form:

$$\max_{y \in \mathcal{Y}} \left[\min_{u \in \mathcal{U}} \left(\max_{z \in \mathcal{Z}(y,u)} \boldsymbol{w}^{\mathrm{T}} \boldsymbol{z} \right) \right] \qquad (9.56)$$

where y, u, and z are first-stage decisions (i.e. MPS pre-positioning and network reconfiguration), uncertainties (i.e. branch damages), and second-stage decisions (i.e. power dispatch), respectively. Their feasible regions are denoted as follows: $\mathcal{Y}$, $\mathcal{U}$, and $\mathcal{Z}(y,u)$, and defined by constraints (9.2)–(9.8), (9.9) and (9.10), and (9.11)–(9.21), respectively, and, $\mathcal{Z}(y,u) = \{z | Az = a, Dz \leq b - Ey - \mathrm{diag}(F_1 y) F_2 u, ||B_n z||_2 \leq y^{\mathrm{T}} H_n u, \forall n = 1, \ldots, |\mathcal{L}|\}$, which depends on the values of y and u. Dualizing the innermost maximization problem and incorporating it with the mid-level minimization problem, a min–max subproblem (SP) in (9.56) becomes:

$$\min_{u \in \mathcal{U}, \pi, \varphi, \mu, \sigma} \boldsymbol{a}^{\mathrm{T}} \pi + (\boldsymbol{b} - \boldsymbol{E}\boldsymbol{y}_* - \mathrm{diag}(F_1 \boldsymbol{y}_*) F_2 \boldsymbol{u})^{\mathrm{T}} \varphi + \sum_{n=1}^{|\mathcal{L}|} \boldsymbol{y}_*^{\mathrm{T}} H_n \boldsymbol{u} \sigma_n \qquad (9.57)$$

$$A^{\mathrm{T}}\pi + D^{\mathrm{T}}\varphi + \sum_{n=1}^{|\mathcal{L}|} B_n^{\mathrm{T}}\mu_n = w \tag{9.58}$$

$$||\mu_n||_2 \leq \sigma_n, \ \forall n = 1, \dots, |\mathcal{L}| \tag{9.59}$$

$$\varphi, \sigma \geq 0 \tag{9.60}$$

where y_* is a given y; π, φ, μ, and σ are dual variables, and the master problem (MP) is as follows:

$$\max_{y \in \mathcal{Y}, \vartheta, z_k} \vartheta \tag{9.61}$$

$$\vartheta \leq w^{\mathrm{T}} z_k, \ \forall k = 1, \dots, K \tag{9.62}$$

$$A z_k = a, \ \forall k = 1, \dots, K \tag{9.63}$$

$$D z_k + Ey + \mathrm{diag}(F_1 y) F_2 u_k \leq b, \ \forall k = 1, \dots, K \tag{9.64}$$

$$||B_n z_k||_2 \leq y^{\mathrm{T}} H_n u_k, \ \forall n = 1, \dots, |\mathcal{L}|, \ \forall k = 1, \dots, K \tag{9.65}$$

where K is the number of iterations. The C&CG algorithm thus is implemented in a two-level MP-SP iterative framework as follows:

Algorithm 9.1 C&CG for solving the MPS pre-positioning problem

---

1: Set the optimality tolerance ϵ; arbitrarily select a feasible y_* from $\mathcal{Y}$ and solve the SP to obtain the optimal solution: $(u_*, \pi_*, \varphi_*, \mu_*, \sigma_*)$ and the optimal value *obj*; set $LB = obj$, $UB = +\infty$, $K = 1$, $u_1 = u_*$

2: Solve the MP to obtain the optimal solution: $(y_*, \vartheta_*, z_{1*}, \dots, z_{K*})$; update $UB = \vartheta_*$; if $UB - LB \leq \epsilon$, return y_* and terminate

3: Solve the SP with the current y_* to obtain $(u_*, \pi_*, \varphi_*, \mu_*, \sigma_*)$ and *obj*; update $LB = \max\{LB, obj\}$

4: If $UB - LB \leq \epsilon$, return y_* and terminate; otherwise, update $K = K + 1$, set $u_K = u_*$, create variables z_K constrained by (9.66)–(9.69), and go to Step 2

$$\vartheta \leq w^{\mathrm{T}} z_K \tag{9.66}$$

$$A z_K = a \tag{9.67}$$

$$D z_K + Ey + \mathrm{diag}(F_1 y) F_2 u_K \leq b \tag{9.68}$$

$$||B_n z_K||_2 \leq y^{\mathrm{T}} H_n u_K, \forall n = 1, \dots, |\mathcal{L}| \tag{9.69}$$

9.5.2 Linearization Techniques

The objective function of the SP (9.57) contains bilinear terms in the form of a binary variable multiplying a continuous variable. Equations (9.46) and (9.47) also contain bilinear terms in this form. These non-linear terms can be equivalently linearized. For example, terms $\beta_{i,t}^m \cdot cp_t^m$ in Eq. (9.46) can be replaced with $CP_{i,t}^m$ constrained by

$$0 \leq CP_{i,t}^m \leq \beta_{i,t}^m \cdot cp_m^{\max} \tag{9.70}$$

$$cp_t^m + (\beta_{i,t}^m - 1) \cdot cp_m^{\max} \leq CP_{i,t}^m \leq cp_t^m \tag{9.71}$$

That is, if $\beta_{i,t}^m = 1$, then $CP_{i,t}^m = cp_t^m$; else if $\beta_{i,t}^m = 0$, then $CP_{i,t}^m = 0$.

After linearization, the involved MPs and SPs when solving the MPS pre-positioning problem by Algorithm 9.1, and the MPS dynamic dispatch problem, all become mixed-integer second-order cone programming (MISOCP) models. If constraints (9.18) and (9.52) are also linearized, the encountered optimization problems become mixed-integer linear programming (MILP) models. Both MISOCP and MILP problems can be efficiently solved by many off-the-shelf solvers such as Gurobi.

9.6 Case Studies

In this section, the proposed MPS routing and scheduling method is demonstrated on two test systems depicted in Figures 9.3 and 9.7. A computer with an Intel i5-4278U processor and 8 GB of memory is used. Involved MISOCP problems are solved by Gurobi 7.5.2.

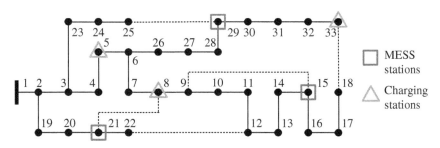

Figure 9.3 The modified IEEE 33-node test system.

9.6.1 Illustration on a Small Test System

Assume that the first test system has three charging stations and three MESS stations. Three MPSs are considered: EV fleet 1 consisting of two electric buses with 150 kW/150 kWh capacity and 0.25 kWh/km energy consumption rate, MESS 1 with 500 kW/776 kWh capacity, and MEG 1 with 800 kW/600 kVar capacity [14, 16, 30, 37, 49]. The randomly generated priority weights of loads are between 0 and 10, lying within or lower than the typical ranges of interruption cost of different kinds of customers [42]. The energy storage levelized cost is set as 0.221\$/kWh [22]. Some other data can be found in [45].

9.6.1.1 Results of MPS Proactive Pre-positioning
Prior to the natural disaster, MPSs are proactively pre-positioned in the DS using the proposed method. Based on the failure probabilities, one-third and two-thirds branches are classified into $\mathcal{L}^{\text{vul.}}$ and $\mathcal{L} \backslash \mathcal{L}^{\text{vul.}}$, respectively. Parameters Γ_1 and Γ_2 are set as 5 and 3, respectively. The problem is solved in 36.8 seconds. Figure 9.4 depicts the convergence curve for Algorithm 9.1.

Table 9.1 lists the solutions. For comparison, the solution without DS proactive network reconfiguration is included, too. Moreover, 10 000 Monte Carlo simulations are conducted to evaluate the effectiveness of proposed proactive actions. That is, DS damage scenarios are repeatedly generated and the amounts of survived loads are calculated. Simulation statistics are summarized in Table 9.2. The

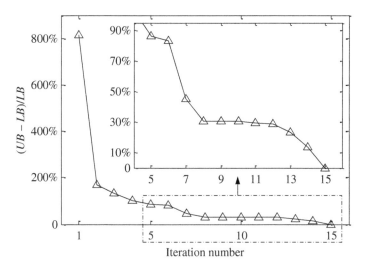

Figure 9.4 Relative optimality gap $(UB - LB)/LB$ in each iteration.

Table 9.1 MPS pre-positioning results and DS topologies: w/ and w/o proactive reconfiguration.

	w/ proactive reconfiguration	w/o proactive reconfiguration
EV fleet 1	Node 33	Node 8
MESS 1	Node 29	Node 29
MEG 1	Node 21	Node 21
Open branches	$(2,3),(9,15),(10,11),$ $(13,14),(23,24)$	$(8,21),(9,15),(12,22),$ $(18,33),(25,29)$ (i.e. the initial topology in Figure 9.3)

Table 9.2 Simulation statistics for survived loads: w/ and w/o proactive reconfiguration.

	Weighted sum of survived loads: $\sum_{i\in\mathcal{N}}\omega_i \cdot p_{i,0}$ (kW)				Box plots				
	Average	Standard	Maximum	Minimum	0	5 000	10 000	15 000	20 000
1: w/ proactive reconfiguration	13 727	3 088	20 672	3 085					
2: w/o proactive reconfiguration	11 012	3 880	20 672	2 837					
3: Benchmark (no actions)	5 641	5 006	20 672	0					

benchmark case neither pre-positions MPSs nor reconfigures the DS. It is indicated that the pre-positioning of MPSs can greatly enhance grid survivability, especially by coordinating with DS reconfiguration to form MGs as a pre-restoration strategy after the natural disaster. As shown in Table 9.2, the method with proactive network reconfiguration attains higher average and higher minimum values of $\sum_{i\in\mathcal{N}}\omega_i \cdot p_{i,0}$, and a smaller standard deviation. Note that the maximum amount of survived loads is the same in three approaches due to the zero-damage scenario in simulations.

9.6.1.2 Results of MPS Dynamic Dispatch

After the natural disaster strikes, MPSs are dynamically dispatched in the DS to assist electric service restoration. For better illustration, in this case, system-level recovery is prioritized over the survivability of individual nodes in the service

Table 9.3 Time sequence of repairing damaged branches.

Time period (t)	3	6	9	13
Repaired branch	$(1, 2)$	$(19, 20)$	$(8, 21)$	$(16, 17)$
Time period (t)	16	20	22	24
Repaired branch	$(32, 33)$	$(29, 30)$	$(28, 29)$	$(24, 25)$

restoration process. Note that the resulting number of damages is not necessarily equal to or less than $\Gamma_1 + \Gamma_2$. Here a scenario with 10 branch damages is considered, so that more time periods are needed for grid recovery, and more dynamic and coordinated dispatch of the involved resources is required. Moreover, the formation of MGs is needed, as the grid is separated into several islands by the damaged branches. Assume that the repair plan in Table 9.3 is adopted ($\Delta t = 0.5$ hours). This section sets $|\mathcal{T}| = 24$, as the electric service can be fully restored with MPSs after the eighth repair. The initial SoC of EV fleet 1 and MESS 1 at $t = 1$ are set as 50% of their capacities.

The post-disaster MPS dynamic dispatch problem is solved in 6.5 seconds. Values of the weighted sum of restored loads over time and the battery lifecycle degradation cost are 10 478.85 kWh and 332.34\$, respectively. The benefit due to prompt service restoration is much higher than the battery degradation cost. Figure 9.5

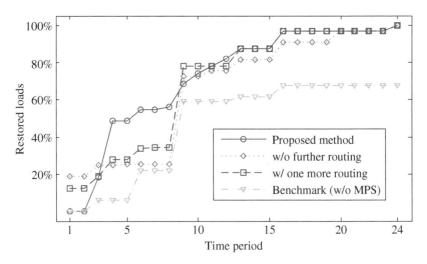

Figure 9.5 Restored loads in each time period for different cases.

depicts the evolution of restored loads. For comparison, curves for the case without further routing (i.e. MPSs stay at their pre-positioned nodes), the case with one further routing (i.e. MPSs can change connection nodes once), and the benchmark case without MPS, are also included. It is indicated that MPSs can greatly improve the performance of restoration in reducing the outage scale and duration, especially by forming MGs. As shown, without MPSs, the benchmark case has the smallest restored loads at all time. With MPSs, the proposed method restores 97% loads at $t = 16$ and 100% loads at $t = 24$. Specifically, the value of $\sum_{t \in T} \sum_{i \in \mathcal{N}} \omega_i \cdot p_{i,t}$ of the proposed method is 45% higher than that of the benchmark case. It is also indicated that the further routing of MPSs facilitates better utilization of MPS resources. Specifically, the value of $\sum_{t \in T} \sum_{i \in \mathcal{N}} \omega_i \cdot p_{i,t}$ of the proposed method is 21% and 9% higher than that of the cases without further routing and with one further routing, respectively.

Table 9.4 lists the further routing of MPSs in the stage of post-disaster MPS dynamic dispatch. The symbol "———→" means that the associated MPS is traveling on the road network. The case with one further routing is also included in the table only for completeness. Table 9.5 lists DS reconfiguration actions that might form MGs in the service restoration process. Note that eight remote-controlled switches are allocated in the DS [50]. Figure 9.6 depicts curves for the SoC of EV fleet 1, the SoC of MESS 1, and the real power output of MEG 1. The system load curve is also included for reference. As battery lifecycle degradation cost is considered, the optimization model tends to increase power outputs of the MEG, and avoid unnecessary charging/discharging of the EV fleet and MESS. In this regard, SoC variations of the EV fleet and MESS in the restoration process' later time periods are flat.

Owing to the co-optimization, different strategies and resources are well coordinated to achieve better post-disaster service restoration. At $t = 4$, EV fleet 1 and MEG 1 are connected to node 8 and node 15, respectively, and branch $(9, 15)$ is

Table 9.4 Connecting nodes of MPSs in each time period: proposed method and the benchmark case w/ one further routing.

		Time period											
		1	2	3	4	5	6	7	8	9	10	11	12~14
Proposed method	EV fleet 1	Node 33	⟶					Node 8			⟶		Node 33
	MESS 1	Node 29	⟶		Node 21		⟶			Node 29			
	MEG 1	Node 21	⟶				Node 15		⟶			Node 29	
w/ one further routing	EV fleet 1	Node 33	⟶				Node 8						
	MESS 1	Node 29	⟶				Node 21						
	MEG 1	Node 21		⟶			Node 29						

Table 9.5 Dynamic network reconfiguration of the DS.

Time period	RCS actions
$t = 4$	Close branch (9, 15)
$t = 20$	Open branch (30, 31)
$t = 21$	Close branch (30, 31)

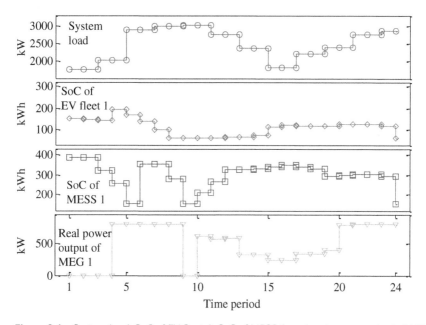

Figure 9.6 System load, SoC of EV fleet 1, SoC of MESS 1, and real power output of MEG 1 in each time period.

closed so that the two MPSs can work cooperatively. EV fleet 1 is charged by MEG 1 at $t = 4$ when the system load is low, and then they together serve the peak load at $t = 5 \sim 8$. After branch (8, 21) is repaired at $t = 9$, the loads they supply get access to the main grid power. MEG 1 thus leaves for node 29, and EV fleet 1 leaves for node 33, to restore loads that are still in outage by forming a MG. As for MESS 1, it arrives at node 21 at $t = 3$ and serves the loads in that MG. After it reaches its minimum SoC at $t = 5$, the loads it supplies get access to the main grid power at $t = 6$. It then gets charged first and leaves for node 29. It again reaches its minimum SoC at $t = 10$, when MEG 1 also arrives at node 29 to continue the electricity supply for restored loads in that MG. For space limit, details of the coordination at $t = 12 \sim 14$ are not elaborated on. In general, EV fleet 1 and MESS 1 tend to charge and discharge when the system load is low and high, respectively, and the power output changes of MEG 1 are positively correlated with the system load levels.

9.6.2 Results on a Large Test System

Assume six MESS stations and six charging stations for the second test system in Figure 9.7. Moreover, two EV fleets, two MESSs, and two MEGs with parameters the same as those in the 33-node system case are considered. Some other data can be found in [50, 51]. In this case, MPSs are pre-positioned and dispatched in a more coordinated manner. That is, the electricity supply to survived loads is uninterrupted in the post-disaster service restoration process.

For the MPS pre-positioning problem, Γ_1 and Γ_2 are set as 10 and 6, respectively. It is solved in 18.3 minutes. Table 9.6 lists the solution, and Table 9.7 lists the simulation statistics of 10 000 Monte Carlo simulations. As indicated by the comparison, survivability of this larger system is also greatly enhanced by pre-positioned MPSs, and the case with proactive DS reconfiguration again has higher average, higher minimum, and smaller standard deviation for the value of $\sum_{i \in \mathcal{N}} \omega_i \cdot p_{i,0}$. That is, coordinating with DS network reconfiguration helps attain better performance.

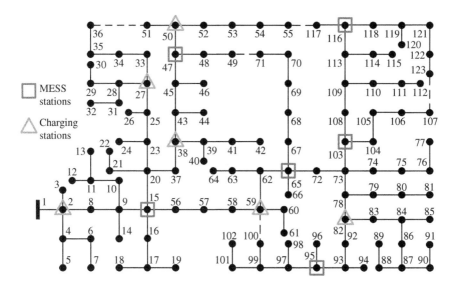

Figure 9.7 The modified IEEE 123-node test system.

Table 9.6 MPS pre-positioning results and DS topologies: w/ and w/o proactive reconfiguration.

	w/ proactive reconfiguration	w/o proactive reconfiguration
EV fleet 1	Node 27	Node 27
EV fleet 2	Node 59	Node 38
MESS 1	Node 116	Node 116
MESS 2	Node 15	Node 15
MEG 1	Node 103	Node 65
MEG 2	Node 47	Node 95
Open branches	$(37, 38), (53, 54), (62, 65),$ $(65, 72), (103, 108)$	$(36, 51), (49, 71), (59, 100),$ $(55, 117), (107, 123)$ (i.e. the initial topology in Figure 9.7)

Table 9.7 Simulation statistics for survived loads: w/ and w/o proactive reconfiguration.

	Weighted sum of survived loads: $\sum_{i \in \mathcal{N}} \varpi_i \cdot p_{i,0}$ (kW)				Box plots				
	Average	Standard	Maximum	Minimum	0	5 000	10 000	15 000	20 000
1: w/ proactive reconfiguration	15 440	2 022	20 862	6 647	1				
2: w/o proactive reconfiguration	13 457	2 138	20 862	5 674	2				
3: Benchmark (no actions)	6 549	5 636	20 384	0	3				

As for the post-disaster MPS dynamic dispatch problem, for the same reasons as in the 33-node system case, a scenario with 20 damaged branches is considered. The damaged branches are repaired in the following order: $(2, 8)$, $(15, 16), (15, 20), (57, 58), (59, 62), (99, 100), (95, 97), (83, 84), (73, 103), (106, 107),$ $(107, 123), (113, 116), (52, 53), (70, 71), (68, 69), (38, 39), (45, 47), (27, 28), (25, 27),$ $(34, 35)$. The damage of branch $(2, 8)$ actually separates almost the whole system from the main grid. The problem is solved in 7.8 minutes. The weighted sum of restored loads over time and the battery degradation cost are 221 453.07 kWh and 498.77$, respectively. Again, the benefit due to prompt service restoration is

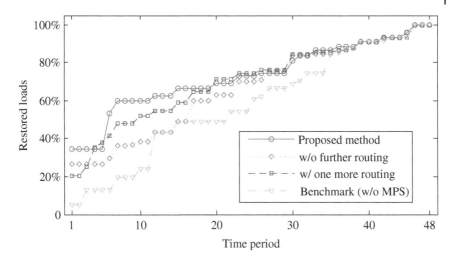

Figure 9.8 Restored loads in each time period for different cases.

Table 9.8 Dynamic network reconfiguration of the DS.

Time period	RCS actions
$t = 6$	Close branch $(62, 72)$
$t = 30$	Open branch $(73, 103)$
$t = 32$	Close branch $(73, 103)$
$t = 46$	Open branch $(78, 82)$
$t = 48$	Open branch $(47, 50)$, close branch $(78, 82)$

much higher than the battery degradation cost. Figure 9.8 depicts the evolution of restored loads. Specifically, the proposed method's value of $\sum_{t \in \mathcal{T}} \sum_{i \in \mathcal{N}} \omega_i \cdot p_{i,t}$ is 35%, 16%, and 6% higher than the benchmark case, the case without further routing, and the case with one further routing, respectively. It is again indicated that the further routing of MPSs facilitates better utilization of MPSs. Table 9.9 lists the further routing of MPSs. Table 9.8 lists DS reconfiguration actions in the MG formation and coordinated restoration process. The figures and tables are self-explanatory. This section does not go into more details for space limit, and note that, for better power quality, reactive power supports provided by MPSs to bound voltage levels within the allowable limits are considered [16].

Table 9.9 Connecting nodes of MPSs in each time period: proposed method and the benchmark case w/ one further routing.

		1	2	3	4	5	6~9	10~11	12	13	14	15	16	17	18~19	20~48
Proposed method	EV fleet 1			Node 27			→			Node 50 →					Node 38	
	EV fleet 2			Node 59				→						Node 27		
	MESS 1		Node 15	→					Node 116							
	MESS 2		Node 103							Node 103					→	Node 116
	MEG 1					→				Node 95				→		Node 103
	MEG 2								Node 47							
w/ one further routing	EV fleet 1	Node 27				→					Node 38			Node 27		
	EV fleet 2			Node 11					→							
	MESS 1		Node 15	→					Node 116							
	MESS 2				→						Node 103					
	MEG 1	Node 103	→									Node 95				
	MEG 2								Node 47							

9.7 Summary and Conclusions

In this chapter, a two-stage framework is proposed, so as to implement resilient routing and scheduling of MPSs before and after a natural disaster. Before the natural disaster, MPSs are pre-positioned in the DS to enable rapid service pre-restoration. After the disaster strikes, MPSs are dynamically dispatched in the DS to coordinate with conventional restoration efforts and novel restoration strategies including the dynamic formation of MGs. Thus, both survivability and recovery performance of the system are enhanced. Case studies demonstrate the proposed method's effectiveness. Future investigation topics include extending the proposed method for other spatial optimization problems, such as resilient dispatch of repair crews, and its co-optimization with MPS dispatch.

Nomenclature

Indices and Sets

i, j	indices of nodes
t, τ	indices of time periods
m	indices of MPSs
$\mathcal{N}$	set of all nodes
$\mathcal{N}^{\text{sub.}}$	set of substation nodes
$\mathcal{N}_m$	set of nodes that MPS m can be connected to
$\mathcal{N}_t^{\text{src.}}$	set of nodes selected as the sources of fictitious flow at time t
$\mathcal{L}$	set of all branches
$\mathcal{L}^{\text{vul.}}$	set of vulnerable branches
$\mathcal{L}^{\text{RCS}}$	set of branches equipped with RCSs
$\mathcal{L}_t^{\text{off}}$	set of branches off (i.e. damaged and unrepaired) at time t
$\mathcal{T}$	set of time periods
$\mathcal{M}, \mathcal{G}, \mathcal{S}, \mathcal{V}$	sets of MPSs, MEGs, MESSs, and EV fleets, i.e. $\mathcal{M} = \{\mathcal{G}, \mathcal{S}, \mathcal{V}\}$
$\mathcal{M}_i$	set of MPSs that can be connected to node i

Parameters

ω_i	priority weight of the power demand at node i
$l_{i,t}$	fictitious load of node i at time t
$\overline{p}_{i,t}, \overline{q}_{i,t}$	real and reactive power demand of node i at time t
$V_i^{\min}, V_i^{\max}$	minimum and maximum voltage values of node i

CAP_i	allowed number of MPSs connected to the station at node i
r_{ij}, x_{ij}	resistance and reactance of branch (i,j)
s_{ij}^{max}	apparent power capacity of branch (i,j)
Δt	duration of one time period
tr_{ij}^m	travel time of MPS m from node i to node j
tp^m	energy consumption rate of EV fleet m when traveling
cp_m^{max}, dp_m^{max}	maximum charging and discharging power of MESS or EV fleet m
η_m^c, η_m^d	charging and discharging efficiency of MESS or EV fleet m
SoC_m^{min}, SoC_m^{max}	minimum and maximum SoC of MESS or EV fleet m
P_m^{max}, Q_m^{max}	maximum real power output and reactive power output of MPS m
I_t	number of islands resulting from damaged and unrepaired branches at time t
Γ_1, Γ_2	estimated maximum number of damaged branches in the sets $\{\mathcal{L}^{vul.}\}$ and $\{\mathcal{L}\backslash\mathcal{L}^{vul.}\}$, respectively
δ_m	transportation cost coefficient of MPS m
K_1, K_2	large enough positive numbers
ζ_m	degradation slope of MESS or EV fleet m
C_m^B	power rating price of MESS or EV fleet m

Variables

α_i^m	binary, 1 if MPS m is prepositioned to node i, 0 otherwise
$\beta_{i,t}^m$	binary, 1 if MPS m is connected to node i at time t, 0 otherwise
γ_t^m	binary, 1 if MPS m is traveling at time t, 0 otherwise
c_t^m	binary, 1 if MESS or EV fleet m is charging at time t, 0 otherwise
d_t^m	binary, 1 if MESS or EV fleet m is discharging at time t, 0 otherwise
cp_t^m	charging power of MESS or EV fleet m at time t
dp_t^m	discharging power of MESS or EV fleet m at time t
SoC_t^m	SoC of MESS or EV fleet m at time t
gp_t^m, gq_t^m	real power output and reactive power output of MPS m at time t
u_{ij}	binary, 1 if branch (i,j) is intact, 0 if damaged

$\lambda_{ij,t}$	binary, 1 if branch (i,j) is closed at time t, 0 if open
$g_{i,t}$	fictitious supply of source node i at time t
$f_{ij,t}$	fictitious flow on branch (i,j) at time t
$p_{i,t}, q_{i,t}$	survived or restored real and reactive power demand of node i at time t
$P_{i,t}, Q_{i,t}$	real power output and reactive power output of node i at time t
$v_{i,t}$	squared voltage magnitude of node i at time t
$pf_{ij,t}, qf_{ij,t}$	real and reactive power flow on branch (i,j) at time t

References

1 U.S. Executive Office of the President and Department of Energy's Office of Electricity and Energy Reliability, "Economic benefits of increasing electric grid resilience to weather outages," 2013.

2 R. J. Campbell and S. Lowry, "Weather-related power outages and electric system resiliency," Congressional Research Service, Washington, DC, USA, 2012.

3 The GridWise Alliance, "Improving electric grid reliability and resilience: Lessons learned from superstorm Sandy and other extremes events," 2013.

4 U.S. Federal Emergency Management Agency, "Mitigation assessment team report: Hurricane Sandy in New Jersey and New York," 2013.

5 Electricity Advisory Committee of U.S. Department of Energy, "Enhancing grid resilience with integrated storage from electric vehicles," 2018.

6 U.S. Electric Power Research Institute, "Enhancing distribution resiliency: Opportunities for applying innovative technologies," 2013.

7 U.S. National Academies of Sciences, Engineering, and Medicine, "Enhancing the resilience of the nation's electricity system," 2017.

8 U.K. House of Lords, "The resilience of the electricity system," 2015.

9 M. Panteli, D. N. Trakas, P. Mancarella, and N. D. Hatziargyriou, "Power systems resilience assessment: Hardening and smart operational enhancement strategies," *Proceedings of the IEEE*, vol. 105, no. 7, pp. 1202–1213, 2017.

10 M. Panteli and P. Mancarella, "The grid: Stronger, bigger, smarter? - Presenting a conceptual framework of power system resilience," *IEEE Power and Energy Magazine*, vol. 13, no. 3, pp. 58–66, 2015.

11 Y. Wang, C. Chen, J. Wang, and R. Baldick, "Research on resilience of power systems under natural disasters–A review," *IEEE Transactions on Power Systems*, vol. 31, no. 2, pp. 1604–1613, 2015.

12 R. Billinton and R. N. Allan, *Reliability Evaluation of Power Systems*. Springer, 2013.

13 B. Zhou, D. Xu, C. Li, Y. Cao, K. W. Chan, Y. Xu, and M. Cao, "Multiobjective generation portfolio of hybrid energy generating station for mobile emergency power supplies," *IEEE Transactions on Smart Grid*, vol. 9, no. 6, pp. 5786–5797, 2017.

14 S. Lei, J. Wang, C. Chen, and Y. Hou, "Mobile emergency generator pre-positioning and real-time allocation for resilient response to natural disasters," *IEEE Transactions on Smart Grid*, vol. 9, no. 3, pp. 2030–2041, 2016.

15 U.S. Electric Power Research Institute, "Transportable energy storage systems project," 2009.

16 H. H. Abdeltawab and Y. A.-R. I. Mohamed, "Mobile energy storage scheduling and operation in active distribution systems," *IEEE Transactions on Industrial Electronics*, vol. 64, no. 9, pp. 6828–6840, 2017.

17 Y. Chen, Y. Zheng, F. Luo, J. Wen, and Z. Xu, "Reliability evaluation of distribution systems with mobile energy storage systems," *IET Renewable Power Generation*, vol. 10, no. 10, pp. 1562–1569, 2016.

18 C. Chen, J. Wang, F. Qiu, and D. Zhao, "Resilient distribution system by microgrids formation after natural disasters," *IEEE Transactions on Smart Grid*, vol. 7, no. 2, pp. 958–966, 2015.

19 K. Schneider, "Microgrids as a resiliency resource," in *2014 International Test Conference*, 2014.

20 R. A. H. El Rashidy, *The resilience of road transport networks redundancy, vulnerability and mobility characteristics*. PhD thesis, University of Leeds, 2014.

21 N. Gordon-Bloomfield, "Nissan, Mitsubishi, Toyota Turn Electric Cars into Backup Batteries," 2014.

22 B. Ozpineci, "Technology development: Updates challenges and opportunities," 2018.

23 S. Iwai, T. Kono, M. Hashiwaki, and Y. Kawagoe, "Use of mobile engine generators as source of back-up power," in *INTELEC 2009 - 31st International Telecommunications Energy Conference*, pp. 1–6, 2009.

24 U.S. Department of Energy, *Global Energy Storage Database*. 2018.

25 H. Farzin, M. Fotuhi-Firuzabad, and M. Moeini-Aghtaie, "Reliability studies of modern distribution systems integrated with renewable generation and parking lots," *IEEE Transactions on Sustainable Energy*, vol. 8, no. 1, pp. 431–440, 2016.

26 V. Monteiro, B. Exposto, J. C. Ferreira, and J. L. Afonso, "Improved vehicle-to-home (iV2H) operation mode: Experimental analysis of the electric vehicle as off-line UPS," *IEEE Transactions on Smart Grid*, vol. 8, no. 6, pp. 2702–2711, 2016.

27 H. Shin and R. Baldick, "Plug-in electric vehicle to home (V2H) operation under a grid outage," *IEEE Transactions on Smart Grid*, vol. 8, no. 4, pp. 2032–2041, 2016.

28 S. M. Mohammadi-Hosseininejad, A. Fereidunian, A. Shahsavari, and H. Lesani, "A healer reinforcement approach to self-healing in smart grid by PHEVs parking lot allocation," *IEEE Transactions on Industrial Informatics*, vol. 12, no. 6, pp. 2020–2030, 2016.

29 N. Z. Xu and C. Y. Chung, "Reliability evaluation of distribution systems including vehicle-to-home and vehicle-to-grid," *IEEE Transactions on Power Systems*, vol. 31, no. 1, pp. 759–768, 2015.

30 H. Gao, Y. Chen, S. Mei, S. Huang, and Y. Xu, "Resilience-oriented pre-hurricane resource allocation in distribution systems considering electric buses," *Proceedings of the IEEE*, vol. 105, no. 7, pp. 1214–1233, 2017.

31 C. Pang, P. Dutta, and M. Kezunovic, "BEVs/PHEVs as dispersed energy storage for V2B uses in the smart grid," *IEEE Transactions on Smart Grid*, vol. 3, no. 1, pp. 473–482, 2011.

32 K. S. A. Sedzro, A. J. Lamadrid, and L. F. Zuluaga, "Allocation of resources using a microgrid formation approach for resilient electric grids," *IEEE Transactions on Power Systems*, vol. 33, no. 3, pp. 2633–2643, 2017.

33 J. Kim and Y. Dvorkin, "Enhancing distribution system resilience with mobile energy storage and microgrids," *IEEE Transactions on Smart Grid*, vol. 10, no. 5, pp. 4996–5006, 2018.

34 M. H. Amirioun, F. Aminifar, and H. Lesani, "Towards proactive scheduling of microgrids against extreme floods," *IEEE Transactions on Smart Grid*, vol. 9, no. 4, pp. 3900–3902, 2017.

35 J. Sun, D. Xie, Y. Lou, M. Yang, and Y. Zhang, "Black-start scheme based on EV's intelligent integrated station," in *2014 International Conference on Power System Technology*, pp. 3118–3123, 2014.

36 S. Yao, P. Wang, and T. Zhao, "Transportable energy storage for more resilient distribution systems with multiple microgrids," *IEEE Transactions on Smart Grid*, vol. 10, no. 3, pp. 3331–3341, 2018.

37 A. Briones, J. Francfort, P. Heitmann, M. Schey, S. Schey, and J. Smart, "Vehicle-to-grid (V2G) power flow regulations and building codes review by the AVTA," 2012.

38 G. Putrus, R. Kotter, Y. Wang, R. Das, G. Obrien, X. Dai, E. Bentley, Y. Cao, R. Heller, and R. Prateek, "Interreg North Sea Region SEEV4-City project: Summary of the state-of-the-art report," 2018.

39 J. M. Conrad, M. E. Randall, G. Vaughn, P. F. Shadwell, and G. W. Randall, "Development and deployment of the IEEE MOVE emergency relief vehicle," in *2016 IEEE Global Humanitarian Technology Conference (GHTC)*, pp. 255–260, 2016.

40 H. Ni, "PJM advanced technology pilots for system frequency control," in *2012 IEEE PES Innovative Smart Grid Technologies Conference (ISGT)*, pp. 1–6, 2012.

41 H. Gao, Y. Chen, Y. Xu, and C.-C. Liu, "Resilience-oriented critical load restoration using microgrids in distribution systems," *IEEE Transactions on Smart Grid*, vol. 7, no. 6, pp. 2837–2848, 2016.

42 O. K. Siirto, A. Safdarian, M. Lehtonen, and M. Fotuhi-Firuzabad, "Optimal distribution network automation considering earth fault events," *IEEE Transactions on Smart Grid*, vol. 6, no. 2, pp. 1010–1018, 2015.

43 M. Lavorato, J. F. Franco, M. J. Rider, and R. Romero, "Imposing radiality constraints in distribution system optimization problems," *IEEE Transactions on Power Systems*, vol. 27, no. 1, pp. 172–180, 2011.

44 W. Yuan, J. Wang, F. Qiu, C. Chen, C. Kang, and B. Zeng, "Robust optimization-based resilient distribution network planning against natural disasters," *IEEE Transactions on Smart Grid*, vol. 7, no. 6, pp. 2817–2826, 2016.

45 M. Baran and F. Wu, "Network reconfiguration in distribution systems for loss reduction and load balancing," *IEEE Transactions on Power Delivery*, vol. 4, no. 2, pp. 1401–1407, 1989.

46 A. Arif, Z. Wang, J. Wang, and C. Chen, "Power distribution system outage management with co-optimization of repairs, reconfiguration, and DG dispatch," *IEEE Transactions on Smart Grid*, vol. 9, no. 5, pp. 4109–4118, 2017.

47 M. A. Ortega-Vazquez, "Optimal scheduling of electric vehicle charging and vehicle-to-grid services at household level including battery degradation and price uncertainty," *IET Generation, Transmission and Distribution*, vol. 8, no. 6, pp. 1007–1016, 2014.

48 B. Zeng and L. Zhao, "Solving two-stage robust optimization problems using a column-and-constraint generation method," *Operations Research Letters*, vol. 41, no. 5, pp. 457–461, 2013.

49 G. Wang, X. Zhang, H. Wang, J.-C. Peng, H. Jiang, Y. Liu, C. Wu, Z. Xu, and W. Liu, "Robust planning of electric vehicle charging facilities with an advanced evaluation method," *IEEE Transactions on Industrial Informatics*, vol. 14, no. 3, pp. 866–876, 2017.

50 S. Lei, J. Wang, and Y. Hou, "Remote-controlled switch allocation enabling prompt restoration of distribution systems," *IEEE Transactions on Power Systems*, vol. 33, no. 3, pp. 3129–3142, 2017.

51 IEEE PES Power System Analysis, Computing and Economics Committee, *IEEE 123 Node Test Feeder*. Feb. 2014.

10

Co-Optimization of Grid Flexibilities in Recovery Logistics

Mobile power sources (MPSs) and repair crews (RCs) are both crucial flexibility resources for distribution grid outage management after a disastrous event. Nevertheless, the optimization of their logistics is not well studied. This chapter proposes a resilient approach for disaster recovery logistics co-optimizing distribution grid restoration with the dispatch of MPSs and RCs. A novel co-optimization model is constructed to coordinately route MPSs and RCs in the transportation network, schedule them in the distribution grid, and reconfigure the distribution grid for forming microgrids, etc. The model incorporates the coupling between power and transportation networks and different timescales of RC/MPS dispatch and distribution grid restoration, etc. To assure radiality of the distribution grid with variable MPS allocation and physical structure, this chapter also formulates topology constraints based on the concept of *spanning forest* introduced in Chapter 8. The model is convexified equivalently and linearized into a mixed-integer linear programming model. To improve its computational efficiency, preprocessing techniques are proposed for pre-assigning a minimum set of repair tasks to depots and reducing the number of candidate nodes for connecting MPSs. Resilient recovery strategies therefore are obtained to improve service restoration, especially by dynamically forming microgrids that are energized by MPSs and topologized by network reconfiguration of the distribution grid and repairing actions of RCs. The proposed methodology is illustrated by case studies.

10.1 Post-Disaster Recovery Logistics of Grids

As aforementioned, recent years have witnessed more frequent disastrous events leading to severe electric outages, which are associated with enormous economic losses, etc. The urgency of improving power grid resilience has been highlighted. Particularly, efficient outage management with effective utilization of flexibility resources is one of the crucial requirements for resilient electric grids.

Power Grid Resilience against Natural Disasters: Preparedness, Response, and Recovery, First Edition.
Shunbo Lei, Chong Wang, and Yunhe Hou.
© 2023 John Wiley & Sons Ltd. Published 2023 by John Wiley & Sons Ltd.

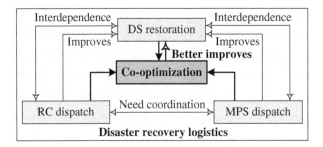

Figure 10.1 Relationships among MPS dispatch, RC dispatch, and DS restoration.

RCs are critical response resources for power grid outage management against disastrous events. It is expected that they repair damaged components in an optimal sequence. MPSs, including truck-mounted mobile energy storage systems (MESSs) and mobile emergency generators (MEGs) [1, 2], are also crucial flexibility resources for grid recovery. They can power critical loads that lose access to the main grid power. Regarding power grids, including distribution systems (DSs) in this chapter, their recovery involves different strategies and many decisions, e.g. network reconfiguration for forming microgrids [3–5]. Incorporating those elements (see Figure 10.1), DS outage management for electric service recovery becomes a *disaster recovery logistics* problem, which involves the dispatch of different flexibility resources in different infrastructures, including the load pick-up and reconfiguration plans of the DS, the scheduling of MPSs' power outputs and RCs' repair actions in the DS, and the routing of MPSs and RCs in the transportation network [6]. The involved decisions are interdependent. As will be elaborated on in later sessions of this chapter, this problem has not been well investigated.

10.1.1 Power Infrastructure Recovery

Utilities normally resolve MPS dispatch, RC dispatch, and DS restoration as three separate problems. To schedule RCs, experience and pre-defined criteria or priorities are considered, while the dependencies of DS operation on the repair plan are commonly neglected [7]. Regarding MPSs, in general, they are simply allocated to nodes with more critical loads [8], without systematic consideration of their utilization. Those approaches do not obtain optimal solutions for the dispatch of flexibilities provided by MPSs and RCs.

As shown in Figure 10.1, in practice those three sub-problems are interrelated and interdependent. For instance, the fast recovery of a critical load may rely on the operability of a line to be repaired and the availability of a MPS at a specific node. Sophisticated coordination among MPSs, RCs, and the DS, can significantly improve service recovery. It is difficult to attain such an optimal outage management plan solely based on experience of the operator. An integrated approach for coordinating available flexibility resources is needed. References [9, 10] review the planning and dispatch of RCs in DSs. The involved interdependencies are often ignored or simplified, e.g. in [11]. Only several works, e.g. [12], co-optimize DS restoration with RC dispatch. Regarding MPSs, in [13, 14], and [15], MESSs are dispatched to decrease DS operation cost, enhance reliability, and improve preparedness against disastrous events, respectively. Reference [16] provides a review on the use of MEGs. Generally, MPS dispatch is seldom co-optimized with DS restoration in a dynamic manner. Besides, few works have considered both MPSs and RCs for DS service recovery.

Approaches with quite less- detailed modeling of the dispatch of involved flexibilities may be especially suitable for mid-term or long-term management of RCs, etc. However, disaster recovery of DSs explored here is a near-term problem. An approach considering the detailed dispatch of available flexibilities is needed. In this regard, interdependencies among the three involved sub-problems need to be taken into account.

A very related problem, i.e. transmission system restoration and repairing, has been investigated in some works. In [17, 18], the routing and scheduling of RCs for post-disaster transmission system recovery is first studied. They decouple the vehicle routing and power recovery sub-problems with the interdependencies characterized by constraints injection and demonstrate the proposed *LPAC* power flow model. A co-optimization method for the same problem is presented in [19]. In [20], an integer program for scheduling repair tasks is developed. In [21], a pre-hurricane restoration model is proposed for pre-allocating RCs, etc. These approaches are not applicable here mainly because of the characteristics of the DS repair and recovery problem and the incorporation of MPSs. Compared to transmission systems, DSs are built with a quite lower level of redundancy and are directly connected with customers. An approach with more specific modeling of the DS recovery and repair process, etc., is needed to enable more coordinated and interactive dispatch of the available flexibilities.

10.1.2 Microgrid-Based Service Restoration

Dynamic network reconfiguration is an essential flexibility of smart DSs. In [22], with a mixed-integer non-linear, multi-period optimal power flow model, both dynamic and static network reconfiguration are adopted for enhancing distributed generation hosting capacity. In [23], for minimizing power losses, a hierarchical decentralized agent-based dynamic network reconfiguration approach is proposed. Reference [24] investigates segmented-time reconfiguration coupled with reactive power control of distributed generators for minimizing the comprehensive cost. In [25], hourly reconfiguration in the presence of renewable energy resources is investigated using mixed-integer second-order cone programming that minimizes daily network losses. To reduce load shedding in islanded operation and power losses in grid-connected operation, microgrid optimal scheduling with dynamic network reconfiguration is explored in [26]. More detailed literature reviews on DS dynamic network reconfiguration can be found in [22, 27, 28], etc. The microgrid formation problem explored in [3–5] is actually static network reconfiguration. This chapter is one of the first to use dynamic network reconfiguration for dynamically forming microgrids.

Using microgrids for smart DS restoration is explored in many works. Reference [29] investigates voltage and frequency controls during the sequential black-start restoration of microgrids. In [30], an algorithm for obtaining the automatic switching time is developed for recovering microgrids. Reference [31] models microgrids as virtual feeders and uses spanning tree search algorithms to find the strategy that minimizes the number of switching actions and maximizes restored loads. Reference [32] proposes to optimally sectionalize the DS into networked self-adequate microgrids that continuously provide reliable electric service for a maximum number of loads. In [33], using the concept of *continuous operating time*, microgrids' availability for critical load recovery and the service time are assessed to improve DS resilience. More detailed literature reviews can be found in [3, 33], etc. In general, those publications assume that microgrids are installed beforehand for providing flexibility resources. Followed by Refs. [4, 5], etc., Ref. [3] is the first to form microgrids with distributed generations for continuing the power supply of critical loads after a natural disaster. As mentioned earlier, this chapter further extends the microgrid formation strategy to be dynamic microgrid formation. Besides, rather than distributed generations, MPSs are used in this process.

10.1.3 A Co-Optimization Approach

This chapter presents a co-optimization approach for disaster recovery logistics of DSs. Available flexibility resources are managed simultaneously and coordinately. The three involved interdependent sub-problems are included in a single optimization problem. Their inter-dependencies are also formulated using relevant constraints. With the proposed approach, decisions of the three sub-problems are co-optimized, so as to obtain an outage management plan optimally coordinating different flexibility resources. Specifically, MPS dispatch and RC dispatch are jointly coordinated with DS restoration for better improving grid resilience. A non-convex mixed-integer non-linear programming (MINLP) model is formulated for co-optimizing the scheduling and routing of MPSs and RCs, and the dynamic microgrid formation of DSs, etc. Issues including the modeling of radiality constraints for a DS with variable MPS allocation and physical structure, a state of charge (SoC) variations of MESSs over time, the coupling between power and transportation networks, and the different timescales of DS restoration and RC/MPS dispatch, are resolved. The model is equivalently convexified as a mixed-integer second-order cone programming (MISCOP) model and further linearized to be a mixed-integer linear programming (MILP) model. Similar to [18], etc., some heuristic pre-processing approaches, e.g. pre-assigning the minimum repair tasks, are also developed to decrease computation time of the co-optimization model.

In the following, Section 10.2 formulates RC dispatch, MPS dispatch, and DS restoration. Section 10.3 develops the co-optimization model. Sections 10.4–10.6 provide the solution method, case studies, and summary, respectively.

10.2 Flexibility Resources in Grid Recovery Logistics

10.2.1 Routing and Scheduling of Repair Crews

After a disastrous event that damages multiple components, patrols first are dispatched for assessing damages. Other flexibility resources, e.g. unmanned aerial vehicles [34], may also be used. Damages are located and assessed. The repairing plan for each damage is also drawn up. For instance, some damages need to be repaired by replacing the pole and some by rejoining cable sections. With rich experience, the repairing time may be estimated. After the damage assessment, RCs are then dispatched, and other recovery efforts are initiated, too.

RC dispatch involves two interdependent sub-tasks, i.e. routing and scheduling. Routing is to choose a route for each RC to travel among damaged components and depots. Scheduling is to set a timetable for repairing and traveling actions of RCs. Let $\mathcal{V} \triangleq \mathcal{V}_1 \cup \mathcal{V}_2$ be the set of damaged components ($\mathcal{V}_1$) and depots ($\mathcal{V}_2$). Let $\mathcal{E} \triangleq \{(m, n), \forall m, n \in \mathcal{V}\}$ be the set of edges for all pairs of vertices. Then, the routing of RCs is to find their optimal paths in graph $\mathcal{G} \triangleq (\mathcal{V}, \mathcal{E})$. As the routing problem can be seen as a generalization of the traveling salesman problem (TSP) [35], its modeling is commonly based on the TSP formulation, e.g. in [12]. This chapter proposes a formulation that is more appropriate and much simpler for the studied problem.

The TSP formulation is based on edge-wise routing variables a_{mn}^k (1 if RC k travels through edge (m, n), 0 otherwise). This chapter uses vertex-wise variables $a_{m,t}^k$ instead (1 if RC k is at vertex m at time t, 0 otherwise). Note that with the subscript t, variables $a_{m,t}^k$ are essentially scheduling variables. This chapter will show that the routing problem can be formulated and incorporated with variables $a_{m,t}^k$ only.

The considered transportation network $\mathcal{G}$ can be formed based on the original transportation network $\mathcal{G}_0$. In general, an edge in $\mathcal{G}$ represents a path in $\mathcal{G}_0$.

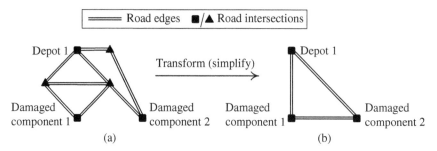

Figure 10.2 An illustration for the construction of $\mathcal{G}$. (a) Graph $\mathcal{G}_0$, the original transportation network. (b) Graph $\mathcal{G}$, the considered transportation network.

Figure 10.2 is self-explaining. If in $\mathcal{G}_0$ a path between a damaged component and another damaged component or the depot exists, then an edge between them can be included in $\mathcal{G}$. Typically, the shortest path can be found by the Floyd–Warshall algorithm [36] and the Dijkstra's algorithm [37], etc. Mostly, $\mathcal{G}_0$ has at least one path between any two vertices considered in $\mathcal{G}$. Thus, $\mathcal{G}$ is commonly a complete graph. In some undesired cases, the transportation network is seriously damaged, leading to a disconnected graph $\mathcal{G}_0$. (And the alternative paths are too distant and not included in $\mathcal{G}_0$.) In those cases, $\mathcal{G}$ can be disconnected. The travel time between disconnected vertices can be set as infinite. To avoid numerical issues, setting it as $|\mathcal{T}|$ is sufficiently large here, where $\mathcal{T}$ is the set of sequential time periods considered in the recovery logistics problem.

In the scheduling of RCs, each RC k can only be visiting at most one vertex at each time step t:

$$\sum_{m \in \mathcal{V}} a^k_{m,t} \le 1, \ \forall k, \ \forall t \tag{10.1}$$

For conciseness of the objective function to be formulated in Section 10.3, auxiliary variables β^k_t are introduced:

$$\beta^k_t = 1 - \sum_{m \in \mathcal{V}} a^k_{m,t}, \ \forall k, \forall t \tag{10.2}$$

That is, if RC k is traveling on the transportation network at time t, $\beta^k_t = 1$; if it is visiting one of the vertices, $\beta^k_t = 0$. As β^k_t is set as binary, Eq. (10.2) essentially includes constraint (10.1).

The routing of RCs enforces traveling time constraints on the scheduling of RCs. For instance, if RC k is at vertex m at time $t = 1$ (i.e. $a^k_{m,1} = 1$), and it takes two time steps to travel from vertex m to n, then RC k can be at vertex n only after $t = 3$ (i.e. $a^k_{n,2} = a^k_{n,3} = 0$). Such constraints are formulated as follows:

$$a^k_{n,t+\tau} + a^k_{m,t} \le 1, \ \forall k, \ \forall m \ne n, \ \forall \tau \le tr_{mn}, \ \forall t \le |\mathcal{T}| - \tau \tag{10.3}$$

where tr_{mn} is the travel time between vertices m and n, and $|\mathcal{T}|$ is the number of time steps. Note that constraint (10.3) actually includes constraint (10.1) implicitly with $\tau = 0$. To decrease the number of constraints, constraint (10.3) can be equivalently transformed into:

$$\sum_{\tau=t}^{\min\{t+tr_{mn},|\mathcal{T}|\}} a_{n,\tau}^k \leq \left(1 - a_{m,t}^k\right) \cdot \min\{tr_{mn}, |\mathcal{T}| - t\}, \forall k, \forall m \neq n, \forall t$$

(10.4)

Actually, constraint (10.3) or (10.4) is sufficient to incorporate the RC routing sub-problem in the co-optimization model. Other routing constraints, e.g. path-flow balance, will be satisfied implicitly. The validity of constraints (10.3) and (10.4) is supported by the proposition as follows:

Proposition 10.1 *For any RC scheduling plan satisfying constraint (10.3) or (10.4), there exists at least one associated feasible path in graph $\mathcal{G}$ representing the transportation network.*

The proof of Proposition 10.1 is straightforward. The key is that variables $a_{m,t}^k$ can be used to retrieve the path of RC k simply by

$$\begin{bmatrix} a_{1,1}^k & \cdots & a_{|\mathcal{V}|,1}^k \\ \vdots & \ddots & \vdots \\ a_{1,|\mathcal{T}|}^k & \cdots & a_{|\mathcal{V}|,|\mathcal{T}|}^k \end{bmatrix} \cdot \begin{bmatrix} 1 \\ \vdots \\ |\mathcal{V}| \end{bmatrix}$$

(10.5)

where $|\mathcal{V}|$ is the cardinality of $\mathcal{V}$. (Hereinafter, $|\cdot|$ denotes the cardinality of the symbol inside it.) For the detailed proof of Proposition 10.1, please see Appendix 10.A.

The advantages of the proposed formulation for the RC routing subproblem are mainly twofold: (i) The issues of transportation-power networks' coupling and their different timescales are resolved in a simpler manner compared to common TSP formulations. (ii) It does not force RCs to visit all vertices of damaged components ($\mathcal{V}_1$), so that the co-optimization model may flexibly select to repair a minimal set of damaged components to recover all loads.

Next, the repair plan, which relies on the flexible routing and scheduling of RCs, is formulated. Let $z_{m,t}^k$ be 1 if a damaged component m is repaired by RC k at time t, 0 otherwise. Let rt_m^k be the required time steps for RC k to repair damaged component m. Then, the following constraints are enforced:

$$z_{m,t}^k \leq \frac{\sum_{\tau=1}^t a_{m,\tau}^k}{rt_m^k}, \forall k, \forall m \in \mathcal{V}_1, \forall t$$

(10.6)

$$z_{m,t}^k \leq z_{m,t+1}^k, \forall k, \forall m \in \mathcal{V}_1, \forall t \leq |\mathcal{T}| - 1$$

(10.7)

Particularly, constraint (10.6) states that $z_{m,t}^k = 1$ only if RC k has spent rt_m^k time steps repairing damaged component m; constraint (10.7) fixes the status of a damaged component in the following time steps once repaired. For instance, if it takes $rt_1^1 = 3$ time steps of RC 1 to repair damaged component 1, and RC 1 repairs it during time steps $t = 1 \sim 3$ (i.e. $\left[a_{1,1}^1, a_{1,2}^1, a_{1,3}^1\right] = [1, 1, 1]$), then according to constraint (10.6), we have $z_{1,1}^1 \leq \frac{1}{3}$, $z_{1,2}^1 \leq \frac{1+1}{3}$, and $z_{1,3}^1 \leq \frac{1+1+1}{3}$. As $z_{m,t}^k$ is binary, and considering constraint (10.7), we can further have $\left[z_{1,1}^1, z_{1,2}^1, z_{1,3}^1, \ldots, z_{1,|\mathcal{T}|}^1\right] = [0, 0, 1, \ldots, 1]$. Note that decisions such as $\left[a_{1,1}^1, a_{1,2}^1, a_{1,3}^1, a_{1,4}^1\right] = [1, 0, 1, 1]$ (idling the RC at $t = 2$ during the repair) and $\left[a_{1,1}^1, a_{1,2}^1, a_{1,3}^1, a_{1,4}^1\right] - [1, 1, 1, 1]$ (idling the RC at $t = 4$ after the repair) arc also feasible for constraint (10.6). As long as the idling of RCs deteriorates the optimality, the to-be-obtained optimal solution will eliminate those situations. However, a solution with insignificant idling of RCs may still be optimal. For instance, if the recovery of the last outage load depends on the repairing of both damaged components 3 and 4, which can be repaired by RC 1 and RC 2 at $t = 11$ and $t = 12$, respectively, it is acceptable to idle RC 1 at $t = 11$ and get damaged component 3 repaired at $t = 12$. Those situations are acceptable, since they do not result in sub-optimality. However, if a solution with idle visiting is obtained, it has to be recast to eliminate the idling of RCs. For instance, if the solution suggests that the scheduling of RC k is $\left[m_1^k, m_2^k, \ldots, m_{|\mathcal{T}|}^k\right] = [1, 0, 2, 0, 2, 0, 0, 3, 0, 3]$, it can be recast to be $[1, 0, 2, 2, 0, 0, 3, 3]$. (Here, m_t^k is the vertex that RC k is visiting at time t; $m_t^k = 0$ means that RC k is not visiting any vertex at time t.)

The following constraint enforces that each damaged component is repaired only once by one of the RCs:

$$\sum_k z_{m,t}^k \leq 1, \ \forall m \in \mathcal{V}_1, \ \forall t \tag{10.8}$$

Actually, constraint (10.8) is redundant. Without it, the co-optimization model also seeks an optimal solution without repeated repairs. Nevertheless, with constraint (10.8), the linear programming relaxation of the MILP co-optimization model is tightened. It helps reduce the solution time using MILP solvers that are generally based on the branch-and-cut framework [38].

The following constraint enforces that RC k's resource capacity suffices the total resources required by its repair tasks:

$$\sum_{m \in \mathcal{V}_1} z_{m,|\mathcal{T}|}^k \cdot rs_m \leq RS^k, \ \forall k \tag{10.9}$$

where rs_m is the number of resources required to repair damaged component m, and RS^k is RC k's resource capacity.

10.2.2 Routing and Scheduling of Mobile Power Sources

The dispatch of MPSs also involves two interdependent subtasks, i.e. MPS routing and MPS scheduling. Routing is to select a route for each MPS to travel among candidate DS nodes for MPS connection. Scheduling is to manage MPSs' traveling behaviors and power outputs (or power inputs for MESSs when charging) over the considered time window. The dispatch of MPSs and RCs have major differences. To name a few, first, a damaged component is repaired by one visit of one RC, while a DS node can be connected with more than one MPSs, and a MPS can be connected to a node more than once. Second, once repaired, a component can be operated in the following time periods, while the availability or unavailability of MPS(s) at a node are interchangeable states in the restoration process. Third, the time required to repair a component is a parameter, while the time of a MPS connecting to a node is a variable. These features contribute to the existing methods' inapplicability here for optimizing the utilization of flexibility provided by MPSs.

Let $\mathcal{N}$ be the set of DS nodes, and $\mathcal{N}' \subset \mathcal{N}$ be the set of candidate nodes for MPS connection. Let $\mathcal{M}_1$ be the set of MEGs, $\mathcal{M}_2$ be the set of MESSs, and $\mathcal{M} \triangleq \{\mathcal{M}_1 \cup \mathcal{M}_2\}$ be the set of MPSs. Let CAP_i be the allowed number of MPSs connected to node i. Similar to the RC dispatch model, let $\alpha_{i,t}^s$ be 1 if MPS s is connected to node i at time t, 0 otherwise. Then, the following constraints are enforced:

$$\sum_{i \in \mathcal{N}'} \alpha_{i,t}^s \leq 1, \ \forall s, \ \forall t \tag{10.10}$$

$$\sum_{\tau=t}^{\min\{t+tr_{ij}, |\mathcal{T}|\}} \alpha_{j,\tau}^s \leq \left(1 - \alpha_{i,t}^s\right) \cdot \min\{tr_{ij}, |\mathcal{T}| - t\}, \ \forall s, \ \forall i \in \mathcal{N}', \ \forall j \in \mathcal{N}'\backslash\{i\}, \ \forall t \tag{10.11}$$

$$\sum_{s \in \mathcal{M}} \alpha_{i,t}^s \leq CAP_i, \ \forall i \in \mathcal{N}', \ \forall t \tag{10.12}$$

Above, constraints (10.10) and (10.11) are essentially the same as constraints (10.1) and (10.4), respectively. Constraint (10.12) limits the number of MPSs connected to each candidate node by its capacity. The routing and traveling behaviors of MPSs are sufficiently formulated by constraints (10.10)–(10.12). Again, for conciseness of the objective function to be modeled in Section 10.3, auxiliary variables β_t^s are introduced:

$$\beta_t^s = 1 - \sum_{i \in \mathcal{N}'} \alpha_{i,t}^s, \ \forall s, \ \forall t \tag{10.13}$$

That is, $\beta_t^s = 1$ if MPS s is traveling on the transportation network at time t, and $\beta_t^s = 0$ if it is connected to the DS.

Next, the power dispatch flexibility of MPSs is formulated. First, for MEGs, let gp_t^s and gq_t^s be the real and reactive power outputs of MEG s at time t, respectively. Let $\overline{gp}^s$ and $\overline{gq}^s$ be the maximum real and reactive power outputs of MEG s, respectively. Then, the following constraints are enforced:

$$0 \le gp_t^s \le \sum_{i \in \mathcal{N}'} \alpha_{i,t}^s \cdot \overline{gp}^s, \forall s \in \mathcal{M}_1, \forall t \tag{10.14}$$

$$0 \le gq_t^s \le \sum_{i \in \mathcal{N}'} \alpha_{i,t}^s \cdot \overline{gq}^s, \forall s \in \mathcal{M}_1, \forall t \tag{10.15}$$

With constraints (10.14) and (10.15), MEGs' power outputs are restricted by their capacities, and are enforced to be zero if a MEG is not connected to the DS. Fuel limits of MEGs are not included, as they can be refueled by tanker trucks in case of long-term blackouts [39]. Still, to use MEGs as flexibility resources against natural disasters, fuel supply is a critical issue. After the disastrous event, efficient fuel transportation may be hindered by communication problems between distributors [40]. Potential risks due to the fuel supply issue can be resolved or relieved by measures as follows:

- Selected candidate nodes for MEG connection are required to be free of threats such as flooding and allow fuel trucks to access them via reliable roadways [16, 39]. Nodes at locations with sub-base or underground storage tanks, which can be filled prior to the disastrous event, are surely preferable [41, 42].
- Before the disastrous event, operators can pre-allocate fuel in the system [15]. For nodes without pre-installed fuel tanks, portable tanks can be used. MEGs can also be equipped with towable fuel tanks, which can greatly extend MEGs' continuous operating time [43].
- After the disastrous event, suppliers may maintain fuel supply by optimally dispatching fuel trucks. The massive requirements for transferring fuel necessitate using the optimal fuel truck routing and sourcing additional vehicles and drivers [40]. Methods and algorithms in [44, 45], etc., can be used.

By the above measures, MEGs' continuous operating time can be sufficiently prolonged. The US Federal Emergency Management Agency and authorities of other countries also instruct utilities to build schemes for maintaining fuel supply in emergency situations [42]. Thus, this chapter follows [8, 39], etc., and assumes adequate fuel for MEGs.

Second, for MESSs, let c_t^s and d_t^s be 1 if MESS s is charging and discharging at time t, respectively, and 0, otherwise. Let cp_t^s and dp_t^s be the charging and discharging power of MESS s at time t, respectively. Let $\overline{cp}^s$ and $\overline{dp}^s$ be the maximum charging and discharging power of MESS s, respectively. Let soc_t^s be the SoC of MESS s at time t. Let η_s^c and η_s^d be the charging and discharging efficiency of MESS s, respectively. Let $\underline{soc}^s$ and $\overline{soc}^s$ be the minimum and maximum SoC of MESS s, respectively. Then, the following constraints are enforced:

$$c_t^s + d_t^s \le \sum_{i \in \mathcal{N}'} \alpha_{i,t}^s, \forall s \in \mathcal{M}_2, \forall t \tag{10.16}$$

$$0 \le cp_t^s \le c_t^s \cdot \overline{cp}^s, \forall s \in \mathcal{M}_2, \forall t \tag{10.17}$$

$$0 \le dp_t^s \le d_t^s \cdot \overline{dp}^s, \forall s \in \mathcal{M}_2, \forall t \tag{10.18}$$

$$0 \le gq_t^s \le (c_t^s + d_t^s) \cdot \overline{gq}^s, \forall s \in \mathcal{M}_2, \forall t \tag{10.19}$$

$$soc_{t+1}^s = soc_t^s + \left(cp_t^s \cdot \eta_s^c - \frac{dp_t^s}{\eta_s^d} \right) \cdot \Delta t, \forall s \in \mathcal{M}_2, \forall t \le |\mathcal{T}| - 1 \tag{10.20}$$

$$\underline{soc}^s \le soc_t^s \le \overline{soc}^s, \forall s \in \mathcal{M}_2, \forall t \tag{10.21}$$

where Δt is the duration of one time period. Specifically, constraint (10.16) ensures that in each time period, charging and discharging are mutually exclusive states of a MESS, and it can neither charge nor discharge if it is not connected to the DS. Constraints (10.17) and (10.18) specify MESSs' charging and discharging power limits, respectively, (if a MESS is not in the charging/discharging state, its charging/discharging power is limited to be zero). Constraint (10.19) is similar to constraint (10.15). Constraint (10.20) expresses MESSs' SoC variations over time. Constraint (10.21) imposes SoC ranges for MESSs.

MESSs are normally delivered by trucks, whose running needs fuel. Nevertheless, this matter does not enforce a fuel constraint here, as the footprint of a DS or several neighboring DSs is commonly much less than the range of trucks. In general, the flexible operation of MESSs is not subject to the fuel supply issue.

Electric vehicles (EVs), especially utility-owned medium-duty EVs, can also be dispatched to facilitate DS restoration using the proposed method in a manner similar to that of MESSs. Therefore, in undesired cases that the fuel supply cannot support MEGs' effective utilization, the proposed method can still work by using flexibility provided by other types of MPSs, e.g. MESSs and EVs.

10.2.3 Grid Reconfiguration and Operation

To coordinate with RC and MPS dispatches, the DS is dynamically reconfigured, e.g. to form microgrids. DS topology has to be radial in this process. As researchers have extensively studied DS reconfiguration, the modeling of radiality constraints is resolved to some extent. However, simply adding some commonly used radiality constraints does not work here, as the formulation also has to allow dynamic and flexible microgrid formation. Moreover, the problem considered in this chapter encounters a new situation. That is, the physical structure of the DS varies with the to-be-optimized repair plan, and the distribution of power source nodes in the DS varies with the to-be-optimized allocation of MPSs. In general, this chapter is reconfiguring a variable DS. For example, a natural disaster splits the DS in Figure 10.3 into four physical islands (PIs). Then, at a future time step t of the recovery process, the number of PIs, and the components in each PI, etc., are all variables that are not only dependent on but also to be co-optimized with RC and

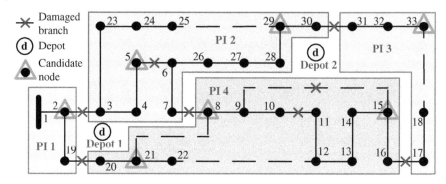

Figure 10.3 IEEE 33-node test system split into multiple PIs.

MPS dispatch decisions. Existing methods of formulating radiality constraints in the literature are not applicable in this case.

Different from common DS reconfiguration problems that seek an optimal *spanning tree* (i.e. a radial topology connecting all nodes without loops), in this chapter, with PIs, each feasible topology is a *spanning forest* (i.e. a radial topology with each sub-graph being a spanning tree; see Figure 10.4). Recent publications [3–5] on microgrids formation are essentially constructing a spanning forest, too. However, their models are neither applicable here. Note that Ref. [12] is essentially also reconfiguring a variable DS. It uses the constraints from Ref. [46] to ensure radiality. However, a topology satisfying such constraints is not necessarily radial [47].

Observing the differences and connections between spanning trees and spanning forests, this chapter uses the approach proposed in Chapter 8 to express radiality constraints for a variable DS, so that the topological flexibility is fully enabled. For the DS network, let λ_t^{ij} be the connection status of branch (i,j) at time t (1 if closed, 0 if open), and let u_t^{ij} be the operable status of branch (i,j) at time t (1 if can be used, 0 if damaged and unrepaired). Another fictitious network that is the same as the DS network but without damages is introduced. Let $e_t \triangleq [e_t^{ij}, \forall(i,j)]$ be the connection status of branches in the fictitious network at time t. Then, radiality

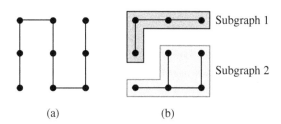

Figure 10.4 (a) A spanning tree. (b) A spanning forest.

constraints of the studied problem are expressed as follows:

$$\lambda_t^{ij} \le e_t^{ij}, \forall(i,j), \forall t \tag{10.22}$$

$$e_t \in \Omega, \forall t \tag{10.23}$$

$$\lambda_t^{ij} \le u_t^{ij}, \forall(i,j), \forall t \tag{10.24}$$

where Ω is the set of spanning tree topologies of the fictitious network. For any time step t, constraint (10.22) restricts the DS to close only a subset of the closed branches in the fictitious network. In other words, constraint (10.22) demands that the closed branches in the fictitious network determined by e_t is a superset of the closed branches in the DS determined by $\lambda_t \triangleq [\lambda_t^{ij}, \forall(i,j)]$. In this regard, constraint (10.23) requires that there exists feasible value(s) of e_t in the set Ω to satisfy constraint (10.22). That is, constraint (10.23) requires e_t to form a fictitious spanning tree in satisfying constraint (10.22). Therefore, constraints (10.22) and (10.23) together enforce the DS to close a subset of the branches in the fictitious spanning tree, so that λ_t forms a spanning forest in each time period. And constraint (10.24) further enforces inoperable branches to be open. The explicit formulation of constraint (10.23) will be presented in a later part of this chapter.

The validity of constraints (10.22)–(10.24) is supported by the following property: Removing $l \ge 0$ edges from a spanning tree results in a spanning forest. In other words, connected components in a sub-graph of a spanning tree are also spanning trees, thus forming a spanning forest.

An example network in Figure 10.5 is used to illustrate constraints (10.22)–(10.24). Apparently, $e_t = [e_t^{12}, e_t^{13}, e_t^{23}]$, $\Omega = \{[0, 1, 1], [1, 0, 1], [1, 1, 0]\}$,

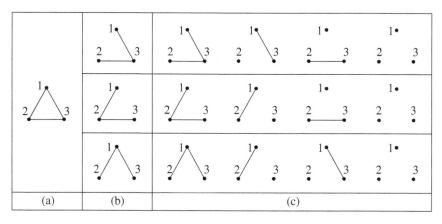

(a)	(b)	(c)

Figure 10.5 (a) An example of three-node network. (b) Possible spanning tree topologies of the example network, i.e. possible values of e_t. (c) Possible spanning forest topologies of the example network, i.e. possible values of λ_t corresponding to different realized values of e_t.

and $\lambda_t = [\lambda_t^{12}, \lambda_t^{13}, \lambda_t^{23}]$. As aforementioned, constraints (10.22) and (10.23) ensure the DS topology determined by λ_t to have closed branches that is a subset of the branches in the fictitious spanning tree determined by e_t. That is, constraint (10.23) enforces $e_t = [0, 1, 1]$, $[1, 0, 1]$ or $[1, 1, 0]$, and constraint (10.22) requires that, in a disjunctive manner, $\lambda_t = [0, 1, 1]$, $[0, 1, 0]$, $[0, 0, 1]$ or $[0, 0, 0]$ if $e_t = [0, 1, 1]$, $\lambda_t = [1, 0, 1]$, $[1, 0, 0]$, $[0, 0, 1]$ or $[0, 0, 0]$ if $e_t = [1, 0, 1]$, and $\lambda_t = [1, 1, 0]$, $[1, 0, 0]$, $[0, 1, 0]$ or $[0, 0, 0]$ if $e_t = [1, 1, 0]$. All valid trees or forests are feasible for the constraints. Constraint (10.24) further eliminates trees and forests that cannot be formed due to branch damages. For example if $u_t^{12} = 0$, constraint (10.24) enforces $\lambda_t^{12} = 0$. Thus, λ_t cannot be $[1, 1, 0]$, $[1, 0, 1]$ or $[1, 0, 0]$.

Note that constraint (10.23) is not to enumerate all spanning trees. It is just expressed conceptually as in constraint (10.23). As mentioned in Chapter 8, the literature contains many kinds of spanning tree constraints [48]. In this work, the extensively used single-commodity flow model is adopted to formulate constraint (10.23) [49]. For a DS with one substation node, its topology is a spanning tree if and only if both of the following conditions are satisfied: (i) It has $|\mathcal{N}| - 1$ closed branches; and (ii) All nodes are connected [50]. Condition (i) is enforced by the following:

$$\sum_{(i,j) \in \mathcal{L}} e_t^{ij} = |\mathcal{N}| - 1, \forall t \tag{10.25}$$

As for Condition (ii), the single-commodity flow model assumes a virtual commodity flow between nodes via branches. Let the substation node and other nodes be the source and sinks for the assumed commodity flow, respectively. Then, this model enforces Condition (ii) by requiring the source to send one unit of the commodity to each sink. Let f_t^{ij} be the commodity flow on branch (i, j) at time t, let $\overline{f}^{ij}$ be the maximum commodity flow allowed on branch (i, j), and let i_g be the substation node. Condition (ii) is formulated as follows:

$$\sum_{(j,i) \in \mathcal{L}} f_t^{ji} - \sum_{(i,j) \in \mathcal{L}} f_t^{ij} = 1, \forall i \in \mathcal{N} \backslash i_g, \forall t \tag{10.26}$$

$$-e_t^{ij} \cdot \overline{f}^{ij} \le f_t^{ij} \le e_t^{ij} \cdot \overline{f}^{ij}, \forall (i, j), \forall t \tag{10.27}$$

Specifically, constraint (10.26) ensures all sinks to satisfy the commodity flow balance condition. Constraint (10.27) restricts the commodity flow on disconnected branches to be zero. Setting $\overline{f}^{ij} = |\mathcal{N}| - 1$ is large enough. For more details on using the single-commodity flow model to construct spanning tree constraints, interested readers can refer to Chapter 8 and Refs. [49, 51], etc. While some other kinds of spanning tree constraints may require additional discrete variables and potentially a combinatorial number of constraints [52, 53], the single-commodity flow-based spanning tree constraints (10.25)–(10.27) here only involve $|\mathcal{L}|$ binary variables, $|\mathcal{L}|$ continuous variables and $|\mathcal{N}| + |\mathcal{L}|$ constraints for each time period.

Next, power dispatch of the DS is formulated. Let δ_t^i be 1 if the load at node i is restored at time t, 0, otherwise. Let p_t^i and q_t^i be the real and reactive power demand of node i at time t, respectively. Let P_t^i and Q_t^i be the real and reactive power output of the MPS(s) connected to node i at time t, respectively. Let v_t^i be the voltage magnitude of node i at time t. Let $\underline{v}^i$ and $\overline{v}^i$ be the minimum and maximum voltage value of node i, respectively. Let pf_t^{ij} and qf_t^{ij} be the real and reactive power flow on branch (i,j) at time t, respectively. Let r_{ij}, x_{ij} and $\overline{S}_{ij}$ be the resistance, reactance and apparent power capacity of branch (i,j), respectively. Then, the following constraints are enforced:

$$\delta_t^i \leq \delta_{t+1}^i, \forall i, \forall t \leq |\mathcal{T}| - 1 \tag{10.28}$$

$$P_t^i - \delta_t^i \cdot p_t^i + \sum_{(j,i) \in \mathcal{L}} pf_t^{ji} - \sum_{(i,j) \in \mathcal{L}} pf_t^{ij} = 0, \forall i, \forall t \tag{10.29}$$

$$Q_t^i - \delta_t^i \cdot q_t^i + \sum_{(j,i) \in \mathcal{L}} qf_t^{ji} - \sum_{(i,j) \in \mathcal{L}} qf_t^{ij} = 0, \forall i, \forall t \tag{10.30}$$

$$(\underline{v}^i)^2 \leq (v_t^i)^2 \leq (\overline{v}^i)^2, \forall i, \forall t \tag{10.31}$$

$$(pf_t^{ij})^2 + (qf_t^{ij})^2 \leq \lambda_t^{ij} \cdot (\overline{S}_{ij})^2, \forall (i,j), \forall t \tag{10.32}$$

$$(v_t^i)^2 - (v_t^j)^2 \leq 2 \cdot \left(r_{ij} \cdot pf_t^{ij} + x_{ij} \cdot qf_t^{ij} \right) + \left(1 - \lambda_t^{ij} \right) \cdot K, \forall (i,j), \forall t \tag{10.33}$$

$$(v_t^i)^2 - (v_t^j)^2 \geq 2 \cdot \left(r_{ij} \cdot pf_t^{ij} + x_{ij} \cdot qf_t^{ij} \right) + \left(\lambda_t^{ij} - 1 \right) \cdot K, \forall (i,j), \forall t \tag{10.34}$$

where $\mathcal{L}$ is the set of DS branches, and K is a large enough positive number. As above, constraint (10.28) prevents de-energizing loads that are already restored. Constraints (10.29) and (10.30) require all nodes to satisfy the real and reactive power balance conditions, respectively. Constraint (10.31) specifies voltage magnitude limits. Constraint (10.32) limits the apparent power on each branch by its capacity and restricts both real and reactive power flows on a branch to be zero if it is open. As for the AC power flow equations of DS circuits, they are often formulated by the DistFlow model, which is first proposed in [54, 55] for radial DSs. However, the DistFlow model is non-convex. This chapter uses its simplified and linearized version, i.e. the LinDistFlow model [54, 56] as described by constraints (10.33) and (10.34), which can be convexified much more easily. Specifically, constraints (10.33) and (10.34) represent the LinDistFlow model based power flow equations for closed branches with $\lambda_t^{ij} = 1$ (that is, K vanishes in this case), and get relaxed for open branches with $\lambda_t^{ij} = 0$ (that is, constraints (10.33) and (10.34) are redundant in this case). Note that K needs to be big enough so that for any open branch (i,j), variables v_t^i and v_t^j do not directly influence each other. It is adequate to set $K = \max_{i \in \mathcal{N}}(\overline{v}^i)^2 - \min_{j \in \mathcal{N}}(\underline{v}^j)^2$. Also, note that constraints (10.31), (10.33), and (10.34) are still non-convex. As the LinDistFlow model is used instead of the DistFlow model, v_t^i only appears in its square form in the optimization problem. Therefore, one can let $y_t^i = (v_t^i)^2$, and substitute $(v_t^i)^2$ with y_t^i. By doing this, constraints (10.31), (10.33), and (10.34) are equivalently convexified. The LinDistFlow model has been extensively used in the literature, e.g. in [3, 57–60] with radially operated DSs. Its use is justified by the fact that, in practice, the non-linear branch loss terms are much smaller than the branch power terms, enabling the LinDistFlow model to produce results almost indistinguishable from the DistFlow model for radial networks [56, 61].

Therefore, with constraints (10.22)–(10.34), dynamic network reconfiguration and power dispatch of the DS are coordinated to achieve prompt service restoration via strategies such as microgrid formation. Flexibility resources are also optimally coordinated.

Although DSs are normally operated radially, in practice, weakly meshed topologies may also be adopted in some cases [62]. Relaxing the radiality requirement and allowing the DS to be weakly meshed, the supply capability may be enhanced by the added flexibility, thus to restore more loads. For example, due to the branches' limited apparent power capacities, there may exist an outage area that can be fully restored only by two or more energizing paths that form a mesh. Besides, more secured voltage profiles and stability, etc., may be achieved. In general, allowing both radial and weakly meshed operating topologies of the DS can enlarge the operating point's feasible region, thus to potentially attain better service restoration. The studies on weakly meshed DSs have adequate methodological supports [62, 63]. In this chapter, to extend the proposed optimization model for meshed, operated DSs, the radiality constraints can be simply discarded. And the DistFlow model should be used instead of the LinDistFlow model, which requires only small changes of constraints (10.33) and (10.34). By such minor modifications, the AC power flow equations can be precisely formulated. Note that the DistFlow model is actually also applicable for meshed networks [64, 65], although it is originally developed for radial networks [54, 55]. However, as aforementioned, the DistFlow model is non-convex. Thus, modifying constraints (10.33) and (10.34) will introduce new non-convexity. This issue can be resolved by convexification methods in [64, 65], etc. Nevertheless, the application of those methods here is non-trivial. Actually, the LinDistFlow model may remain satisfactorily approximating the DistFlow model for meshed networks here. Still, extensive numerical experiments need to be conducted for accuracy examination. The above topics, including the meshed operation of DSs in the studied problem, are left for future investigation.

10.3 Co-Optimization of Flexibility Resources

In this section, DS service restoration is co-optimized with RC dispatch and MPS dispatch to enhance DS resilience. Involved flexibility resources are also coordinated in this process. Let R be the resilience metric. Let $F(t)$ be a function indicating the system performance. As in [33, 66], etc., an extensively used evaluation of R is the integral of $F(t)$ over the concerned time duration:

$$R = \int_0^{|\mathcal{T}|} F(t)dt \tag{10.35}$$

During system restoration, $F(t)$ is commonly a non-decreasing function, as the system performance is gradually improved by restoration efforts. For example, it is normally disallowed to de-energize restored loads. Therefore, the faster $F(t)$ achieves higher values, the better R is attained. That is, Eq. (10.35) indicates both system performance levels and the speed to reach those levels [67].

Let ω^i be the priority weight of the power demand at node i. Let $p^i(t)$ be the served power demand at node i at time t. As in [15, 33], etc., $F(t)$ is typically specified as the priority-weighted served loads as follows:

$$R = \int_0^{|\mathcal{T}|} F(t)dt = \int_0^{|\mathcal{T}|} \sum_i \omega^i \cdot p^i(t)dt \approx \sum_t \sum_i \omega^i \cdot p_t^i \cdot \Delta t \tag{10.36}$$

where the integral is approximated by the discrete sum. To optimize R, the constant Δt can be ignored, resulting in an extensively used objective function, e.g. in [15, 33]. Based on that, the objective function of the co-optimization problem is modeled as follows:

$$\max \sum_t \left[\sum_i \omega^i \cdot \delta_t^i \cdot p_t^i - \varepsilon \cdot \left(\sum_k \beta_t^k + \sum_s \beta_t^s \right) \right] \tag{10.37}$$

which maximizes the weighted sum of restored loads over time and minimizes the total number of travels of RCs and MPSs. Relative weights of the two objectives are adjusted by the parameter ε, which is set as a small value so that the first objective is generally dominating. It seeks solutions that optimally use flexibility resources to reduce both outage scale and outage time and eventually achieve 100% restoration. That is, it targets for both served load maximization and rapid recovery. The second objective is added for two main reasons:

- If all loads are restored at a time step $t < |\mathcal{T}|$, it is no longer necessary to transport RCs or MPSs during time periods $t \sim |\mathcal{T}|$. The second objective restricts the transportation of RCs and MPSs during time $t \sim |\mathcal{T}|$ in this case.
- There may exist multiple RC and MPS dispatch strategies that can achieve the optimal service restoration. As redundant travels of RCs and MPSs cause extra cost, the second objective is added to select a dispatch strategy attaining the optimal recovery effect by a minimum number of travels of RCs and MPSs.

In general, the two objectives in (10.37) have conflicts. It is critical to set an appropriate weight (i.e. ε) to preserve the hierarchical nature of the two objectives with different priorities. Too small values of ε can lead to the second objective's insignificance in optimization, while too large values of ε may deteriorate the DS service restoration due to the expensive and thus limited dispatch of RCs and MPSs. Its selection can depend on the preference and knowledge of decision-makers. To avoid decision-makers' subjective influence on its selection, the analytic hierarchy process method [68] can be applied. In this chapter, load restoration is prioritized over travel limitation in utilizing the flexibility resources. Still, decision-makers may set ω^i and ε (which can be further specified as ε^k and ε^s for each MEG and MESS, respectively) based on the customer interruption cost and RC/MPS dispatch cost, respectively, or derive the Pareto frontier by solving the problem with different values of the weights.

Next, the interdependencies among RC dispatch, MPS dispatch, and DS service restoration are formulated. First, the relationships between RC dispatch and DS restoration are modeled. Let branch (i_m, j_m) denote the associated damaged component m. Let $\mathcal{L}_1 \triangleq \{(i_m, j_m), \forall m \in \mathcal{V}_1\}$ be the set of damaged branches. Let $\mathcal{L}_2$ be the set of branches with switches. Then, the following constraints are enforced:

$$u_t^{ij} = 1, \forall (i,j) \in \mathcal{L} \backslash \mathcal{L}_1, \forall t \tag{10.38}$$

$$u_{t+1}^{i_m j_m} \leq \sum_k z_{m,t}^k, \forall m \in \mathcal{V}_1, \forall t \leq |\mathcal{T}| - 1 \tag{10.39}$$

$$\lambda_t^{ij} = 1, \forall (i,j) \in \mathcal{L} \backslash \{\mathcal{L}_1 \cup \mathcal{L}_2\}, \forall t \tag{10.40}$$

As mentioned above, constraint (10.38) states that an intact branch is always operable. Constraint (10.39) indicates that a damaged branch is operable only if it is repaired by one of the RCs in a previous time period. That is, the to-be-repaired components' operability and flexibility in DS restoration depend on RC dispatch. Constraint (10.40) enforces undamaged branches without switches to be closed. As in the example used to explain constraints (10.6) and (10.7), one can have $\left[u_1^{i_1 j_1}, u_2^{i_1 j_1}, u_3^{i_1 j_1}, u_4^{i_1 j_1}, \ldots, u_{|\mathcal{T}|}^{i_1 j_1}\right] = [0, 0, 0, 1, \ldots, 1]$. That is, damaged component 1 can be used at time periods $t = 4 \sim |\mathcal{T}|$, i.e. one time period after it is repaired at $t = 3$.

Second, the relationships between MPS dispatch and DS service restoration are incorporated by the following equations:

$$P_t^i = \sum_{s \in \mathcal{M}_1} \alpha_{i,t}^s \cdot gp_t^s + \sum_{s \in \mathcal{M}_2} \alpha_{i,t}^s \cdot (dp_t^s - cp_t^s), \forall i \in \mathcal{N}', \forall t \tag{10.41}$$

$$Q_t^i = \sum_{s \in \mathcal{M}} \alpha_{i,t}^s \cdot gq_t^s, \forall i \in \mathcal{N}', \forall t \tag{10.42}$$

$$P_t^i = Q_t^i = 0, \forall i \in \mathcal{N} \backslash \mathcal{N}', \forall t \tag{10.43}$$

With Eqs. (10.41) and (10.42), P_t^i and Q_t^i are derived by summing the real and reactive power outputs of MPSs connected to node i at time t, respectively. Equation (10.43) enforces P_t^i and Q_t^i to be zero for nodes that are not for MPS connection. That is, DS restoration relies on MPSs' dispersion, connection status, and power/energy states, etc., which are determined by the flexible dispatch of MPSs.

Note that although constraints (10.38)–(10.43) do not show direct relationships between RC dispatch and MPS dispatch, they are intrinsically inter-related in the studied problem. Their flexibilities need to be coordinated, i.e. co-optimized, to attain better service restoration of the DS.

Therefore, the co-optimization model can be formulated as follows:

$$
\begin{aligned}
\text{Objective}: \quad & (10.37) \\
\text{Constraints}: \quad & (10.1), (10.2), (10.4), (10.6) - (10.34), (10.38) - (10.43) \\
\text{Variables}: \quad & a_{m,t}^k, \beta_t^k, z_{m,t}^k, \\
& \alpha_{i,t}^s, \beta_t^s, gp_t^s, gq_t^s, c_t^s, d_t^s, cp_t^s, dp_t^s, soc_t^s, \\
& \delta_t^i, P_t^i, Q_t^i, v_t^i, \lambda_t^{ij}, u_t^{ij}, e_t^{ij}, f_t^{ij}, pf_t^{ij}, qf_t^{ij}
\end{aligned}
$$

The above mentioned list of variables is also provided for clarity. The co-optimization model incorporates the sub-problems of DS restoration, RC dispatch, and MPS dispatch. It is solved to simultaneously optimize decisions of the three involved sub-problems so that the involved flexibility resources are fully coordinated.

The inherent relationships among the three sub-problems incorporated in the co-optimization model are discussed as follows:

- DS restoration depends on damaged components' operability determined by the repair plan. To attain the optimal RC dispatch, the distribution grid's state and operation, load priorities, and flexibilities, etc., need to be considered. For example, if the DS has a part isolated by a damaged branch and there exists a normally open branch between this island and the main grid, the repair of this damaged branch is optional, as simply closing the normally open branch can reconnect the isolated part to the main grid.
- DS restoration also relies on MPSs' availability decided by MPS dispatch. On the other hand, MPSs' power and energy states rely on DS operation. For example, the charging and discharging of MESSs are impacted by the power output of substation(s) and the pick-up of loads, etc. To attain the optimal MPS dispatch, the distribution grid's state and operation, load priorities, and flexibilities, etc., need to be considered, too. For example, normally MPSs are first dispatched to PIs of the DS.

- RCs and MPSs should be dispatched coordinately to enhance DS restoration. Sometimes the flexibilities provided by RCs and MPSs should aid different parts of the DS. For example, if a PI already has an MPS, it may be better to dispatch RCs for restoring other PIs by repairing relevant components. Sometimes RCs and MPSs should aid the same part or the adjacent parts of the DS (e.g. by repairing a line to extend the supply area of an MPS). DS operation should also be tuned to enable sophisticated coordination between RCs and MPSs.

The above model is a deterministic one. It can produce practical solutions for coordinating flexibility resources. Still, solutions more robust to uncertain parameters are also desired. In that case, the proposed model needs to be extended. First, the uncertain parameters need to be characterized, e.g. using lognormal distribution to feature the uncertain repair time [69]. Second, based on the proposed model, stochastic optimization or robust optimization models need to be formulated to deal with uncertain parameters [70, 71]. A rolling horizon scheme may be necessary, too. Third, algorithms need to be developed to efficiently solve the extended models. In general, the uncertainty issue here is a challenging topic that is worth investigating. It is beyond the scope of this book and left for future investigation.

10.4 Solution Method

10.4.1 Pre-assigning Minimal Repair Tasks

To reduce the MILP co-optimization model's computational complexity, this chapter follows the idea of clustering damaged components in [12], and pre-assigns repair tasks to depots using a method majorly different from Ref. [12] in terms of the considered issues and problem formulation, etc. Note that at this stage, selected repair tasks are not yet assigned to RCs, which are accommodated in depots. The repair tasks are further assigned and scheduled to RCs when solving the co-optimization problem. Let ψ_n^m be 1 if damaged component m is assigned to depot n, 0, otherwise. Let l_{mn} be the distance between damaged component m and depot n. Let Φ_n be the set of RCs accommodated in depot n. Then, the pre-assignment of repair tasks to depots is determined by a small-scale MILP problem as follows:

$$\min_{\substack{\psi_n^m, u^{ij}, y^{ij}, e^{ij}, \\ p f^{ij}, q f^{ij}, v^i}} \sum_{m \in \mathcal{V}_1} \sum_{n \in \mathcal{V}_2} \psi_n^m \cdot l_{mn} \tag{10.44}$$

$$\text{s.t.} \quad \sum_{m \in \mathcal{V}_1} \psi_n^m \cdot rs_m \leq \sum_{k \in \Phi_n} RS^k, \forall n \in \mathcal{V}_2 \tag{10.45}$$

$$\sum_{n \in \mathcal{V}_2} \psi_n^m \leq 1, \forall m \in \mathcal{V}_1 \tag{10.46}$$

$$u^{i_m j_m} \leq \sum_{n \in \mathcal{V}_2} \psi_n^m, \forall m \in \mathcal{V}_1 \tag{10.47}$$

$$\sum_{(j,i) \in \mathcal{L}} pf^{ji} - \sum_{(i,j) \in \mathcal{L}} pf^{ij} - p^i = 0, \forall i \tag{10.48}$$

$$\sum_{(j,i) \in \mathcal{L}} qf^{ji} - \sum_{(i,j) \in \mathcal{L}} qf^{ij} - q^i = 0, \forall i \tag{10.49}$$

$$(10.22) - (10.24), (10.31) - (10.34), (10.38), (10.40) \text{ (without time subscript } t) \tag{10.50}$$

Specifically, the objective function (10.44) minimizes the sum of distances between each damaged component and its assigned depot, so as to reduce the transportation time of RCs and thus enhance DS restoration. Constraint (10.45) requires each depot to have adequate resources to complete its assigned repair tasks. Constraint (10.46) states that each damaged component is assigned to at most one depot. Note that, different from Ref. [12], here the proposed model does not demand all repair tasks to be assigned. The above model only seeks and assigns *a minimal set of repair tasks* that are able to fully restore all DS loads without using MPSs. Thus, constraints (10.47)–(10.50) are added. Constraint (10.47) states that a damaged branch will be operable only if it is assigned to be repaired. Constraints (10.48) and (10.49) represent the power balance conditions to fully supply all loads without using MPSs. Constraints (10.50) include the other DS operational constraints. Parameters p^i and q^i can be set as the maximum power demand of each DS node, and note that the substation power outputs are implicitly modeled as $pf^{i_g j}$ and $qf^{i_g j}$, where i_g denotes the substation node. Although without the time subscript t, other involved variables and constraints are introduced hereinbefore. Operational constraints of the DS are considered to ensure a selection of repair tasks that can fully restore all DS loads without using MPSs. Therefore, the above model also instructs the DS operation at the time with all pre-assigned repair tasks finished. By selecting and assigning a minimal set of repair tasks rather than all tasks to depots, the computational complexity of the proposed co-optimization model is further reduced. Another advantage is that, after all loads are restored by a minimal set of repair tasks, the flexibilities provided by RCs can be better utilized to help recover other DSs.

10.4.2 Selecting Candidate Nodes to Connect Mobile Power Sources

The co-optimization model's computation time also partially depends on the number of candidate nodes for MPS connection in the DS. Here, one candidate node is selected from each PI.

First, the set of candidate nodes in each PI is found. Let $\mathcal{I}$ be the set of all candidate nodes, i.e. nodes meeting the requirements (e.g. facility requirements [16]) for connecting MPSs. Let $\mathcal{I}_\sigma$ be the set of candidate nodes in PI σ. Let $v2s$ be an operation transforming a vector into a set, e.g. $v2s([0, 1, 2]) = \{0, 1, 2\}$. Let $\odot$ be the symbol of element-wise multiplication for obtaining the Hadamard product. Let $\mathcal{A}$ be the adjacency matrix of the DS (that is $\mathcal{A}_{ij} = 1$ if there is an intact branch (i, j), whether closed or open, and $\mathcal{A}_{ij} = 0$, otherwise). Let $\mathcal{A}_i$ be the ith row of $\mathcal{A}$. Algorithm 10.1 as follows is proposed to find $\mathcal{I}_\sigma$ for each PI σ:

Algorithm 10.1 Find the set of candidate nodes in each PI

Input: $\mathcal{N}, \mathcal{A}, \mathcal{I}$

Output: $\mathcal{I}_\sigma, \forall \sigma$

1: $\sigma \leftarrow 1, \tilde{\mathcal{N}} \leftarrow \emptyset$
2: **while** $\mathcal{N} \setminus \tilde{\mathcal{N}} \neq \emptyset$ **do**
3:     randomly select $i \in \mathcal{N} \setminus \tilde{\mathcal{N}}, \mathcal{I}_\sigma \leftarrow \{i\}, \tilde{\mathcal{I}}_\sigma \leftarrow \{i\}$
4:     $C_i \leftarrow v2s(\mathcal{A}_i \odot [1, \ldots, N]), \mathcal{I}_\sigma \leftarrow \{\mathcal{I}_\sigma \bigcup C_i\} \setminus \{0\}$
5:     **if** $\mathcal{I}_\sigma \setminus \tilde{\mathcal{I}}_\sigma \neq \emptyset$ **then**
6:         randomly select $i \in \mathcal{I}_\sigma \setminus \tilde{\mathcal{I}}_\sigma, \tilde{\mathcal{I}}_\sigma \leftarrow \tilde{\mathcal{I}}_\sigma \bigcup \{i\}$
7:         **go to** *Step 4*
8:     **end if**
9:     $\tilde{\mathcal{N}} \leftarrow \tilde{\mathcal{N}} \bigcup \mathcal{I}_\sigma, \mathcal{I}_\sigma \leftarrow \mathcal{I}_\sigma \bigcap \mathcal{I}, \sigma \leftarrow \sigma + 1$
10:     **go to** *Step 3*
11: **end while**

Then, the final selection of candidate nodes is determined by a small-scale integer programming (IP) problem as follows:

$$\min_{\phi_i} \frac{1}{2} \sum_\sigma \sum_{i \in \mathcal{I}_\sigma} \sum_{j \in \mathcal{I} \setminus \mathcal{I}_\sigma} \phi_i \cdot \phi_j \cdot l_{ij} \tag{10.51}$$

$$\text{s.t.} \quad \sum_{i \in \mathcal{I}_\sigma} \phi_i = 1, \forall \sigma \text{ with } |\mathcal{I}_\sigma| \neq 0 \tag{10.52}$$

where $\phi_i = 1$ if candidate node i is selected, $\phi_i = 0$ otherwise, and l_{ij} is the distance between nodes i and j. The objective function (10.51) minimizes the sum of distances between all pairs of selected candidate nodes, in order to reduce the transportation time of MPSs and thus enhance DS restoration. Constraint (10.52) enforces one node to be selected from each PI that has at least one candidate node. Again, by the McCormick envelopes [72], terms $\phi_i \cdot \phi_j$ can be equivalently linearized.

In short, the problem (10.44)–(10.50) finds a minimal set of repair tasks that can restore all loads without using MPSs, and pre-assigns them to depots. The problem

(10.51) and (10.52) selects one candidate node in each PI for MPS connection. Their solutions serve as parameters for the co-optimization model. They decrease the number of possible combinations of repair task assignments/scheduling and the number of candidate nodes for MPS connection, respectively, for the co-optimization model. In general, the search space of the co-optimization problem is narrowed without deteriorating the optimality of its solution. The optimal coordination of flexibility resources is neither deteriorated.

After the above preprocessing, the co-optimization model is solved by advanced solvers based on state-of-the-art branch-and-cut methods [73]. Note that the optimal solutions of NP-hard MILP and IP problems are difficult to obtain. Sometimes, it is even difficult to find a feasible solution. In this regard, the proposed preprocessing methods reduce the computational complexity and, thus, enhance the capability of attaining optimality. It may still end up with suboptimal solutions, e.g. with RCs' idling deteriorating the optimality. Some other methods, e.g. genetic algorithms [74], may be applied to solve the studied problem. It is beyond the scope of this book and left for future study.

10.4.3 Linearization Techniques

The proposed co-optimization model is a non-convex MINLP. First, constraints (10.31), (10.33), and (10.34) are non-convex. As aforementioned, since the model only involves the square form of v_t^i, those constraints can be equivalently convexified by replacing $(v_t^i)^2$ with y_t^i. Second, constraints (10.41) and (10.42) have non-linear and non-convex terms as follows: $\alpha_{i,t}^s \cdot gp_t^s$, $\alpha_{i,t}^s \cdot gq_t^s$, $\alpha_{i,t}^s \cdot cp_t^s$, and $\alpha_{i,t}^s \cdot dp_t^s$. They can be linearized by the McCormick envelopes [72]. For example, to linearize $\alpha_{i,t}^s \cdot gp_t^s$, one can let $GP_{i,t}^s = \alpha_{i,t}^s \cdot gp_t^s$, and add constraints as follows:

$$0 \le GP_{i,t}^s \le \alpha_{i,t}^s \cdot \overline{gp}^s \tag{10.53}$$

$$gp_t^s + (\alpha_{i,t}^s - 1) \cdot \overline{gp}^s \le GP_{i,t}^s \le gp_t^s \tag{10.54}$$

As $\alpha_{i,t}^s \in \{0, 1\}$, constraints (10.53) and (10.54) indicate that $GP_{i,t}^s = 0$ when $\alpha_{i,t}^s = 0$, and $GP_{i,t}^s = gp_t^s$ when $\alpha_{i,t}^s = 1$. The other similar terms can be linearized in the same manner. In general, owing to the facts that $\alpha_{i,t}^s$ is binary and the other involved continuous variables have explicit lower and upper bounds, the linearization is equivalent. By doing this, the co-optimization model is equivalently convexified into a mixed-integer second-order cone programmings (MISOCP) model. Constraint (10.32) is further linearized using the technique in [60]. The co-optimization model thus becomes a MILP problem, which can be solved by off-the-shelf solvers such as Gurobi. As the involved linearizations are straightforward, this section does not further elaborate on the detailed reformulations.

10.5 Case Studies

In this section, the proposed co-optimization method for disaster recovery logistics is demonstrated on two systems based on per-phase analysis. A computer with an Intel i5-4278U processor and 8GB memory is used. Involved MILP and IP problems are solved using Gurobi 7.5.2 with the parameter *MIPGap* set as 0.01%.

10.5.1 Illustration on a Small Test System

For this DS, a scenario with eight branches damaged by the natural disaster is considered (see Figure 10.3). The system has two depots and two RCs, i.e. RC 1 accommodated in depot 1 and RC 2 accommodated in depot 2. The resource capacity of RCs is set as 8. Both the number of resources and time periods required to repair different damaged components vary from 1 to 4. The DS also has two MPSs, i.e. 500 kW/400 kVar MEG 1, and 300 kW/300 kWh MESS 1. Due to the lack of relevant data, priority weights of the loads and the travel time/distance data are randomly generated in *Matlab*. Branches $(9, 10)$, $(9, 15)$, $(12, 22)$, $(14, 15)$, $(18, 33)$, $(25, 29)$, $(28, 29)$, and $(30, 31)$ are equipped with remote-controlled switches [75]. For both this small system case (i.e. Case I) and the later large system case (i.e. Case II), the minimum and maximum voltage values are set as 0.95 and 1.05 of the nominal voltage, respectively. And, for this 12.66 kV system, setting $K = 33$ is big enough. Some other data can be found in [56].

First, a minimal set of repair tasks is sought and pre-assigned to depots. The pre-assignment model (10.44)–(10.50) is solved within 0.13 seconds. Table 10.1 lists the results. In virtue of the flexibility provided by back-up branches, i.e. normally open branches, repairing four damaged branches is sufficient to fully restore all loads. The repair of the other four un-assigned damaged branches can be arranged later considering an objective other than maximizing the sum of restored loads.

Second, candidate nodes for MPS connection are selected. Let $\mathcal{N}' = \{2, 5, 8, 15, 21, 29, 33\}$ be the set of all candidate nodes. Algorithm 10.1, which takes 0.14 seconds, accurately detects that the system is split into four PIs containing candidate nodes $\{2\}$, $\{5, 29\}$, $\{8, 15, 21\}$, and $\{33\}$, respectively. Note that although these results are obvious in this case, Algorithm 10.1 is necessary to

Table 10.1 Preassignment of a minimal set of repair tasks (Case I).

	Depot 1	Depot 2
Preassigned repair tasks	Branches $(2, 3)$, $(5, 6)$	Branches $(7, 8)$, $(30, 31)$
Unassigned repair tasks	Branches $(10, 11)$, $(16, 17)$, $(19, 20)$, $(9, 15)$	

automatically generate such results for both small and large systems. Then, the candidate node selection problem (10.51) and (10.52) is solved in 0.04 seconds. Nodes {2, 8, 29, 33} are selected.

Third, after pre-assigning repair tasks and selecting candidate nodes, the co-optimization problem is solved. Let $|\mathcal{T}| = 12$ and $\Delta t = 0.5$ h, let $CAP_i = 2$ for all candidate nodes of MPS connection, and let $\alpha_{2,1}^s = 1$ for both MPSs. The co-optimization model is solved in 2.26 seconds. Tables 10.2 and 10.3 list the dispatch of RCs and MPSs, respectively. Symbols "→" or "×" mean that this RC/MPS is being transported or has stopped working, respectively. MPSs change their locations only once for this small system. More dynamic dispatch of MPSs will be seen in Case II with a larger system. Figure 10.6 depicts MPSs' real-power

Table 10.2 Routing and scheduling of RCs (Case I).

RC 1	Time period	0		1	2~4		5~7	8~10		11~12
	Dispatch	Depot 1	→		Branch (2, 3)	→		Branch (5, 6)		×
RC 2	Time period	0		1	2~3		4~5	6~7		8~12
	Dispatch	Depot 2	→		Branch (7, 8)	→		Branch (30, 31)		×

Table 10.3 Routing and scheduling of MPSs (Case I).

MEG 1	Time period	0		1~3	4~10		11~12
	Dispatch	Node 2	→		Node 33		×
MESS 1	Time period	0		1~2	3~10		11~12
	Dispatch	Node 2	→		Node 29		×

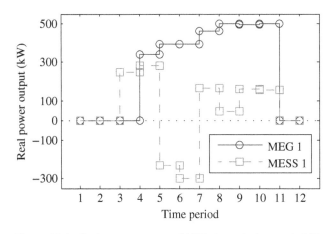

Figure 10.6 Real power outputs of MPSs in each time period (Case I).

Table 10.4 Dynamic network reconfiguration of the DS (Case I).

Time period	3	4	11
Switch actions	Close (25, 29)	Close (8, 21), (12, 22), (18, 33)	Open (28, 29)

outputs. Table 10.4 lists switch actions of the DS. The weight ε is set as 10. To check if this value is appropriate here, the results are compared with a two-step optimization process. The first step solves the co-optimization problem without the second objective in (10.37), and the second step solves the co-optimization problem without the first objective in (10.37) but with a constraint enforcing the weighted sum of restored loads over time equal to that of the first step. The solutions of the proposed model and the two-step optimization process are essentially the same with negligible objective value differences in the order of 10^{-2}. Thus, the value of the weight ε is appropriate here.

The proposed co-optimization method for disaster recovery logistics is also compared to other recovery strategies to demonstrate the proposed method's optimal coordination of flexibility resources. As shown in Figure 10.7, the proposed method co-optimizing both RC dispatch and MPS dispatch with DS restoration has the best performance.

The recovery process based on the proposed co-optimization model is described in details by Figure 10.8. At $t = 3$, MESS 1 arrives at and is connected to node 29. Closing branch (25, 29), microgrid 1 is formed to restore nodes 3, 4, 6, and 30. At $t = 4$, with branch (7, 8) repaired, microgrid 1 can further restore nodes 13 and 22

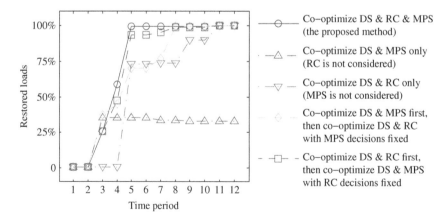

Figure 10.7 Restored loads over time for different restoration strategies (Case I).

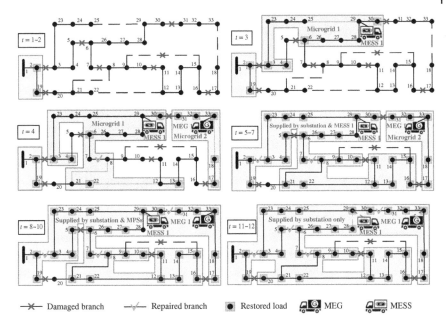

—✕— Damaged branch   —✓— Repaired branch   ● Restored load   🔋🟢 MEG   🔋📼 MESS

Figure 10.8 DS service restoration process co-optimized with both RC dispatch and MPS dispatch (Case I).

by closing branches (8, 21) and (12, 22). At $t = 4$, MEG 1 arrives at and is connected to node 33. Closing branch (18, 33), microgrid 2 is formed to restore all loads in it. At $t = 5$, with branch (2, 3) repaired, microgrid 1 gets connected to the substation. Thus, MESS 1 can get charged at $t = 5 \sim 6$ to supply future peak loads. At $t = 8$, with branch (30, 31) repaired, microgrid 2 also gets connected to the main grid. However, due to operational constraints (low voltage of node 16 and large power flow on branch (24, 25)), nodes 20 and 28 cannot be restored until branch (5, 6) is repaired at $t = 11$, when branch (28, 29) is opened to avoid forming a loop, and both MPSs are disconnected from the DS as they are no longer necessary. As indicated by the recovery process, the proposed method optimally coordinates the involved flexibility resources.

10.5.2 Results on a Large Test System

A scenario with 20 branches damaged by the natural disaster is considered for the second test system in Figure 10.9. It has 2 depots (each with 2 RCs), 4 MPSs (2 MEGs and 2 MESSs), and 12 branches equipped with remote-controlled switches [75]. Let $|\mathcal{T}| = 16$. Setting $K = 8$ is large enough for this case. Some other data can be found in Case I or in [76]. For this case with a large system,

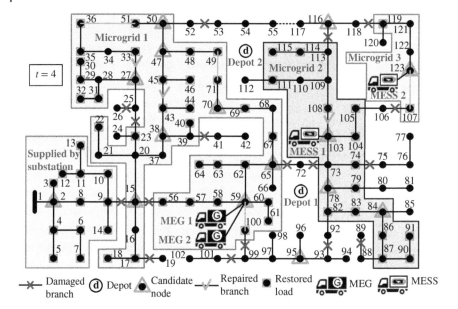

Figure 10.9 IEEE 123-node test system and its restoration state at $t = 4$.

the weight ε is again set as 10, which is appropriate here, as the solutions of the proposed model and the two-step optimization process also only have negligible objective value differences in the order of 10^{-2}.

First, the pre-assignment model (10.44)–(10.50) is solved in 0.43 seconds. Next, Algorithm 10.1 is run for 0.45 seconds, and the candidate node selection problem (10.51) and (10.52) is solved in 0.28 seconds. It selects nodes $\{2, 27, 38, 59, 95, 103, 116, 123\}$. At last, the co-optimization problem is solved in 1183 seconds. For space limit, the recovery process is not elaborated on. Only the DS restoration progress at $t = 4$ is depicted in Figure 10.9 for illustration. Tables 10.5 and 10.6 list the dispatch of RCs and MPSs, respectively. Table 10.5 also implicitly includes the results for the pre-assignment of a minimal set of repair tasks. Figure 10.10 depicts real-power outputs of MPSs. Table 10.7 lists the DS switch actions in its dynamic network reconfiguration. Figure 10.11 again demonstrates the effectiveness and superiority of the proposed co-optimization method for disaster recovery logistics. The tables and figures indicate that the proposed method is effective in coordinating RC dispatch and MPS dispatch to restore loads, especially by dynamically forming microgrids in the DS. The microgrids are powered by MPSs and reconfigured and extended by switch actions of the DS and repair actions of RCs. In general, the flexibility resources are optimally coordinated.

Table 10.5 Routing and scheduling of RCs (Case II).

RC 1	Time period	0	1	2~4	5~6	7~9	10~11	12~13	14	15	16
	Dispatch	Depot 1	→	Branch (9, 15)	→	Branch (72, 73)	→	Branch (88, 89)	→	Branch (106, 107)	×
RC 2	Time period	0	1	2~3	4	5	6	7	8~9	10~11	12~16
	Dispatch	Depot 1	→	Branch (103, 108)	→	Branch (39, 41)	→	Branch (15, 56)	→	Branch (65, 72)	×
RC 3	Time period	0	1	2	3	4~5	6	7~8	9	10~11	12~16
	Dispatch	Depot 2	→	Branch (27, 33)	→	Branch (92, 93)	→	Branch (99, 101)	→	Branch (17, 19)	×
RC 4	Time period	0	1	2~3	4	5	6	7~8	9~10	11~12	13~16
	Dispatch	Depot 2	→	Branch (43, 45)	→	Branch (74, 75)	→	Branch (25, 26)	→	Branch (113, 116)	×

Table 10.6 Routing and scheduling of MPSs (Case II).

MEG 1	Time period	0		1~2	3~4	5	6~11	12~16		—
	Dispatch	Node 2	→		Node 59	→	Node 103	×		
MEG 2	Time period	0		1~2	3~4	5	6~11	12~13	14~15	16
	Dispatch	Node 2	→		Node 59	→	Node 103	→	Node 123	×
MESS 1	Time period	0~1		2~3	4~5	6~7	8	9	10~12	13~16
	Dispatch	Node 123	→		Node 103	→	Node 38	→	Node 116	×
MESS 2	Time period	0~14		15~16				—		
	Dispatch	Node 123	×							

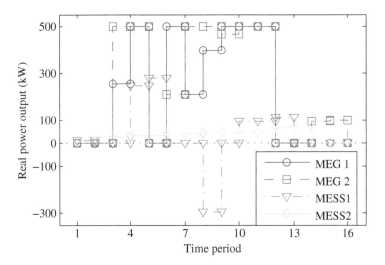

Figure 10.10 Real-power outputs of MPSs in each time period (Case II).

10.5.3 Computational Efficiency

Table 10.8 reports the computation time of different methods for solving MISOCP and MILP models of the co-optimization problem. For Case I with the small system, all methods attain an optimal solution the same as the full model without any preprocessing. That is, pre-processing of the proposed co-optimization model does not result in sub-optimality. The MISOCP and MILP models also attain the same solution. Specifically, integer variables have the same values, and some continuous variables have negligible differences. For Case II with the large system, only the proposed method attains the optimal solution of the MILP

Table 10.7 Dynamic network reconfiguration of the DS (Case II).

Time period	1	3	8	10
Switch actions	Close (107, 123)	Close (36, 51), (49, 71), (59, 100)	Open (49, 71)	Close (55, 117)

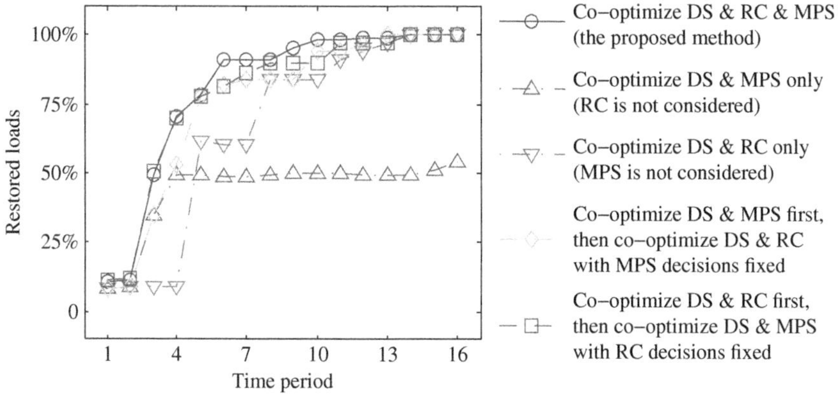

Figure 10.11 Restored loads over time for different restoration strategies (Case II).

Table 10.8 Computation time of the co-optimization model (Cases I & II).

Solution method	Case I		Case II (MILP)
	MISOCP	MILP	
Proposed method	11.24 s	2.26 s	1183 s
Without any pre-processing	242.65 s	66.20 s	gap = 97.20% at 2 h
Only with $\mathcal{N}'$ reduced	114.99 s	37.59 s	gap = 59.92% at 2 h
Only with all repairs pre-assigned	73.15 s	17.69 s	gap = 32.60% at 2 h
Only with minimal repairs pre-assigned	29.83 s	6.67 s	gap = 5.34% at 2 h

Note: For the MISOCP model of Case II, all solution methods cannot find a feasible solution in two hours.

co-optimization model in two hours. Cases I and II demonstrate that the proposed solution method can reduce the computation time without weakening the solution optimality. Specifically, it is indicated that, compared with pre-assigning all repair tasks, pre-assigning a minimal set of repair tasks can better improve the computational efficiency. It is also confirmed that the solutions obtained by the MILP approximation model satisfy all constraints of the original problem. In general, compared with the MISOCP model, the MILP model has much lower-computational complexity and thus much shorter computation time, and it does not lead to sub-optimal or infeasible solutions.

10.5.4 LinDistFlow Model Accuracy

It is well justified in the literature that the LinDistFlow model can provide accurate results. Still, this section uses the DistFlow model to evaluate the LinDistFlow model-based solutions. After solving the co-optimization problem, the following validation problem is also solved:

$$\text{Objective : } \min \ \Delta v \tag{10.55}$$

$$\text{Constraints : } (10.14) - (10.24), (10.29) - (10.32), (10.38) - (10.43)$$

$$(v_t^i)^2 - (v_t^j)^2 = 2 \cdot \left(r_{ij} \cdot pf_t^{ij} + x_{ij} \cdot qf_t^{ij} \right)$$
$$- \left(r_{ij}^2 + x_{ij}^2 \right) \cdot \frac{(pf_t^{ij})^2 + (qf_t^{ij})^2}{(v_t^i)^2}, \quad \forall t, \ \forall (i,j) \text{ with } \check{\lambda}_t^{ij} = 1 \tag{10.56}$$

$$\Delta v \geq |v_t^i - \check{v}_t^i|, \ \forall i, \ \forall t \tag{10.57}$$

$$\text{Variables : } gp_t^s, gq_t^s, cp_t^s, dp_t^s, soc_t^s, P_t^i, Q_t^i, v_t^i, pf_t^{ij}, qf_t^{ij}$$

Above, $\check{\lambda}_t^{ij}$ and $\check{v}_t^i$ are the values of λ_t^{ij} and v_t^i attained by solving the co-optimization problem, respectively. For other symbols that are variables in the co-optimization problem but not in the validation problem, here they are also seen as parameters using their values attained by solving the co-optimization problem. The validation problem finds an operating point of the DS to realize the restoration plan obtained by the co-optimization model. Specifically, it finds the operating point with the least distance to the operating point provided by the co-optimization model. The "distance" here is assessed by the maximum voltage difference. Actually, as long as the feasible region of the validation problem is non-empty, the restoration plan attained by the co-optimization model is feasible. If the validation problem finds an operating point that is virtually indistinguishable from the one provided by the co-optimization model, it further proves that the operating point provided by the co-optimization model is feasible. The validation problem is solved by IPOPT [77]. For Case I and Case II, the validation problem finds operating points with Δv equal to 0.002 and 0.003 pu, respectively. That is, the solutions based on the LinDistFlow model here are physically plausible.

10.6   Summary and Conclusions

To enhance DS resilience, this chapter builds a co-optimization method for disaster recovery logistics. RC dispatch, MPS dispatch, and DS operation are jointly cooperated for electric service restoration so that the involved flexibility resources are optimally coordinated. A MINLP model is formulated to attain resilient strategies that involve the routing and scheduling of RCs and MPSs, and dynamic network reconfiguration of the DS, etc. The model is transformed into a MILP, and pre-processed to reduce the computational complexity. Case studies demonstrate that the co-optimization method for disaster recovery logistics efficiently improves DS service restoration, especially by the dynamic formation of microgrids that are powered by MPSs and topologized by repair actions of RCs and reconfiguration of the DS.

Specifically, many issues need to be addressed for stable operation of the formed microgrids. To name a few, first, voltage and frequency regulations need to be particularly dealt with [78]. Second, intelligent control schemes enabling the system's plug-and-play operation are required [79]. Third, the system is expected to be capable of using localized information and nearest-neighbor communication to collectively perform control actions [80]. These issues can be investigated in future works.

Appendix 10.A

10.A.1   Proof of Proposition 10.1

Proof: Let m_t^k be the vertex that RC k is visiting at time t. As $\mathcal{V} = \{1, 2, \ldots, |\mathcal{V}|\}$, m_t^k can be obtained by

$$m_t^k = \sum_{m \in \mathcal{V}} a_{m,t}^k \cdot m = \sum_{m=1}^{|\mathcal{V}|} a_{m,t}^k \cdot m \tag{10.A.1}$$

As constraints (10.A.3) and (10.A.4) implicitly include constraint (10.A.1), one can have $m_t^k \in \{0, 1, 2, \ldots, |\mathcal{V}|\}$. If $m_t^k = 0$, RC k is traveling on the transportation network at time t, and if $m_t^k \in \mathcal{V}$, RC k is at vertex m_t^k at time t, i.e. $a_{m_t^k,t}^k = 1$.

RC k's scheduling is expressed as follows: $[m_1^k, m_2^k, \ldots, m_{|T|}^k]$. Its path can be easily retrieved by its scheduling. For example, if its scheduling is $[1, 0, 3, 3, 0, 0, 2, 2, 2]$, its path is $1 \xrightarrow{\text{1 time period}} 3 \xrightarrow{\text{2 time periods}} 2$, which also indicates the time reserved for traveling. The retrieved path is feasible if and only if each reserved travel time is adequate for RC k to travel through the associated edge.

That is, for any RC k, the following constraint has to be satisfied:

$$\tau \geq tr_{m_t^k m_{t+\tau}^k}, \quad \forall t, \ \forall \tau \leq |\mathcal{T}| - t, \ m_t^k \neq 0, \ m_{t+\tau}^k \neq 0, \ m_t^k \neq m_{t+\tau}^k \quad (10.A.2)$$

Assume that, for some RC k, there exist some t and $\tau \leq |\mathcal{T}| - t$ with $m_t^k \neq 0$, $m_{t+\tau}^k \neq 0$ and $m_t^k \neq m_{t+\tau}^k$ to violate constraint 10.A.2:

$$\tau < tr_{m_t^k m_{t+\tau}^k} \quad (10.A.3)$$

As $m_t^k \neq 0$ and $m_{t+\tau}^k \neq 0$, one can have $m_t^k \in \mathcal{V}$ and $m_{t+\tau}^k \in \mathcal{V}$, which further indicate $a_{m_t^k,t}^k = 1$ and $a_{m_{t+\tau}^k,t+\tau}^k = 1$, respectively. Thus, for RC k, one can have some $\tau < tr_{m_t^k m_{t+\tau}^k}, t \leq |\mathcal{T}| - \tau$ and $m_t^k \neq m_{t+\tau}^k$ that result in the following equation:

$$a_{m_{t+\tau}^k,t+\tau}^k + a_{m_t^k,t}^k = 2 > 1 \quad (10.A.4)$$

which violates constraint (10.A.3). Moreover, for this RC k, one can also have some t and $m_t^k \neq m_{t+\tau}^k$ that result in the following equation:

$$\sum_{\tau'=t}^{\min\left\{t+tr_{m_t^k m_{t+\tau}^k}, |\mathcal{T}|\right\}} a_{m_{t+\tau}^k,\tau'}^k$$

$$= a_{m_{t+\tau}^k,t+\tau}^k + \sum_{\tau'=t}^{t+\tau-1} a_{m_{t+\tau}^k,\tau'}^k + \sum_{\tau'=t+\tau+1}^{\min\left\{t+tr_{m_t^k m_{t+\tau}^k}, |\mathcal{T}|\right\}} a_{m_{t+\tau}^k,\tau'}^k$$

$$\geq 1 \quad (10.A.5)$$

which violates constraint (10.A.4), as $\left(1 - a_{m_t^k,t}^k\right) \cdot \min\left(tr_{m_t^k m_{t+\tau}^k}, |\mathcal{T}| - t\right) = 0$.

That is, if the retrieved path violates constraint 10.A.2, the scheduling also violates constraints (10.A.3) and (10.A.4). Thus, if the scheduling satisfies constraint (10.A.3) or constraint (10.A.4), which are equivalent, the retrieved path also satisfies constraint 10.A.2. This completes the proof. □

References

1 B. Zhou, D. Xu, C. Li, Y. Cao, K. W. Chan, Y. Xu, and M. Cao, "Multiobjective generation portfolio of hybrid energy generating station for mobile emergency power supplies," *IEEE Transactions on Smart Grid*, vol. 9, no. 6, pp. 5786–5797, 2018.

2 B. Norris, "Transportable energy storage systems project," Electric Power Research Institute, Palo Alto, CA, USA, Tech. Rep. 1017818, Oct. 2009.

3 C. Chen, J. Wang, F. Qiu, and D. Zhao, "Resilient distribution system by microgrids formation after natural disasters," *IEEE Transactions on Smart Grid*, vol. 7, no. 2, pp. 958–966, 2016.

4 T. Ding, Y. Lin, G. Li, and Z. Bie, "A new model for resilient distribution systems by microgrids formation," *IEEE Transactions on Power Systems*, vol. 32, pp. 4145–4147, 2017.

5 K. S. A. Sedzro, A. J. Lamadrid, and L. F. Zuluaga, "Allocation of resources using a microgrid formation approach for resilient electric grids," *IEEE Transactions on Power Systems*, vol. 33, no. 3, pp. 2633–2643, 2018.

6 P. Toth and D. Vigo, *The Vehicle Routing Problem*. Philadelphia, PA USA: SIAM, 2002.

7 C. J. Zapata, S. C. Silva, H. I. Gonzalez, O. L. Burbano, and J. A. Hernandez, "Modeling the repair process of a power distribution system," in *IEEE/PES Transmission and Distribution Conference and Exposition - Latin America*, pp. 1–7, Bogota, Colombia, Aug. 2008.

8 J. Shang, X. Sheng, J. Zhang, and W. Zhao, "The optimized allocation of mobile emergency generator based on the loads importance," in *Asia-Pacific Power and Energy Engineering Conference*, pp. 1–4, Wuhan, China, Mar. 2009.

9 N. Perrier, B. Agard, P. Baptiste, J.-M. Frayret, A. Langevin, R. Pellerin, D. Riopel, and M. Trépanier, "A survey of models and algorithms for emergency response logistics in electric distribution systems - Part I: Reliability planning with fault considerations," *Computers & Operations Research*, vol. 40, no. 7, pp. 1895–1906, 2013.

10 N. Perrier, B. Agard, P. Baptiste, J.-M. Frayret, A. Langevin, R. Pellerin, D. Riopel, and M. Trépanier, "A survey of models and algorithms for emergency response logistics in electric distribution systems - Part II: Contingency planning level," *Computers & Operations Research*, vol. 40, no. 7, pp. 1907–1922, 2013.

11 K. G. Zografos, C. Douligeris, and P. Tsoumpas, "An integrated framework for managing emergency-response logistics: The case of the electric utility companies," *IEEE Transactions on Engineering Management*, vol. 45, no. 2, pp. 115–126, 1998.

12 A. Arif, Z. Wang, J. Wang, and C. Chen, "Power distribution system outage management with co-optimization of repairs, reconfiguration, and DG dispatch," *IEEE Transactions on Smart Grid*, vol. 9, no. 5, pp. 4109–4118, 2018.

13 H. H. Abdeltawab and Y. A.-R. I. Mohamed, "Mobile energy storage scheduling and operation in active distribution systems," *IEEE Transactions on Industrial Electronics*, vol. 64, no. 9, pp. 6828–6840, 2017.

14 Y. Chen, Y. Zheng, F. Luo, J. Wen, and Z. Xu, "Reliability evaluation of distribution systems with mobile energy storage systems," *IET Renewable Power Generation*, vol. 10, no. 10, pp. 1562–1569, 2016.

15 H. Gao, Y. Chen, S. Mei, S. Huang, and Y. Xu, "Resilience-oriented pre-hurricane resource allocation in distribution systems considering electric buses," *Proceedings of the IEEE*, vol. 105, pp. 1214–1233, 2017.

16 S. Lei, J. Wang, C. Chen, and Y. Hou, "Mobile emergency generator pre-positioning and real-time allocation for resilient response to natural disasters," *IEEE Transactions on Smart Grid*, vol. 9, no. 3, pp. 2030–2041, 2018.

17 P. V. Hentenryck, C. Coffrin, and R. Bent, "Vehicle routing for the last mile of power system restoration," in *17th Power System Computation Conference*, pp. 1–8, Stockholm, Sweden, Aug. 2011.

18 C. Coffrin and P. V. Hentenryck, "Transmission system restoration with co-optimization of repairs, load pickups, and generation dispatch," *International Journal of Electrical Power & Energy Systems*, vol. 72, pp. 144–154, 2015.

19 A. Arif, S. Ma, and Z. Wang, "Optimization of transmission system repair and restoration with crew routing," in *North American Power Symposium*, pp. 1–6, Denver, CO, USA, Sep. 2016.

20 N. Xu, S. D. Guikema, R. A. Davidson, L. K. Nozick, Z. Çağnan, and K. Vaziri, "Optimizing scheduling of post-earthquake electric power restoration tasks," *Earthquake Engineering and Structural Dynamics*, vol. 36, no. 2, pp. 265–284, 2007.

21 A. Arab, A. Khodaei, S. K. Khator, K. Ding, V. A. Emesih, and Z. Han, "Stochastic pre-hurricane restoration planning for electric power systems infrastructure," *IEEE Transactions on Smart Grid*, vol. 6, no. 2, pp. 1046–1054, 2015.

22 F. Capitanescu, L. F. Ochoa, H. Margossian, and N. D. Hatziargyriou, "Assessing the potential of network reconfiguration to improve distributed generation hosting capacity in active distribution systems," *IEEE Transactions on Power Systems*, vol. 30, no. 1, pp. 346–356, 2015.

23 F. Ding and K. A. Loparo, "Hierarchical decentralized network reconfiguration for smart distribution systems–Part II: Applications to test systems," *IEEE Transactions on Power Systems*, vol. 30, no. 2, pp. 744–752, 2015.

24 S. Chen, W. Hu, and Z. Chen, "Comprehensive cost minimization in distribution networks using segmented-time feeder reconfiguration and reactive power control of distributed generators," *IEEE Transactions on Power Systems*, vol. 31, no. 2, pp. 983–993, 2016.

25 M. R. Dorostkar-Ghamsari, M. Fotuhi-Firuzabad, M. Lehtonen, and A. Safdarian, "Value of distribution network reconfiguration in presence of renewable energy resources," *IEEE Transactions on Power Systems*, vol. 31, no. 3, pp. 1879–1888, 2016.

26 A. Kavousi-Fard, A. Zare, and A. Khodaei, "Effective dynamic scheduling of reconfigurable microgrids," *IEEE Transactions on Power Systems*, vol. 33, no. 5, pp. 5519–5530, 2018.

27 T. Thakur and Jaswanti, "Study and characterization of power distribution network reconfiguration," in *IEEE/PES Transmission and Distribution Conference and Exposition - Latin America*, pp. 1–6, Caracas, Venezuela, Aug. 2006.

28 S. Lei, Y. Hou, F. Qiu, and J. Yan, "Identification of critical switches for integrating renewable distributed generation by dynamic network reconfiguration," *IEEE Transactions on Sustainable Energy*, vol. 9, no. 1, pp. 420–432, 2018.

29 C. L. Moreira, F. O. Resende, and J. A. P. Lopes, "Using low voltage microgrids for service restoration," *IEEE Transactions on Power Systems*, vol. 22, no. 1, pp. 395–403, 2007.

30 E. Zare and M. Shahabi, "Microgrid restoration after major faults in main grid with automatic and constant time switching," *International Journal of Intelligent Systems and Applications*, vol. 5, no. 10, pp. 50–58, 2013.

31 J. Li, X. Y. Ma, C. C. Liu, and K. P. Schneider, "Distribution system restoration with microgrids using spanning tree search," *IEEE Transactions on Power Systems*, vol. 29, no. 6, pp. 3021–3029, 2014.

32 Z. Wang and J. Wang, "Self-healing resilient distribution systems based on sectionalization into microgrids," *IEEE Transactions on Power Systems*, vol. 30, no. 6, pp. 3139–3149, 2015.

33 H. Gao, Y. Chen, Y. Xu, and C. C. Liu, "Resilience-oriented critical load restoration using microgrids in distribution systems," *IEEE Transactions on Smart Grid*, vol. 7, no. 6, pp. 2837–2848, 2016.

34 G. J. Lim, S. Kim, J. Cho, Y. Gong, and A. Khodaei, "Multi-UAV pre-positioning and routing for power network damage assessment," *IEEE Transactions on Smart Grid*, vol. 9, pp. 3643–3651, 2018.

35 G. B. Dantzig and J. H. Ramser, "The truck dispatching problem," *Management Science*, vol. 6, no. 1, pp. 80–91, 1959.

36 T. H. Cormen, C. Stein, R. L. Rivest, and C. E. Leiserson, *Introduction to Algorithms*. McGraw-Hill Higher Education, 2001.

37 E. W. Dijkstra, "A note on two problems in connexion with graphs," *Numerische Mathematik*, vol. 1, no. 1, pp. 269–271, 1959.

38 A. J. Miller and L. A. Wolsey, "Tight formulations for some simple mixed integer programs and convex objective integer programs," *Mathematical Programming*, vol. 98, no. 1, pp. 73–88, 2003.

39 S. Iwai, T. Kono, M. Hashiwaki, and Y. Kawagoe, "Use of mobile engine generators as source of back-up power," in *IEEE 31st International Telecommunications Energy Conference*, pp. 1–6, Incheon, South Korea, Oct. 2009.

40 U.S. Federal Emergency Management Agency, *2017 Hurricane Season FEMA After-Action Report*. 2018.

41 C. R. Nightingale, "The design of mobile engine driven generating sets and their role in the British telecommunications network," in *5th International Telecommunications Energy Conference*, pp. 144–150, Tokyo, Japan, Oct. 1983.

42 U.S. Federal Emergency Management Agency, *Mitigation Assessment Team Report: Hurricane Sandy in New Jersey and New York*. 2013.

43 Global Power Supply Ltd., *Mobile Diesel Generators*. 2018.

44 W. Ng, S. Leung, J. Lam, and S. Pan, "Petrol delivery tanker assignment and routing: A case study in Hong Kong," *Journal of the Operational Research Society*, vol. 59, no. 9, pp. 1191–1200, 2008.

45 G. G. Brown, C. J. Ellis, G. W. Graves, and D. Ronen, "Real-time wide area dispatching of mobil tank trucks," *Interfaces*, vol. 17, no. 1, pp. 107–120, 1987.

46 R. A. Jabr, R. Singh, and B. C. Pal, "Minimum loss network reconfiguration using mixed-integer convex programming," *IEEE Transactions on Power Systems*, vol. 27, no. 2, pp. 1106–1115, 2012.

47 H. Ahmadi and J. R. Martí, "Mathematical representation of radiality constraint in distribution system reconfiguration problem," *International Journal of Electrical Power & Energy Systems*, vol. 64, pp. 293–299, 2015.

48 P. C. Pop, "A survey of different integer programming formulations of the generalized minimum spanning tree problems," *Carpathian Journal of Mathematics*, vol. 25, no. 1, pp. 104–118, 2009.

49 P. C. Pop, *The generalized minimum spanning tree problem*. Ph.D. Dissertation, Twente University Press, Enschede, The Netherlands, 2002.

50 R. K. Ahuja, T. L. Magnanti, and J. B. Orlin, *Network Flows: Theory, Algorithms, and Applications*. Prentice Hall, 1993.

51 M. Lavorato, J. F. Franco, M. J. Rider, and R. Romero, "Imposing radiality constraints in distribution system optimization problems," *IEEE Transactions on Power Systems*, vol. 27, no. 1, pp. 172–180, 2012.

52 E. R. Ramos, A. G. Expósito, J. R. Santos, and F. L. Iborra, "Path-based distribution network modeling: Application to reconfiguration for loss reduction," *IEEE Transactions on Power Systems*, vol. 20, no. 2, pp. 556–564, 2005.

53 J.-Y. Fan, L. Zhang, and J. D. McDonald, "Distribution network reconfiguration: Single loop optimization," *IEEE Transactions on Power Systems*, vol. 11, no. 3, pp. 1643–1647, 1996.

54 M. E. Baran and F. F. Wu, "Optimal sizing of capacitors placed on a radial distribution system," *IEEE Transactions on Power Delivery*, vol. 4, no. 1, pp. 735–743, 1989.

55 M. E. Baran and F. F. Wu, "Optimal capacitor placement on radial distribution systems," *IEEE Transactions on Power Delivery*, vol. 4, no. 1, pp. 725–734, 1989.

56 M. E. Baran and F. F. Wu, "Network reconfiguration in distribution systems for loss reduction and load balancing," *IEEE Transactions on Power Delivery*, vol. 4, no. 2, pp. 1401–1407, 1989.

57 K. Turitsyn, P. Sulc, S. Backhaus, and M. Chertkov, "Options for control of reactive power by distributed photovoltaic generators," *Proceedings of the IEEE*, vol. 99, no. 6, pp. 1063–1073, 2011.

58 J. A. Taylor and F. S. Hover, "Convex models of distribution system reconfiguration," *IEEE Transactions on Power Systems*, vol. 27, no. 3, pp. 1407–1413, 2012.

59 H. G. Yeh, D. F. Gayme, and S. H. Low, "Adaptive VAR control for distribution circuits with photovoltaic generators," *IEEE Transactions on Power Systems*, vol. 27, no. 3, pp. 1656–1663, 2012.

60 X. Chen, W. Wu, and B. Zhang, "Robust restoration method for active distribution networks," *IEEE Transactions on Power Systems*, vol. 31, no. 5, pp. 4005–4015, 2016.

61 K. Turitsyn, P. Sulc, S. Backhaus, and M. Chertkov, "Distributed control of reactive power flow in a radial distribution circuit with high photovoltaic penetration," in *IEEE Power and Energy Society General Meeting*, pp. 1–6, Minneapolis, MN, USA, Jul. 2010.

62 D. Issicaba, *Power Flow Analysis for Radial and Weakly Meshed Distribution Systems*. Novas Ediç oes Acadêmicas, 2016.

63 D. Shirmohammadi, H. W. Hong, A. Semlyen, and G. X. Luo, "A compensation-based power flow method for weakly meshed distribution and transmission networks," *IEEE Transactions on Power Systems*, vol. 3, no. 2, pp. 753–762, 1988.

64 M. Farivar and S. H. Low, "Branch flow model: Relaxations and convexification–Part I," *IEEE Transactions on Power Systems*, vol. 28, pp. 2554–2564, 2013.

65 M. Farivar and S. H. Low, "Branch flow model: Relaxations and convexification–Part II," *IEEE Transactions on Power Systems*, vol. 28, pp. 2565–2572, 2013.

66 Z. Bie, Y. Lin, G. Li, and F. Li, "Battling the extreme: A study on the power system resilience," *Proceedings of the IEEE*, vol. 105, no. 7, pp. 1253–1266, 2017.

67 M. Panteli, D. N. Trakas, P. Mancarella, and N. D. Hatziargyriou, "Power systems resilience assessment: Hardening and smart operational enhancement strategies," *Proceedings of the IEEE*, vol. 105, no. 7, pp. 1202–1213, 2017.

68 M. J. Sharma, I. Moon, and H. Bae, "Analytic hierarchy process to assess and optimize distribution network," *Applied Mathematics and Computation*, vol. 202, no. 1, pp. 256–265, 2008.

69 C. J. Zapata, S. C. Silva, and O. L. Burbano, "Repair models of power distribution components," in *IEEE/PES Transmission and Distribution Conference and Exposition - Latin America*, pp. 1–6, Bogota, Colombia, Aug. 2008.

70 A. Arif, S. Ma, Z. Wang, J. Wang, S. M. Ryan, and C. Chen, "Optimizing service restoration in distribution systems with uncertain repair time and demand," *IEEE Transactions on Power Systems*, vol. 33, no. 6, pp. 6828–6838, 2018.

71 A. Ben-Tal, A. Goryashko, E. Guslitzer, and A. Nemirovski, "Adjustable robust solutions of uncertain linear programs," *Mathematical Programming*, vol. 99, no. 2, pp. 351–376, 2004.

72 G. P. McCormick, "Computability of global solutions to factorable nonconvex programs: Part I–Convex underestimating problems," *Mathematical Programming*, vol. 10, no. 1, pp. 147–175, 1976.

73 M. Jünger, T. M. Liebling, D. Naddef, G. L. Nemhauser, W. R. Pulleyblank, G. Reinelt, G. Rinaldi, L. A. Wolsey, *50 Years of Integer Programming 1958-2008: From the Early Years to the State-of-the-Art*. Berlin, Germany: Springer, 2010.

74 K. Deep, K. P. Singh, M. Kansal, and C. Mohan, "A real coded genetic algorithm for solving integer and mixed integer optimization problems," *Applied Mathematics and Computation*, vol. 212, no. 2, pp. 505–518, 2009.

75 S. Lei, J. Wang, and Y. Hou, "Remote-controlled switch allocation enabling prompt restoration of distribution systems," *IEEE Transactions on Power Systems*, vol. 33, no. 3, pp. 3129–3142, 2018.

76 IEEE PES Power System Analysis, Computing and Economics Committee, *IEEE 123 Node Test Feeder*. Feb. 2014.

77 A. Wächter and L. T. Biegler, "On the implementation of an interior-point filter line-search algorithm for large-scale nonlinear programming," *Mathematical Programming*, vol. 106, no. 1, pp. 25–57, 2006.

78 H. Qi, X. Wang, L. M. Tolbert, F. Li, F. Z. Peng, P. Ning, and M. Amin, "A resilient real-time system design for a secure and reconfigurable power grid," *IEEE Transactions on Smart Grid*, vol. 2, no. 4, pp. 770–781, 2011.

79 F. Dörfler, J. W. Simpson-Porco, and F. Bullo, "Breaking the hierarchy: Distributed control and economic optimality in microgrids," *IEEE Transactions on Control of Network Systems*, vol. 3, no. 3, pp. 241–253, 2016.

80 J. W. Simpson-Porco, Q. Shafiee, F. Dörfler, J. C. Vasquez, J. M. Guerrero, and F. Bullo, "Secondary frequency and voltage control of islanded microgrids via distributed averaging," *IEEE Transactions on Industrial Electronics*, vol. 62, no. 11, pp. 7025–7038, 2015.

Index

a

adaptability 5
adaptive multi-microgrids 122
allocation decisions 105
analytic hierarchical process 86
ant colony system-based method 86
approximate dynamic programming
    (ADP)
  forward dynamic algorithm 174–175
  iteration process 177
  post-decision states 174
  proposed model reformulation
    175–176
  solution challenges 172–173
automation systems 86–87

b

backward induction 44
Bayesian Additive Regression Trees 68
BlackEnergy 8
branch-and-cut (B&C) algorithms 198

c

candidate nodes for MPS connection
    281–283
capacity utilization rate (CUR) 73–75
carbon emissions 3, 10
CIC-oriented grid automation
  optimization model 108

CIC-oriented model 102–103
climate change 4
column-and-constraint generation
    (C&CG) algorithm 227, 243–244
compactness 198, 205–207
component-level deterioration transition
    probability 38–40
component-level resilience enhancement
    13–14
component reliability enhancement
    9–10
computational efficiency 76–77
conceptual resilience curve 60–61, 231
condition-based maintenance scheduling
    13
condition monitoring technology 52
continuous operating time (COT) 122
controlled separation 16
conventional recovery process 17–18
co-optimization model 267, 268, 279
co-optimization problem 277
customer damage function (CDF) 92
customer interruption cost (CIC) 85
cyber-physical systems 11
cybersecurity-related events 8–9

d

data description 45
decentralized architecture 16

Power Grid Resilience against Natural Disasters: Preparedness, Response, and Recovery, First Edition.
Shunbo Lei, Chong Wang, and Yunhe Hou.
© 2023 John Wiley & Sons Ltd. Published 2023 by John Wiley & Sons Ltd.

Dijkstra's algorithm   266
Dijkstra's shortest-path algorithm   69
directed cutset formulation   200
directed multi-commodity flow-based
    formulation   202
DistFlow model   275, 276
DistFlow model-based power flow
    equations   238
distributed generations (DGs)   12, 14,
    18, 19, 110–111. *see also*
    microgrids (MGs)
  flexibility resources   67
  operational constraints   64–65
distribution grid automation   6
distribution grid operators   97
distribution grid planners   106
distribution grid planning   111
distribution systems (DSs)   228, 262
  damages   58
  Dijkstra's algorithm   69
  network   272
  reconfiguration constraints   64
  restoration   262, 265
  service restoration   278
  survivability   235
  topology   271
dynamic network reconfiguration   250,
    263–264, 288
dynamic state estimation   15

e
economic sustainability   6–7
electric buses   229, 233
electricity market   93
electric service recovery   18
electric vehicles (EVs)   227, 271
emergency resources   55
  electric service recovery   56
  experimental studies   56–57
  grid survivability   58–61

mobile emergency generators (*see*
    mobile emergency generators
    (MEGs))
  preallocation and real-time allocation
    57
  system design   56–57
  utilization of emergency resources
    56
environmental sustainability   6–7
expected costs   51
external conditions prediction   52
extreme weather events   85, 125, 228

f
Federal Emergency Management Agency
    (FEMA)   56
Floyd–Warshall algorithm   69, 266
forward dynamic algorithm   174–175
Fukushima disaster   232

g
Gaussian distributions   10
greenhouse gas emissions   10
grid automation system   85–101
  distribution system automation and
    restoration   87–89
  MICP models   85
  prompt restoration   89–91
  remote-controlled switch allocation
    models   91–95
  smart grid and automation systems
    85–87
  solution method   95–101
grid reconfiguration   271–276

h
harsh external conditions   48
HAVEX   8–9
hierarchical architecture   16
historical data-based regression models
    122

i

IEEE 30-bus system   152–156
  data description   152–153
  effectiveness   155–156
  influences of state reduction   156
  mapping states   153–155
IEEE 123-bus system   181–183
IEEE 33-bus system, MDP-based model
    177–181
  data description   177–178
  estimations of post-decision states
    178–180
  state-based strategies   180–181
IEEE 33-node test system   212–218,
    271–272
immune algorithm   86
integer variables   95
integrated resilience response framework
    122
inter-dependencies   264
interior-point based algorithm   15
isolation capability malfunction   110
iteration process   177

k

Karush–Kuhn-Tucker (KKT) condition
    121

l

Lagrangian function   132
large test system   106–109
LinDistFlow model   204, 219, 275–276,
    292
linearization techniques   100, 283
linear max–min optimization model
    131
linear programming (LP) relaxations
    198
linear scalarization method   150–152
line capacity constraint   171
loop-eliminating model   200

m

maintenance cost   37
maintenance scheduling   51
maintenance strategies   37
Markov-based model   51
Markov-based operation strategy   122
Markov chain   40
Markov decision process (MDP)-based
    model   165–184
  ADP   172–177
    forward dynamic programming
      algorithm   174–175
    iteration process   177
    post-decision states   174
    proposed model reformulation
      175–176
    solution challenges   172–173
  disasters' trajectories   167–168
  IEEE 33-bus system   177–181
    data description   177–178
    estimations of post-decision states
      178–180
    state-based strategies   180–181
  IEEE 123-bus system   181–183
  operational constraints   170–172
    line capacity constraint   171
    power balance constraint   171
    power flow constraints   171
    radiality constraint   170
    repair constraint   170
    voltage constraint   172
  real-time response   165–167
  recursive model   169–170
  transition probability   169
Markovian method   145
Markov process   37
mathematical model, without harsh
      external conditions   40–41
McCormick envelope method   129, 130,
    138
medium-duty EVs   230, 234

microgrids (MGs)   18, 227, 264. *see also*
   mobile emergency generators
   (MEGs)
  construction   122, 123
  dynamic formation   194–198
   flexible boundaries   194–195
   radiality constraints   195–198
   topological flexibility   195–198
  flexibility resources   67
  formation   59
  large test system   215–219
  radiality constraints   198–202
   incidence/characteristic vectors of
    spanning forests   199–200
   incidence/characteristic vectors of
    spanning trees   199–200
   loop eliminating model   200
   MIP model   198–199
   parent–child node relation-based
    model   201
   path-based model   200
   primal and dual graph-based model
    201
   single-commodity flow-based model
    200
   spanning forest-based model
    201–202
  in service restoration   192–194,
   202–211
   applicability and application
    207–211
   compactness and tightness
    205–207
   features   194
   formulation and validity   202–205
   LinDistFlow model   219
   network reconfiguration   191–192
   radiality constraints   192–193
   resilience   192
  small test system   211–215
  minimal set of repair tasks   284

mixed-integer linear programming
  (MILP) method   86
mixed-integer non-linear programming
  (MINLP)   264
mixed-integer programming (MIP)
  198–199
mixed-integer second-order cone
  programming (MISCOP)   245,
  265
mobile emergency generators (MEGs)
  227, 262
  availability of mobile emergency
   generators   66
  capacity utilization rate   73–75
  characteristics   58–59
  computational efficiency   76–77
  connection of mobile emergency
   generators   66–67
  conventional restoration procedures
   60–61
  Dijkstra's shortest-path algorithm   69
  dispatch approach   71–73
  flexibility resources   67
  microgrid formation   59
  preallocation and real-time allocation
   59–60
  scenario decomposition algorithm
   69–70
  scenario generation and reduction
   68–69
  task modules and data sets   67–68
  test system   70–71
  traffic issue and preallocation   75–76
  two-stage stochastic optimization
   model   61–65
mobile emergency resource
  allocation/dispatch   123
mobile energy storage preallocation
  122
mobile energy storage systems (MESSs)
  227

mobile generation resources   6
mobile power sources (MPSs)   261
  C&CG algorithm   243–244
  grid survivability and recovery
    227–230
  large test system   251–254
  linearization techniques   245
  routing and scheduling   230–233
  small test system   246–251
  supporting facilities   233–235
    availability   233–234
    cost-effectiveness   234–235
    grid-forming functions   234
  two-stage dispatch framework
    235–243
    dynamic routing and scheduling
      239–243
    proactive pre-dispatch   235–238
model transformation   99
Monte Carlo method   138
Monte Carlo simulations   46–47, 251
MPS dispatch   262, 278
MPS dynamic dispatch   239
MPS pre-positioning   251
MPSs' dynamic dispatch   227
multiarea networks   11–12
multiarea state estimation   16
multilinear functions   130
multiple energy systems   10–11

n
network hardening   122
network reconfiguration   18–20, 89
North American Electric Reliability
  Corporation   6, 121
NP-hard partition problem   16

o
operational enhancement approach
  122

operational point analysis resilience
  enhancement   126–132
  bilevel optimization model   130–131
  mapping system topology changes
    129–130
  sequential security region   126–129
  solution process   131–132
optimal maintenance activities   47–48
optimal maintenance strategy model   37
optimization-based maintenance
  scheduling   14
optimization model   130
optimization variables   128

p
parallel computation algorithm   15
parent–child node relation-based model
  201
phasor measurement units (PMUs)   10,
  15, 16
physical islands (PIs)   271–272
planned maintenance scheduling   13
post-disaster MG formation models
  209–211
post-disaster MPS dynamic dispatch
  239, 248
post-disaster service restoration strategy
  234
power balance constraint   149, 171
power dispatch flexibility   269
power flow constraints   149–150, 171
power grid resilience, definition   4–5
practical candidate restoration strategies
  (PCRSs)   87, 91, 95–98
practical power grid system   156–159
pre-assigning repair tasks   284–285
pre-hurricane restoration planning   122
preventive maintenance   41–44
  activity vectors   42–43
  expected cost   43–44
  harsh weather   41

preventive maintenance (*contd.*)
  models and methodologies 42
  probabilistic model of repair delays
      42
primal and dual graph-based model
    201
private-owned EVs 233
proactive resilience enhancement
      strategy 143–159
  component failure rate 146
  disasters' trajectories 146–147
  extreme weather events 143–145
  Markov process 143
  practical power grid system 156–159
  sequentially proactive operation
      strategy 148–159
    constraints 148–150
    decision processes 148
    IEEE 30-bus system 152–156
    linear scalarization method
        150–152
  transition probabilities 147–148
probability distribution 8
prompt restoration 90
proposed model 45–46

r
radiality constraints 170
  fictitious 197
  graph theory 196–198
  incidence/characteristic vectors of
      spanning forests 199–200
  incidence/characteristic vectors of
      spanning trees 199–200
  loop eliminating model 200
  MIP model 198–199
  parent–child node relation-based
      model 201
  path-based model 200
  primal and dual graph-based model
      201

single-commodity flow-based model
    200
spanning forest-based model
    195–196, 201–202
spanning tree 195–196
ramping rates of generators 149
real-time allocation constraints 63–64
recovery logistics 261–294
  computational efficiency 290–292
  flexibility resources 265–276
    co-optimization of flexibility
        resources 277–280
    routing and scheduling of MPS
        268–271
    routing and scheduling of repair
        crews 265–268
  large test system 287–290
  LinDistFlow model 292
  linearization techniques 283
  post-disaster 261–265
    co-optimization approach 264–265
    microgrid-based service restoration
        263–264
    power infrastructure recovery
        262–263
  pre-assigning minimal repair tasks
      280–281
  proof of proposition 293–294
  selecting candidate nodes 281–283
  small test system 284–287
recursive model 169–170
reliability 50, 51
reliability-centered maintenance
      scheduling 13–14
reliability investments 95
remote-controlled switches (RCSs) 85,
    249
  allocation models 113
    customer interruption cost 91–93
    system average interruption
        duration index 93–94

system restoration capability 94–95
  malfunction 109–110
  restoration models 112
renewable energy sources (RESs) 7–8
renewable energy uncertainty 11
repair constraint 170
repair crews (RCs) 261
  dispatch 262, 265, 278
  routing subproblem 267
resilience 95, 192
resilience-constrained unit commitment (RCUC) model 122
resilience enhancement strategies 12–20
  controlled separation 16
  coordination and co-optimization 20–21
  preparedness prior to disasters 12–14
  recovery after disasters 17–20
  state acquisition 15–16
resilience-enhancing effects 228
resilience-oriented operational strategies 121–123
resilient electric grids 261
resilient recovery strategies 261
resourcefulness 5
restoration strategies 91, 96
RL-oriented model 105
road networks (RNs) 58, 60, 63, 69, 71, 72
robustness 5

s

SAIDI-oriented model 103–105
Sandworm 9
scenario decomposition (SD) algorithm 69–70, 77
search space reduction method 44–45

sensitivity analyses 104
sequential steady-state security region (SSSR)
  sequential security region 123–125
  uncertain varying system topology changes 125–126
simplification techniques 100
simulation statistics 246
single-commodity flow-based model 200, 214, 274
small-scale MILP problem 280
small test system 102–106
smart grid 86–87
social sustainability 6–7
solution method 95–101
  linearization and simplification techniques 100
  model transformation 99
  overall solution process 100–101
  practical candidate restoration strategies 95–98
spanning forest-based model 195–196, 201–202
spanning tree 195–196, 272–274
static state estimation 15
stochastic optimization 8
subtour elimination formulation 200
system-level deterioration transition probability 40
system-level resilience enhancement 14
system state acquisition 10, 15–16

t

tightness 199, 205–207
topology reconfiguration 18–20
transition matrix 40
transition probabilities 40
transmission system restoration and repairing 263

transmission systems   123
trilevel optimal hardening strategy
    122
truck-mounted mobile energy storage
    systems (MESSs)   262
two-stage decomposition   86
two-stage robust optimization model
    235, 237
two-stage stochastic optimization model
    61–65

u

United States Electric Power Research
    Institute   6, 121
United States National Research Council
    (NRC)   121

US Federal Emergency Management
    Agency   270
U.S. National Security Agency   8–9

v

vehicle routing (VR)   59, 69
vehicle-to-grid's capability   234
voltage constraint   172

w

weather-related disastrous events   6–7
weighted least squares algorithm   15
wind power   7–8, 11

z

Zhejiang Electric Power Grid   48–51